VOLUME ONE HUNDRED AND NINE

CURRENT TOPICS IN DEVELOPMENTAL BIOLOGY

Mouse Models of The Nuclear Envelopathies and Related Diseases

CURRENT TOPICS IN DEVELOPMENTAL BIOLOGY

"A meeting-ground for critical review and discussion of developmental processes"
A.A. Moscona and Alberto Monroy (Volume 1, 1966)

VOLUME ONE HUNDRED AND NINE

CURRENT TOPICS IN DEVELOPMENTAL BIOLOGY

Mouse Models of The Nuclear Envelopathies and Related Diseases

Edited by

COLIN L. STEWART

Development and Regenerative Biology Group,
Institute of Medical Biology,
Immunos, Singapore, Singapore

ELSEVIER

AMSTERDAM • BOSTON • HEIDELBERG • LONDON
NEW YORK • OXFORD • PARIS • SAN DIEGO
SAN FRANCISCO • SINGAPORE • SYDNEY • TOKYO
Academic Press is an imprint of Elsevier

Academic Press is an imprint of Elsevier
525 B Street, Suite 1800, San Diego, CA 92101-4495, USA
225 Wyman Street, Waltham, MA 02451, USA
The Boulevard, Langford Lane, Kidlington, Oxford, OX5 1GB, UK
32 Jamestown Road, London NW1 7BY, UK
Radarweg 29, PO Box 211, 1000 AE Amsterdam, The Netherlands

First edition 2014

For information on all Academic Press publications
visit our website at store.elsevier.com

ISBN: 978-0-12-397920-9
ISSN: 0070-2153

14 15 16 17 11 10 9 8 7 6 5 4 3 2 1

CONTENTS

CONTRIBUTORS

Roger L. Albin
Department of Neurology; Michigan Alzheimer Disease Center, University of Michigan, and Geriatrics Research, Education, and Clinical Center, VAAAHS, Ann Arbor, Michigan, USA

Brian Burke
Nuclear Dynamics and Architecture Group, Institute of Medical Biology, Immunos, Singapore, Singapore

William T. Dauer
Department of Neurology, and Department of Cell and Developmental Biology, University of Michigan, Ann Arbor, Michigan, USA

Lorna R. Fiedler
British Heart Foundation Centre of Research Excellence, National Heart and Lung Institute, Imperial College London, London, UK

Henning F. Horn
Development and Regenerative Biology, Institute of Medical Biology, Singapore, Singapore

Brian K. Kennedy
Buck Institute for Research on Aging, Novato, California, USA

Daniel K. Leventhal
Department of Neurology, University of Michigan, Ann Arbor, Michigan, USA

Chen-Yu Liao
Buck Institute for Research on Aging, Novato, California, USA

Evie Maifoshie
British Heart Foundation Centre of Research Excellence, National Heart and Lung Institute, Imperial College London, London, UK

Samuel S. Pappas
Department of Neurology, University of Michigan, Ann Arbor, Michigan, USA

Justin J. Rochford
Rowett Institute of Nutrition and Health, University of Aberdeen, Foresterhill, Aberdeen, UK

Michael D. Schneider
British Heart Foundation Centre of Research Excellence, National Heart and Lung Institute, Imperial College London, London, UK

Colin L. Stewart
Development and Regenerative Biology Group, Institute of Medical Biology, Immunos, Singapore, Singapore

PREFACE

Our concept of the cell's nucleus has undergone significant reevaluation. Primarily, it had been regarded a largely amorphous porous sphere, containing the cell's genome. Macromolecules (protein–RNA) would enter and exit the nucleus, as part of the gene regulation–translation process, via the nuclear pores that traverse the nuclear membranes.

This view has changed over the past decade. One aspect in this altered perspective has been (unsurprisingly) that there is an extensive network of subnuclear structures/organelles, such as gems, speckles, and Cajal bodies, among others, that appear to largely function in regulating the synthesis of different types of RNA. Firstly, chromosomes are confined to specific locations or territories, with the spatial location of genes on these chromosomes changing as these genes are transcribed/suppressed, so implicating the position of genes being important for their regulation. Secondly, the nuclear periphery, comprised of the nuclear envelope (NE) and underlying nuclear lamina, contains a range of proteins whose expression varies according to cell type and that also show dynamic changes during embryonic development. Among the NE proteins are those of the LINC complex which functions to tether nuclei to the various components of the cytoskeleton—again in a cell-specific manner. This connection between the interphase nucleus and the cytoskeleton is particularly intriguing since it may function as a direct physical link between the cell membrane/extracellular matrix and nuclear interior, that is, nucleoplasm/chromatin. Such a link implicates nuclear structure being a key component in the mechanosignaling/transduction process. Whether such a link has a role in regulating gene expression/chromatin organization in response to mechanical stimuli is a tantalizing prospect.

However, the greatest stimulus to initiating interest in nuclear structure (its functional architecture) has been the finding that some 30 different inherited diseases are caused by mutations in genes encoding many of the NE proteins and above all the lamins. Paramount in generating this interest has been the lamin A gene of which more than 450 mutations have been mapped that result in some 6–8 diseases, the so-called laminopathies. Within the field of human genetics, no other gene is known that when mutated in different ways results in so many different, seemingly tissue-specific diseases.

Linking so many diseases to the proteins of the lamina and NE has stimulated a major reassessment of how the structural organization of the nucleus is important to its function. At the clinical level, despite these diseases being relatively rare, they may potentially provide novel insights into their more common variants, some of which include some of the more pressing issues in twenty-first century "developed world" health issues. These include obesity/diabetes, cardiovascular disease, neurological decline, and aging; the latter being the major catalyst for the other diseases, hence this rather eclectic collection of review articles which cover nuclear structure, the LINC complex, and some of the diseases associated with lamin/NE protein mutations and how they may shed light on their more prevalent equivalents.

In Chapter 1, Burke and Stewart provide an up-to-date summary of the different mouse lines that have been derived with mutations in the three lamin genes and a selection of some of the better-studied NE proteins. The extent to which the various pathologies resulting from these mutations in mice can vary between the mouse and human patient is discussed; however, some of these mice have provided valuable insights into the (continuing) development of therapeutics that may treat the human disease. Some of these mutants have also exemplified how genetics can reveal unexpected interactions between the various NE proteins and lamins, indicating that the different NE proteins function as an integrated network.

In Chapter 2, lipodystrophy, a group of rare diseases resulting from the loss or redistribution of white fat, is covered. Paradoxically, a loss of white fat results in many of the same pathologies associated with obesity, including diabetes, altered triglyceride/cholesterol levels, and steatotic liver. In this chapter, Rochford reviews what insights, together with their strengths and weaknesses, mouse models of the so far 10 generalized and partial congenital lipodystrophies have provided. Many of these diseases arise from mutations in proteins associated with forming or maintaining the intracellular lipid droplets, where others arise from, including the lamins, nuclear proteins.

In Chapter 3, Dauer and colleagues cover mouse models of diseases arising from neurological defects in the basal ganglia and associated circuits in the subcortical region of the brain. These structures are involved in coordinating movement and learning-related motor functions. Congenital diseases that affect these neural circuits include Tourette syndrome and obsessive compulsive disorder, Rett's syndrome, and primary dystonia, with the last arising due to disruption of the NE protein Torsin.

Fiedler and colleagues in Chapter 4 cover the extensive literature relating to the contribution of cell death to heart failure and the extent to which defects in signaling pathways are responsible. They review a wide range of mouse models of heart failure and how some of these models have been used in developing potential therapeutics with varying degrees of success.

Overshadowing all of these congenital diseases is the influence that the aging process has on their development, since aging is the greatest risk factor to developing cancer, cardiovascular disease, and many neurodegenerative diseases. Since the discovery that life-span can be genetically regulated, this has stimulated an enormous interest into what factors/genes influence longevity, either in prolonging or shortening (progeria) life-span. This is now an enormously diverse area of research and Kennedy and Chen in Chapter 5 have accomplished the not insignificant task in reviewing current studies on mouse longevity and how they related to other animal models, as well as suggesting possible avenues that would promote *healthy* human aging, rather than increasing longevity.

The last chapter (Chapter 6) by Horn provides the first comprehensive review of the LINC complex of proteins in many different organisms and their importance in various different tissue functions in mice, fertility, skeletal–cardiac muscle function, and hearing, among others.

From these chapters, it is apparent that the mouse genetics and models of some of the various diseases have had and still make an important contribution to understanding the complexity of the molecular pathology underlying the diseases covered in this issue. At present, existing mouse models are coming under increasingly critical focus as having been often unsatisfactory in developing therapeutics for the equivalent human disease. This situation is echoed in the chapters by Schneider, Kennedy, and Stewart, where it is clear that if mice are to be useful models then greater care has to be used in both developing and maintaining the model in question (Perrin, S. 2014. *Nature* 507, 423–425). Given the enormous level of knowledge gained and the extent of mouse genetic resources developed over the past quarter of a century, it would a shame to squander such a resource. On top of these issues, new techniques of gene manipulation (Talens, Crispr) will help in surpassing some of these concerns (e.g., the genetic heterogeneity of current ES lines to used to manipulate mice). These techniques are also making it easier and economically feasible to develop alternative animal models (rats, pigs, even primates) to model these diseases. Mouse genetics is still an important tool in biomedical research. However, with the advent of new techniques of gene manipulation, these other species will become increasingly important in both modeling various human diseases and in searching for novel therapeutics and may well eclipse mice.

Colin L. Stewart

CHAPTER ONE

Functional Architecture of the Cell's Nucleus in Development, Aging, and Disease☆

Brian Burke*, Colin L. Stewart†,1

*Nuclear Dynamics and Architecture Group, Institute of Medical Biology, Immunos, Singapore, Singapore
†Development and Regenerative Biology Group, Institute of Medical Biology, Immunos, Singapore, Singapore
[1]Corresponding author: e-mail address: colin.stewart@imb.a-star.edu.sg

Contents

☆Dedicated to Boris Joffe, 1953–2014.

Current Topics in Developmental Biology, Volume 109
ISSN 0070-2153
http://dx.doi.org/10.1016/B978-0-12-397920-9.00006-8

Abstract

In eukaryotes, the function of the cell's nucleus has primarily been considered to be the repository for the organism's genome. However, this rather simplistic view is undergoing a major shift, as it is increasingly apparent that the nucleus has functions extending beyond being a mere genome container. Recent findings have revealed that the structural composition of the nucleus changes during development and that many of these components exhibit cell- and tissue-specific differences. Increasing evidence is pointing to the nucleus being integral to the function of the interphase cytoskeleton, with changes to nuclear structural proteins having ramifications affecting cytoskeletal organization and the cell's interactions with the extracellular environment. Many of these functions originate at the nuclear periphery, comprising the nuclear envelope (NE) and underlying lamina. Together, they may act as a "hub" in integrating cellular functions including chromatin organization, transcriptional regulation, mechanosignaling, cytoskeletal organization, and signaling pathways. Interest in such an integral role has been largely stimulated by the discovery that many diseases and anomalies are caused by defects in proteins of the NE/lamina, the nuclear envelopathies, many of which, though rare, are providing insights into their more common variants that are some of the major issues of the twenty-first century public health. Here, we review the contributions that mouse mutants have made to our current understanding of the NE/lamina, their respective roles in disease and the use of mice in developing potential therapies for treating the diseases.

1. INTRODUCTION

1.1. The nuclear envelope and lamina

In most eukaryotic cells, the nucleus is the most prominent organelle. The traditional view of the nucleus's function was that it served as the container for the cell's genome. Here, we discuss recent findings revealing that the nucleus has additional, nongenomic, functions that go beyond being the genome repository and impact on the whole cell, particularly cytoskeletal organization and function. When these functions breakdown this often leads to disease.

Due to improved imaging techniques, it is apparent that the nucleus is not a homogeneous structure, and there is substantial subnuclear organization (Bickmore, 2013). Such organization, besides the nucleolus, includes nuclear structures such as Gems, Paraspeckles, PML bodies, and Cajal bodies, all involved in RNA processing (Mao, Zhang, & Spector, 2011). Furthermore, chromosomes are organized into distinct territories, with their chromatin being distributed into the tightly packed, densely staining heterochromatin at the nuclear periphery and around the nucleolus, and the more loosely packed lighter staining euchromatin occupying the rest of the nucleus (Cremer & Cremer, 2010). The structural organization of chromatin, with regard to its role in regulating gene expression, is currently an area of intense investigation.

Enclosing all these structures is the nuclear envelope (NE). The envelope was originally thought to largely function as a selective barrier regulating the entry and exit of macromolecules via the nuclear pores. This view is changing. Increasing evidence is revealing that in addition to regulating nuclear transport, the NE acts as a cellular "hub" integrating many cellular functions including chromatin organization, signaling pathways, transcriptional regulation, and cytoskeletal organization (Dauer & Worman, 2009; Gruenbaum, Margalit, Goldman, Shumaker, & Wilson, 2005).

The NE consists of the inner and outer nuclear membranes (INM and ONM, respectively) that are separated by the 40–50-nm diameter perinuclear space (PNS). Both membranes connect with each other where they are traversed by the nuclear pore complexes (NPCs) (Grossman, Medalia, & Zwerger, 2012). Since the ONM also connects at multiple points with the cytoplasmic endoplasmic reticulum (ER), this makes the ER, INM, and ONM one continuous membrane system with lumen of the ER being contiguous with the PNS (Stewart, Roux, & Burke, 2007).

Despite the INM, ONM, and ER being one continuous system, each membrane is characterized by its' association with a unique set of proteins. For instance, the peripheral ER contains the reticulon and DP1/Yop1p families of proteins that are required for the ER's assembly and maintenance as a tubular structure (Shnyrova, Frolov, & Zimmerberg, 2008). Both the ONM and ER, but not the INM, are associated with ribosomes. In contrast to the ONM and ER, proteomic studies have identified at least 70 transmembrane proteins that are found in the INM, some of which have already been extensively characterized (see below) and many that vary significantly in their expression between different cell types (Korfali et al., 2012; Schirmer, Florens, Guan, Yates, & Gerace, 2003).

In metazoans, another component of the nuclear periphery is the lamina that underlies the INM (Dwyer & Blobel, 1976; Fig. 1.1). The nuclear lamina is a thin proteinaceous meshwork, which in most cells is some 10–20 nm thick, although the thickness can significantly increase in different cell types (Hoger, Grund, Franke, & Krohne, 1991). The principal components of the

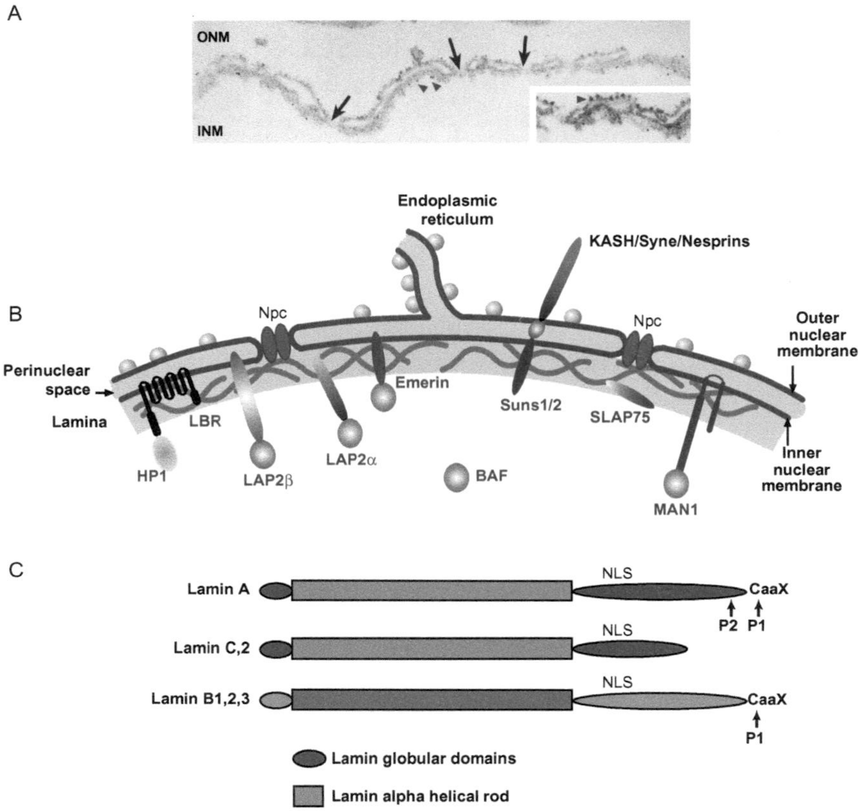

Figure 1.1 The nuclear envelope (NE), lamina, and some of the more extensively characterized proteins associated with the lamina and NE. Panel A shows an electron micrograph of isolated rat liver NEs. The NE consists of the contiguous inner and outer nuclear membranes (INM and ONM, respectively), which are connected where the nuclear pore complexes traverse the two membranes. The arrows show the location of nuclear pore complexes. The red (dark gray in print) arrowheads indicate immunogold-labeled antibodies to lamin A decorating the underside of the INM. Panel B is a diagram of the NE/lamina with the localization of some on the proteins associated with the NE. Underlying the inner nuclear membrane is the 20–50-nm thick lamina, comprised of the lamins. Associated with the ONM, INM, and lamina are various proteins, with many connecting the INM to chromatin and DNA, for example, BAF and HP1. Panel C illustrates the structures of the mammalian lamins consisting of a central rod domain flanked at either end by globular domains. Only lamin A undergoes a second proteolytic cleavage step (P2) that removes the farnesylated cysteine. P, protease cleavage site; NLS, nuclear localization sequence. (See the color plate.)

lamina are the type V intermediate filament proteins—the lamins, which are found exclusively in the nucleus (Gerace & Huber, 2012). The lamins consist of a central α-helical rod domain flanked by a head and tail globular domain (Fisher, Chaudhary, & Blobel, 1986; McKeon, Kirschner, & Caput, 1986). The large carboxy-terminal globular domain of about 200–300 amino acid residues has at its core an immunoglobulin (Ig)-like β-fold, together as well as a nuclear localization sequence (Dhe-Paganon, Werner, Chi, & Shoelson, 2002; Krimm et al., 2002). In mammals, the lamins are grouped into two classes, A- (A, AΔ10, and C) and the B-types (B1, B2, and B3) (Peter et al., 1989; Vorburger, Lehner, Kitten, Eppenberger, & Nigg, 1989). Most adult mammalian somatic cells contain four major lamin proteins, A, B1, B2, and C. A single gene, *LMNA*, encodes the A-type lamins, which are generated by alternate splicing of a common pre-mRNA (Lin & Worman, 1993, 1995; Machiels et al., 1996). A minor spiced *LMNA* variant, lamin C2, is also produced in the testis (Furukawa, Inagaki, & Hotta, 1994). In mammals, most cells express the A-type lamins, both lamin A and C proteins at roughly equimolar amounts. However, in neurons of the central nervous system (CNS) lamin A protein is largely absent due translational inhibition by the microRNA (miR-9) that that binds to the 3′ end of the longer lamin A transcript, and not to the shorter lamin C transcript (Jung, Tu, et al., 2014). Separate genes encode lamins B1 and B2 with lamin B3 being produced as a minor spliced variant of lamin B2 and as with lamin C2 is found in the testis (Furukawa & Hotta, 1993; Lin & Worman, 1995).

During translational processing, lamins A, B1, and B2 undergo a series of sequential modifications (Fig. 1.2). At the C-termini of lamin A and both the B-type lamins is a CaaX motif (where "C" is cysteine, "a" is an aliphatic

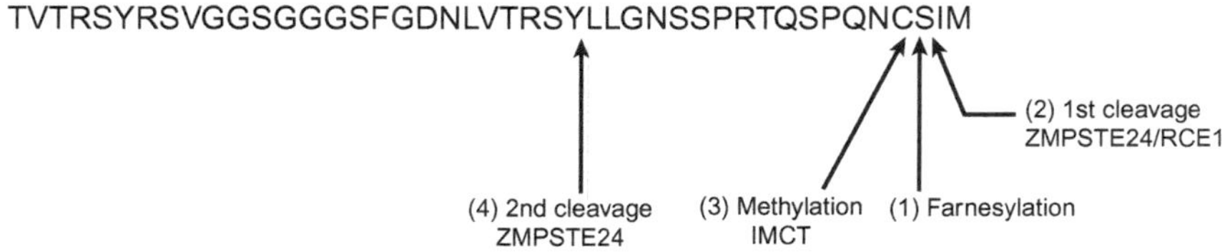

Figure 1.2 The sequential posttranslational processing of the carboxy terminal of prelamin A. The C-terminal amino acid sequence is shown, as is the order of each step with farnesylation of the cysteine in the CaaX domain being the first step. The second step involves cleavage of the terminal—SIM peptide by either Rec1 or Zmpste24. Thirdly, the terminal cysteine is methylated by IMCT. The fourth and last step involves cleavage of the terminal, farnesylated and methylated 15 amino acid peptide by Zmpste24 to produce the mature lamin A protein.

amino acid, and "X" can be many different amino acids, although usually is a Met) that undergoes farnesylation and carboxy methylation. C-terminal processing of lamins A, B1, and B2 is initiated soon after their translation (Beck, Hosick, & Sinensky, 1988; Holtz, Tanaka, Hartwig, & McKeon, 1989; Kitten & Nigg, 1991; Sinensky et al., 1994; Wolda & Glomset, 1988). Posttranslational modification of the lamins begins with the addition of a farnesyl lipid to the cysteine by the cytosolic enzyme protein farnesyl transferase. Subsequently, the –aaX amino acids are removed at the ER surface by specific endoproteases, ZMPSTE24 (FACE1) and/or FACE2 (RCE1). The isoprenylated cysteine is then methylated at the ER by a membrane methyltransferase, isoprenylcysteine carboxyl methyltransferase (ICMT). These modifications result in the carboxyl-terminal domains of the lamins becoming hydrophobic, supposedly to enhance their association with the INM, and are thought to be required for the assembly of lamins A, B1, and B2 into the lamina (Holtz et al., 1989). Prelamin A, but not the B-type lamins, then undergoes a final endoproteolytic cleavage, by ZMPSTE24, that removes an additional 15 amino acids, including the farnesylated and methylated cysteine, to produce the mature non-farnesylated lamin A (Gerace, Comeau, & Benson, 1984; Pendas et al., 2002; Weber, Plessmann, & Traub, 1989; Fig. 1.2).

Why there are 2 A-type lamins and why lamins undergo extensive posttranslational processing is unclear, as mice engineered to only produce either mature lamin A or C are overtly normal revealing a high degree of redundancy between the 2 A-type lamins (Davies, Fong, Yang, Coffinier, & Young, 2009; Fong, Ng, et al., 2006). Furthermore, the direct production of mature lamin A (completely bypassing all prelamin A processing) results in no discernable pathology in mice, as does exclusive production of non-farnesylated prelamin A, although these latter mice eventually develop cardiomyopathy (Coffinier et al., 2010). Similarly, farnesylation of lamin B2 is not essential for normal development, although it is essential for lamin B1. Embryos expressing non-farnesylated lamin B1 have lower levels of laminB1 protein. In the nuclei of the migrating neurons, lamin B1 farnesylation is required to retain chromatin at the NE, and lack of chromatin attachment is associated with perinatal death (Jung, Nobumori, et al., 2013).

Within mammalian cells the 3D organization of the interphase nuclear lamina remains to be established. In *Xenopus* oocytes, the lamina is comprised of a single oocyte-specific B-type lamin, lamin LIII, and appears as a highly structured filament lattice (Aebi, Cohn, Buhle, & Gerace, 1986).

In somatic cells, the structure of the lamina is uncertain due to chromatin masking the nuclear face of the NE. Although the somatic cell lamina appears to be filamentous, recent high-resolution light microscopy studies have suggested that the A- and B-type lamins form their own separate but interacting networks (Shimi et al., 2008). Consistent with this notion, ectopic expression of lamin A and lamin B2 in *X. laevis* oocytes produces filaments with distinct morphologies (Goldberg, Huttenlauch, Hutchison, & Stick, 2008). The organization of the lamina in mammalian somatic cells may therefore be more complex and less regular than that suggested by the frog oocyte lamina.

1.2. The physical functions of the lamins

The derivation of mice deficient in the different lamins has provided some significant insights to understanding lamin function. We can now be certain that lamins, in particular the A-type lamins, have a structural role in that they maintain nuclear shape and rigidity, as well a conferring cellular resistance to mechanical strain (Lammerding et al., 2006, 2004). Loss of the A-type lamins, or the inclusion of specific mutations, results in the typically oval nuclear shape becoming distorted to varying degrees (Raharjo, Enarson, Sullivan, Stewart, & Burke, 2001; Sullivan et al., 1999). An extreme form of this distortion, which occurs naturally, is found in granulocytes lacking lamin A (Olins et al., 2008). The lack of A-type lamins makes nuclei more flexible, a feature that is of importance in highly migratory cells, such as granulocytes and macrophages, that have to squeeze through narrow gaps in the extracellular matrix (ECM) and between other cells (Rowat et al., 2013; Shin, Mendez-Lopez, et al., 2013). In fibroblasts that are deficient in A-type, but not B-type lamins, the nuclei are more deformable and less resistant to mechanical strain, such that repetitive stretching results in the cells becoming physically weaker and more susceptible to mechanically induced necrosis or apoptosis (Lammerding et al., 2006). Different tissues express A-type lamins at levels in direct proportion the overall stiffness of the tissue, with softer tissues such as brain and fat expressing lower lamin A levels, whereas stiff tissues, such as cartilage and bone express higher levels. The tissue variation in lamin levels may also prevent nuclear distortion and disruption that is associated with the degree of physical stress to which different tissues are subjected (Swift et al., 2013). The correlation between A-type lamin levels and the stiffness of the extracellular environment may also influence cell differentiation (e.g., of mesenchymal stem cells), with a soft environment supporting fat differentiation and a stiffer substrate favoring bone/cartilage formation (Swift et al., 2013).

An intriguing aspect of the mechanical role of the lamins is that mutation or loss of *LMNA* reduces cytoplasmic stiffness/compliance (Lammerding & Lee, 2005; Lee et al., 2007). Although the precise molecular basis to this remains unclear, it may well be due to disruption of NE-cytoskeletal connections (Maniotis, Chen, & Ingber, 1997). These connections are mediated, at least in part, by the LInker of the Nucleoskeleton and Cytoskeleton (LINC) complex, which tethers the nucleus to various cytoplasmic organelles, and the three different cytoskeletal networks (see Horn, Chapter 6 for a detailed review of the LINC complex in this volume).

1.3. Lamins in regulating chromatin organization and gene expression

Another important function of the lamins is to act as a scaffold for the correct localization of nuclear proteins, either by concentrating nucleoplasmic proteins to the nuclear lamina, for example, Slap75 (Roux, Kim, Raida, & Burke, 2012) and possibly the transcription factors c-fos, Oct-1, and Lmo7 (Columbaro et al., 2013; Dedeic, Cetera, Cohen, & Holaska, 2011; Ivorra et al., 2006), or to the INM by selective retention (Soullam & Worman, 1995). Some, but by no means all, INM proteins such as emerin (EMD) and certain Lap1 isoforms, concentrate in the INM by virtue of their interaction with the A-type lamins. In the absence of A-type lamins, these proteins redistribute from the INM to the other contiguous membranes systems; the ONM and ER and are therefore lost from the nuclear interior (Powell & Burke, 1990; Sullivan et al., 1999).

In metazoans, it has long been recognized that the electron dense form of chromatin (heterochromatin) tends to localize to the nuclear periphery, whereas the lighter less dense form (euchromatin) is positioned more to the nuclear interior. Heterochromatin has generally been regarded as containing transcriptionally silent parts of the genome and is associated with repressive chromatin marks, whereas euchromatin is more transcriptionally active. Consequently much effort has centered on determining whether the localization of these two forms of chromatin is an important determinant of their transcriptional activity (Bickmore, 2013).

The nuclear lamins may indirectly have a role in the tethering of transcriptionally silent parts of the genome, as a well as heterochromatin to the nuclear periphery. Lamina-associated domains (LADs) are stretches of DNA, which can be many hundreds of kilobases long, and share a physical proximity to the lamins (Guelen et al., 2008). Within these stretches, which are often gene poor, many of the genes are transcriptionally repressed although this repression may itself, at least in cells of the early mouse embryo, be independent of lamins *per se* (Kim et al., 2011). Although LADs were first identified in mammalian cells using DamID genomic mapping in conjunction with lamin B1, it appears unlikely the LADs actually interact directly with the B-type lamins, rather their interaction may be through other Lamin/NE associated proteins (Burke & Stewart, 2013).

Some of the better characterized INM proteins, including the LEM domain group (LEMD1-4) that include EMD, Man1, and Lap2 interact with chromatin/DNA by binding to the small protein BAF (barrier-to-autointegration factor). BAF in turn crosslinks and compacts DNA in a sequence independent manner and therefore has been implicated in regulating the formation and localization of heterochromatin (Margalit, Brachner, Gotzmann, Foisner, & Gruenbaum, 2007). Other NE proteins such as the lamin B receptor, mediate DNA/chromatin attachment to the NE by binding to heterochromatin via the chromobox protein HP1 (Ye, Callebaut, Pezhman, Courvalin, & Worman, 1997), indicating the existence of multiple mechanisms, with some redundancy, in regulating heterochromatin localization. In mice, tethering of heterochromatin to the periphery is shared between the LBR (via HP1) and A-type lamins, with the latter, probably to a large extent, mediating their role through the LEM domain group of proteins. Loss of both LBR1 and lamin A in the same nuclei results in chromatin inversion, with heterochromatin dissociating from the nuclear periphery and collapsing into the center of the nucleus (Solovei et al., 2013).

In yeast, and the worm *C. elegans*, interactions between the nuclear periphery, including NPCs and chromatin are important in regulating transcription (Mattout et al., 2011; Steglich, Sazer, & Ekwall, 2013). However, in mammalian nuclei, the situation is less clear. Attempts to determine whether the NE/lamina has a role in transcriptional regulation in mammalian cells has met with mixed results (Finlan et al., 2008; Kumaran, Thakar, & Spector, 2008; Reddy, Zullo, Bertolino, & Singh, 2008). Artificially tethering some genes to the nuclear periphery results in their repression, but with other genes repression was not observed. Similarly, transcriptional repression does not require relocation to the NE (Kim et al., 2011). However, tethering of chromatin to the NE maybe required for DNA repair (Mekhail & Moazed, 2010; Oza, Jaspersen, Miele, Dekker, & Peterson, 2009) and during mammalian spermatogenesis telomeres are attached to the LINC complex at the nuclear periphery during prophase 1 of meiosis (Horn, Kim, et al., 2013).

1.4. Developmental and tissue-specific variation of the lamins and NE protein expression

In mammals, lamin expression is developmentally regulated. All nucleated cells express at least one B-type lamin, whereas A-type lamins are absent during the early pre- and postimplantation embryonic stages and in embryonic stem (ES) cells, with these types expressing high levels of Lamins B1 and B2 (Rober, Sauter, Weber, & Osborn, 1990; Stewart & Burke, 1987). Initially, A-type lamin expression is transiently expressed in fertilized eggs, disappears during preimplantation and early postimplantation development, but then reappears, as different tissues form in the postimplantation embryo, with some tissues not expressing A-type lamins until after birth, for example, the CNS and others seemingly never expressing the A-type lamins (e.g., hematopoietic cells). Surprisingly, the expression of lamins in ES cells appears to be nonessential, as ES cells lacking all three lamins proliferate normally in culture, maintain euploidy and differentiate into multiple cell types, including those with fibroblastic, cardiac, and neuronal phenotype (Kim, Zheng, & Zheng, 2013).

The levels of lamins A and C vary between different somatic cell types and tissues (Cutler, Sullivan, Marcus-Samuels, Stewart, & Reitman, 2002; Swift et al., 2013; Zhang et al., 2011). In certain stem cell populations in adult mice, in particular the nucleated cells of the hematopoietic system, and the crypt cells in the gut epithelium A-type lamins are absent (Solovei et al., 2013; Fig. 1.3). The lamina composition and NE changes during embryogenesis and varies between different adult tissues, particularly those that are maintained by stem cells (Korfali et al., 2012; Rober et al., 1990; Rober, Weber, & Osborn, 1989). During murine development, A-type lamin expression is apparently dispensable, with mice lacking lamin A being overtly indistinguishable from normal siblings at birth, possibly because of the shared redundancy with the LBR receptor that is expressed, or even reexpressed in cells engineered to lack lamin A (Solovei et al., 2013; Fig. 1.3). Problems arising from the loss of, or as a consequence of specific mutations, appear after birth with loss of lamin A resulting in early (by 3 weeks) postnatal death, which correlates to some extent with the normal silencing of LBR expression in many postnatal tissues. There is some redundancy between LBR and the A-type lamins, since *Lmna*$^{-/-}$/*Lbr*$^{-/-}$ double

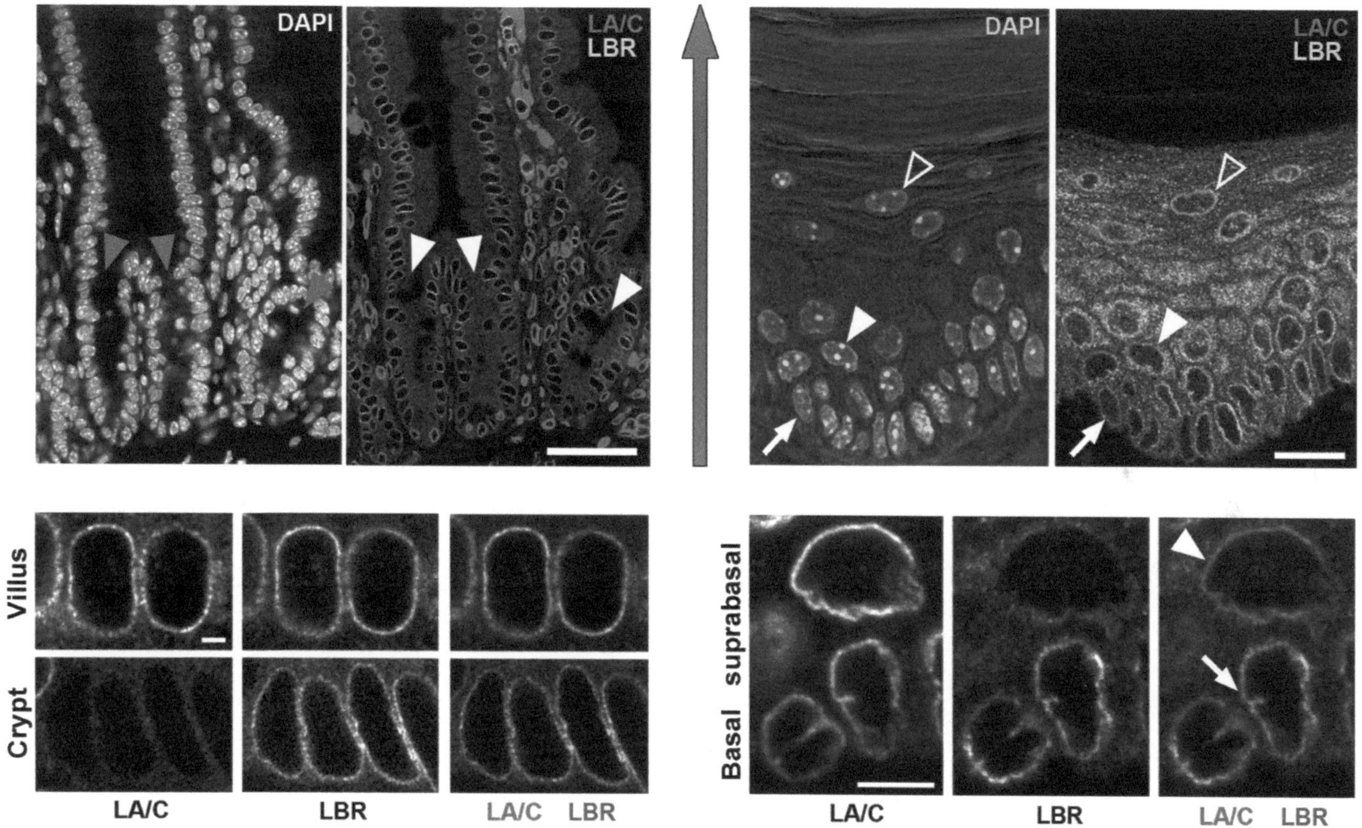

Figure 1.3 Sequential changes in expression of the LBR and lamin A proteins in the nuclei of the gut villous epithelium epidermal layer of the skin. In the left panels, the nuclei in the intestinal crypt epithelial cells express the lamin B receptor (LBR). As the cells move up the villus the epithelial nuclei coexpress LBR and lamin A and then further up the villus LBR expression is lost and with lamin A being exclusively expressed. Similarly, in the skin epidermis (right panels), the nuclei in the basal layer of the epidermis only express the LBR. In the suprabasal layer, the nuclei express both LBR and lamin A and the most distal nuclei express only lamin A (Solovei et al., 2013). (See the color plate.)

nulls die at birth (Solovei et al., 2013). In contrast, human fibroblasts lacking lamin A are difficult to propagate and in human development, loss of lamin A may have severe consequences during embryogenesis, as a fetus homozygous for a *LMNA* mutation carrying a premature stop codon, was a preterm lethal (van Engelen et al., 2005).

Less is known about cell and tissue variation of the B-type lamins. Human and murine fibroblasts with reduced levels of lamin B1 undergo proliferative arrest, and senescence (Dreesen et al., 2013; Freund, Laberge, Demaria, & Campisi, 2012; Shah et al., 2013; Shimi et al., 2011). *In vivo* laminB1 levels decline as keratinocytes mature in the skin epidermis (Dreesen et al., 2013; Shimi et al., 2008; Venables et al., 2001). In contrast, lamin B2 levels do not change as cells senesce, and loss of laminB2 has no effect on fibroblast proliferation (Dreesen et al., 2013). Unlike the A-type lamins, the B-type lamins are not mutually redundant during embryogenesis, as both are required for normal development of the CNS (Coffinier et al., 2011; Vergnes, Peterfy, Bergo, Young, & Reue, 2004). Loss of lamin B1 reduces cell number in and layering of the cerebral cortex, together with hypothalamic abnormalities and a reduction in cerebellar size due to abnormal neural migration. Additional defects were noted in other tissues such as the lungs and skeletal system. Lamin B2 is also necessary for neuronal migration during CNS embryogenesis, with the layering of neurons within the cerebral cortex becoming abnormal, resulting in a smaller cerebellum devoid of folds (Young, Jung, Coffinier, & Fong, 2012). However, loss of lamin B2 does not result in a smaller body size or have effects in other tissues as noted for lamin B1. In adults, B-type lamin expression appears to be nonessential in some tissues, such as the skin epidermis and liver, which can dispense with both B-type lamins without the development of any overt pathology (Yang et al., 2011). Together, these findings reveal that an absolute dependence on Lamin expression in mammals varies between different cell types and that early embryos, including pluripotent cells derived from them and some of their differentiated derivatives, do not require any of the lamins (Kim et al., 2013). Postimplantation development can proceed, independent of lamin A, to birth, at least in mice, whereas development of the CNS is dependent on the B-type lamins. Only in postnatal mice, do cells and tissues develop a significant dependency on the A-type lamins. Which specific tissues have an absolute dependence on the A-type lamins should, in the near future, become apparent with the derivation of mouse lines carrying conditional (floxed) alleles of *Lmna* (Kim & Zheng, 2013; Solovei et al., 2013).

2. THE LAMINOPATHIES

The principal stimulus that led to an increased interest into the functions of the NE/lamina were the discoveries that nearly 30 inherited diseases and anomalies, ranging from muscular dystrophies, premature aging-like syndromes, white fat, and skeletal homeostasis, are caused by mutations in the *LMNA* gene, as wells as in genes encoding some of the NE associated proteins (Worman & Bonne, 2007). The largest of this group of these diseases, the primary laminopathies, is associated with *LMNA* mutations and to a much lesser extent mutations in *LMNB1*. The secondary laminopathies are caused by mutations in the gene encoding the enzyme *ZMPSTE24*, the endoprotease essential for the posttranslational maturation of prelamin A to mature lamin A (Agarwal, Fryns, Auchus, & Garg, 2003; Bergo et al., 2002; Corrigan et al., 2005; Moulson et al., 2005; Navarro et al., 2005; Pendas et al., 2002).

The primary *LMNA* laminopathies can be classified into three groups: the first and largest consists of diseases affecting striated muscle. These include the autosomal dominant form of Emery-Dreifuss muscular dystrophy (AD-EDMD), dilated cardiomyopathy (DCM), and limb-girdle muscular dystrophy 1B (LGMD1B). Mutations causing these diseases are mostly missense mutations and are found throughout the 11 exons of the *LMNA* gene. Included in this group is a rare peripheral neuropathy (R298C), a recessive axonal form of Charcot–Marie–Tooth disease (AR-CMT2A), resulting in demyelination of motor nerves and muscle wasting in the limbs (De Sandre-Giovannoli et al., 2002).

The second group of laminopathies has minimal, if any effects on muscle, but mainly influences white fat distribution and skeletal development. The two diseases are Dunnigan-type familial partial lipodystrophy (FPLD) and mandibuloacral dysplasia (MAD) (Cao & Hegele, 2000; Novelli et al., 2002; Speckman et al., 2000; Vigouroux et al., 2000). The third group of diseases is the premature aging or progeroid syndromes, Hutchinson-Gilford progeria syndrome (HGPS) and a few cases of atypical Werner's syndrome (Bonne & Levy, 2003; De Sandre-Giovannoli et al., 2003; Eriksson et al., 2003; Oshima, Garg, Martin, & Kennedy, 2003). To date, over 460 mutations have been identified in the *LMNA* gene with a database on the "nuclear envelopathies" can be found at (http://lmna.umd.be/*LMNA*/). This has raised the issue as to how do these different mutations that all affect the same ubiquitously expressed protein, result in such a range of tissue-specific diseases?

3. MOUSE MODELS FOR THE A-TYPE LAMINOPATHIES

3.1. Laminopathies affecting striated muscle

The first mutations linked to the *LMNA* gene were those causing the AD-EMD2 (Bonne et al., 1999). As with the X-linked form, which results from mutations in the INM transmembrane protein emerin (*EMD*– see below) (Bione et al., 1994), AD-EMD is characterized by progressive wasting of specific muscle groups, particularly within the limbs, as well as cardiac conduction defects that distinguish it from other muscular dystrophies. AD-EDMD *LMNA* mutations display wide variation in their severity, with the conduction defects, that are potentially fatal, being more acute than those observed arising from the X-linked form of the disease (Morris, 2001). Some *LMNA* mutations result in dilated cardiomyopathy with conduction system disease (DCM-CD1) with no apparent effect on skeletal muscle (Fatkin et al., 1999). Since DCM-CD1 conduction defects are similar to those in patients with AD-EDMD, this has suggested that *LMNA* linked cardiomyopathy represents one extreme of a phenotypic continuum in which skeletal muscle involvement is not observed. Dilation of the heart chambers, hypertrophy, arrhythmic conduction defects, and cardiac arrest are all associated with *LMNA* cardiomyopathy mutations. At the other end of the *LMNA* mutation spectrum, limb-girdle muscular dystrophy 1B (LGMD1B), is associated with tendon contractures, with fewer or no cardiac complications (Muchir et al., 2000). Muscle wasting in the proximal limbs is the main clinical feature of LGMD1B. The heterogeneity in the disease phenotypes, even among members of a single family carrying the same *LMNA* mutation, suggests that genetic or environmental modifiers influence the pathological outcome (Brodsky et al., 2000; Granger et al., 2011). Diseases affecting striated muscle comprise about 60% of the laminopathies. However, only ~50% of patients diagnosed with AD-EDMD or EDMD have EMD or *LMNA* mutations, indicating that mutations in other genes, including other NE proteins (e.g., LAP2, FHL1, the SYNE1 and two genes, and potentially LAP1), may account for the remaining 50% of patients (Granger et al., 2011; Gueneau et al., 2009; Shin, Spinler, et al., 2013; Taylor et al., 2005; Zhang, Bethmann, et al., 2007).

To date, seven lines of mice, each with a different type of *Lmna* mutation, that result in early postnatal lethality, muscular dystrophy/immaturity, and/or cardiomyopathy have been derived. The first line of mice (*Lmna*$^{Sul/Sul}$), develop normally to birth, but then show postnatal growth retardation starting at 2 weeks and by 6–7 weeks of age are dead. Their demise is associated with retarded muscle development, with smaller diameter muscle fibers, muscular dystrophy, and cardiomyopathy (Cohen et al., 2013; Nikolova et al., 2004; Sullivan et al., 1999). However, a recent reexamination of these mice indicated that the induced mutation (deletion of exons 8–11) may not be a true null, as very low levels of a truncated form of lamin A protein, lacking most of the C-terminal globular domain, may still be expressed (Jahn et al., 2012). Although nearly all the identified human AD-EDMD mutations function in a dominant manner, it is still unclear as to why many of the mouse *Lmna* mutations need to bred as homozygotes for a phenotype to manifest. However, 1-year-old heterozygous *Lmna*$^{+/Sul}$ mice do develop atrioventricular (AV) conduction defects with atrial and ventricular arrhythmias, analogous to those in human *LMNA* mutations. *Lmna*$^{+/Sul}$ cardiomyocytes have impaired cell and sarcomere contractility, with the AV node cells undergoing apoptosis and fibrosis (Wolf et al., 2008).

Several other additional mutant lines including, *Lmna*$^{\Delta K32/+}$ (Bertrand et al., 2012), *Lmna*$^{Gt/-}$ (Kubben et al., 2011), Disheveled Hair and Ear (*Dhe*)-a spontaneous *Lmna* L52R mutation (Odgren et al., 2010), and two conditional floxed lines *Lmna*$^{Flx/Flx}$ (Kim & Zheng, 2013; Solovei et al., 2013), that when constitutively deleted, appear to be a true nulls, have been described. All of these additional lines, that as homozygotes, develop overtly normally to birth. However, within the first 2 weeks of postnatal development, the homozygotes are growth retarded, with retardation being associated with defective skeletogenesis, skeletal hypomineralization (Li et al., 2011), immature muscle and heart development, together with metabolic complications associated with cachexia (Cutler et al., 2002). By 3 weeks, the homozygotes are dead, with the *Lmna*$^{Sul/Sul}$ dying around 5–6 weeks, and which therefore express the mildest phenotype. In comparison with humans, there is no report of a viable patient completely lacking lamins A/C, apart from an individual that was probably haploinsufficient for *LMNA* (Bonne et al., 1999), and a fetus homozygous for a premature stop codon in *LMNA* that died late in gestation (van Engelen et al., 2005). Overall, these different lines of mice support the notion that uterine development of mice depleted in *Lmna* can seemingly proceed normally. But once born, postnatal development/maturation becomes acutely dependent on the A-type lamins. It remains to be determined whether the adult laminopathies arise as a consequence of the *LMNA* mutations having nonrecognizable effects on embryonic development but having latent effects on postnatal/adult stages, or whether the pathologies arise *de novo* in adults. Again, the conditional *Lmna* alleles may help in addressing this issue.

In addition to the above mutations that have severe early acting pathologies, two other lines of mice were established, each carrying a missense mutation *H222P* (Arimura et al., 2005) and *N195K* (Mounkes, Kozlov, Rottman, & Stewart, 2005). The *H222P* mutation was originally identified in families with AD-EDMD, whereas the *N195K* mutation was identified independently in two different patients, both diagnosed with DCM (Fatkin et al., 1999; van Tintelen et al., 2007). Adult male mice homozygous for the *H222P* mutation, develop a stiff walking posture, with cardiac fibrosis, chamber dilation, with conduction defects, and die by 9 months of age. Female homozygotes also exhibit these pathologies but survive for longer. The $Lmna^{H222P/H222P}$ mice represent a good model for studying laminopathies affecting striated muscles as they develop a dystrophic condition in both skeletal and cardiac musculature that is similar to the human disease.

The missense mutation *LMNA N195K* acts in an autosomal-dominant manner and causes DCM in humans. Mice homozygous for this mutation, develop cardiac conduction defects consistent with DCM-CD1, with death occurring at three months due to arrhythmia. These mice were largely free of muscular dystrophy. The transcription factor Hf1b/Sp4 and the gap junction proteins connexin 40 and connexin 43 were misexpressed and/or mislocalized in the mutant hearts. Desmin staining revealed a loss of organization in the sarcomeres and intercalated disks, suggesting that *LMNA* mutations causing cardiomyopathy, do so by disrupting the internal organization and cellular connectivity of the cardiomyocyte and/or altering the expression of transcription factors essential to normal cardiac development and function. One such factor could be the mechanosensitive transcription factor megakaryoblastic leukemia 1 (MKL1). Intracellular localization of MKL1 is regulated via changes in actin polymerization. Normally, MKL1 is localized in the cytoplasm. Mechanical stimulation frees MKL1 from its association with G actin resulting in increased nuclear import. In cells from the $Lmna^{Sul/Sul}$, $Lmna^{N195K/N195K}$ and $Emd^{-/Y}$ mice, MLK nuclear translocation is impaired, indicating that functional lamina and the NE protein EMD are required for MLK nuclear translocation (Ho, Jaalouk, Vartiainen, & Lammerding, 2013). However to what extent aberrant

MLK localization influences muscle and heart pathologies is unclear as the *Lmna*$^{N195K/N195K}$ and *Emd*$^{-/Y}$ mouse lines have minimal to no detectable muscle pathology and in the *Emd*$^{-/Y}$ mice no cardiac defects (Melcon et al., 2006; Mounkes et al., 2005).

3.2. Using mutant lines to identify potential therapies for the cardiac laminopathies

One of the objectives in deriving mouse models of disease is to explore potential avenues that may offer therapies for patients. Correcting *LMNA* mutations by gene therapy has not been possible. Even if it were, ablation of the mutant dominant allele that is the basis for the majority of the muscle/cardiac laminopathies, may not be effective as it would render the patient *LMNA* haploinsufficient, and the (limited) evidence suggests that haploinsufficiency itself is pathological. Consequently, attention has focused on identifying the more tractable molecular consequences arising from the laminopathies (e.g., dysregulation of signaling pathways) that may underlie the diseases.

An informatics based gene expression analysis on the hearts from the H222P mouse line, as they developed cardiomyopathy, revealed increased activation of the extracellular signal-regulated kinase1/2 (ERK1/2) pathway. Abnormal ERK1/2 activation was also detected in heart tissue from human laminopathy patients. Treatment of the mutant mice with the MAP kinase (MEK) inhibitor markedly improved cardiac function, the associated skeletal muscle pathology and reduced fibrosis, resulting in a modest but significant prolongation in survival (Muchir, Pavlidis, Bonne, Hayashi, & Worman, 2007; Muchir, Pavlidis, Decostre, et al., 2007; Muchir, Shan, Bonne, Lehnart, & Worman, 2009; Wu, Iwata, Homma, Worman, & Muchir, 2014). Both *Lmna*$^{H222/H222P}$ and *Lmna*$^{Sul/Sul}$ mutations, activate the AKT signaling pathway, which has been implicated in contributing to the pathogenesis of cardiomyopathy (Choi et al., 2012). Among the targets of AKT activation is the mammalian target of rapamycin complex 1 (mTORC1) that positively regulates cell growth, protein synthesis, and metabolism. Both the *Lmna*$^{H222P/H222P}$ and the *Lmna*$^{Sul/Sul}$ mutations exhibit increased activation of the mTOR pathway. Inhibition of mTORC, by rapamycin, is a classic means to enhancing longevity in invertebrate models of aging and has the same effect on outbred mice (Harrison et al., 2009). Autophagy, the lysosome-mediated degradation and recycling of cellular components, is a protective mechanism against stress and is reduced by mTORC1. Drug-mediated inhibition of mTORC1, with temsirolimus or rapamycin, enhanced autophagy in the *Lmna* mutants and significantly increased their mean lifespans and heart function (Ramos et al., 2012). Although activation of both the ERK1/2 and AKT pathways converge on mTORC1, simultaneous inhibition of both pathways did not result in an enhanced improvement in the $Lmna^{H222/H222P}$ mice, although the number of mice tested was quite limited (Choi et al., 2012).

The $Lmna^{Sul/Sul}$ line, as well as another mutant *Lmna* mice (Progeroid Δ9), have increased protein levels of the LINC complex component Sun1 and the lamin associated protein, Lap2α. A similar increase in Sun1 levels was also noted in HGPS fibroblasts. When either Sun1 or Lap2α (neither of which when deleted result in lethality) were reduced, by deleting allele, in the lamin mutant lines, the reduction increased the survival, body weight and reduced the dystrophy of the $Lmna^{Sul/Sul}$ mice, with loss of both Sun1 alleles having an even more marked improvement (Chen et al., 2012). Deletion of Sun1 in other *Lmna* mutants increased their longevity and siRNA knockdown of Sun1 enhanced the proliferation of HGPS fibroblasts. Deletion of Lap2α in $Lmna^{Sul/Sul}$ mice also increases longevity and postnatal growth, but not to the same extent as loss of Sun1 (Cohen et al., 2013). Lap2α has so far only been effective at improving the pathology of the $Lmna^{Sul/Sul}$ line, as deletion did not enhance the growth or viability of the $Lmna^{\Delta K32/+}$ line (Pilat et al., 2013). How these *Lmna* mutations result in Sun1 and Lap2α accumulation is unclear, although reduced protein turnover maybe one explanation. Nevertheless, it is intriguing that rapamycin treatment, that increases the lifespan and growth of the $Lmna^{Sul/Sul}$ mice, does so to an extent comparable to deletion of one Sun1 allele, suggesting some shared mechanism. In the $Lap2\alpha^{-/-}$ mice the TGFβ signaling was hyperactivated, resulting in increased nuclear levels of Smad2/3; however, the effects of Sun1 deficiency has on the TGFβ signaling pathway is not known. Together, these findings reveal that focusing on signaling pathways such as the Tgfβ pathway and regulating Sun1/Lap2α levels may be potential therapeutic alternatives to ameliorating some of the pathologies resulting from *LMNA* mutations.

3.3. Laminopathies affecting adipose and skeletal tissues

The second group of laminopathies, FPLD and MAD do not affect muscle tissue (Cao & Hegele, 2000; Novelli et al., 2002; Shackleton et al., 2000). FPLD is inherited as an autosomal-dominant trait, with about 85% of the cases (from ~200 affected individuals) being associated with a missense mutation at Arg482. FPLD is characterized by the loss of subcutaneous white adipose tissue from the limbs, gluteal region, and areas of the trunk, with a concomitant accumulation of white adipose tissue in the neck, face, and abdominal regions. These changes begin at puberty, suggesting a possible hormonal influence on the initiation of disease phenotypes. Progressive insulin resistance, resulting in type 2 diabetes, accompanies the remodeling and redistribution of adipose tissues. The patients are hyperlipidemic and exhibit an increased susceptibility to atherosclerotic heart disease (Vigouroux et al., 2000).

The aberrant adipose tissue redistribution in lipodystrophy may arise due to an autonomous defect in subsets of mesenchymal or adipocyte precursors, since white fat is a heterogeneous tissue (Fruhbeck, 2008). In lipodystrophy, the body may be attempting to compensate for this loss of fat in some areas by an accumulation of fat in others. Neither the levels of *Lmna* expression nor the ratio of lamins A and C vary significantly in the subcutaneous, omental, and neck fat depots of normal individuals, suggesting that an intrinsic fat depot-specific pattern of A-type lamin expression does not underlie the fat depot abnormalities associated with FPLD (Cutler et al., 2002; Lelliott et al., 2002). Although $Lmna^{Sul/Sul}$ mice exhibit reduced stores of white fat, as do the many of the other postnatal lethal *Lmna* mutants, the $Lmna^{Sul/Sul}$ are cachexic and do not exhibit the hallmark insulin and plasmid lipid alterations found in humans with FPLD (Cutler et al., 2002).

A rare autosomal recessive mutation in the carboxy-terminal globular domain of A-type lamins is responsible for MAD (Novelli et al., 2002; Simha, Agarwal, Oral, Fryns, & Garg, 2003), with 94% cases having a missense mutation at residue 527 (R527H). MAD is a disease with many of the metabolic and fat depot redistribution phenotypes of lipodystrophy, but with an expanded set of skeletal abnormalities, including osteolytic lesions in the bones. The primary sites of skeletal malformations involving bone reabsorption in MAD are the craniofacial region, termini of the digits, and clavicles.

There have been no reports of these diseases being modeled in mice by knock-in of the same mutations. However a transgenic line of mice expressing a *lmna*R482W mutant cDNA was derived and suggested that FPLD maybe due to a failure in adipose regeneration, although the affect of an overall increase in lamin A levels, and the fact that the AP2 promoter used to drive expression of the cDNA, which is now recognized to be widely expressed in the CNS, were not considered (Wojtanik et al., 2009).

3.4. Laminopathies affecting axonal myelination

A single autosomal recessive mutation in the rod domain of *LMNA* results in the peripheral neuropathy Charcot–Marie–Tooth syndrome Type 2A (AR-CMT2A) (Chaouch et al., 2003; De Sandre-Giovannoli et al., 2002). Families homozygous for the R298C variant have no deep-tendon reflexes and exhibit distal amyotrophy, motor deficits, and a loss of large myelinated nerve fibers. The muscular weakening associated with this disease, though maybe a consequence of defective neuromuscular junction function and subsequent muscle atrophy is unknown. Although neurons in the sciatic nerve of the $Lmna^{Sul/Sul}$ mice showed extensive demyelination (De Sandre-Giovannoli et al., 2002), a mouse line homozygous for the R298C mutation exhibited no overt neurological defects, despite *Lmna* transcript levels being reduced. Increased levels of the Peripheral myelin protein 22 (Pmp22) in sciatic nerves were noted that maybe of significance, since Pmp22, when amplified, is associated with another neurological condition, CMT-1A (Poitelon et al., 2012).

3.5. Progeria

HGPS is a rare dominantly inherited disease in which the affected infants show growth retardation, loss of subcutaneous fat, alopecia, reduction in bone density, and weakened muscle function (Gordon et al., 2007; Merideth et al., 2008). The average age of death in HGPS is in the mid-teens, due to cardiovascular failure or stroke caused by vascular occlusion and arthrosclerosis. Atherosclerosis in the HGPS patients is not linked to abnormal systemic lipids levels (Gordon, Harten, Patti, & Lichtenstein, 2005), but can be associated with smooth muscle depletion in sclerotic vessels and disruption to the microvasculature in various tissues (Gordon et al., 2007; Olive et al., 2010; Stehbens, Delahunt, Shozawa, & Gilbert-Barness, 2001). Individuals with HGPS do not show any increase in tumor susceptibility, cataract formation, or cognitive degeneration, features often associated with normal aging. Consequently, HGPS is regarded as a segmental progeroid syndrome, as it only partially reproduces some aspects of the normal aging processes (Ershler, Ferrucci, & Longo, 2008; Martin, 1989).

Another premature aging condition, Werner's syndrome, is inherited as an autosomal recessive trait, due to mutations in *WRN*, a 3′–5′ RecQ DNA helicase-exonuclease. Werner's syndrome patients exhibit a high incidence of cancers, early-onset cataracts, arthrosclerosis, diabetes, premature graying of hair, and early death, usually in the late 1940s, from myocardial infarction (Fry, 2002; Hickson, 2003; Oshima et al., 2003).

Approximately 15% of Werner's patients, however, were found to have mutations in *LMNA* instead of WRN (Chen & Oshima, 2002). These atypical Werner's patients have short stature, alopecia, osteoporosis, lipodystrophy, diabetes, and muscle atrophy and tend to live longer than patients with the most common HGPS mutation (G608G). Additional recessive missense mutations have been identified in *LMNA* and *ZMPSTE24* resulting in syndromes resembling both progeria and MAD (Csoka et al., 2004; Plasilova et al., 2004; Shackleton et al., 2005). The best characterized of these mutations is the *E145K* mutation that arose in one progeric individual. This missense mutation occurs in the α-helical rod domain, thus far removed from the most common C-terminal globular domain mutations. The *E145K* mutation results in proliferative arrest, profound nuclear changes including, lobulation, chromatin reorganization, and centrosome clustering. To what extent, this individual shares pathologies with the most common and better characterized HGPS mutation is unclear (Taimen et al., 2009).

Recessive mutations resulting in the complete absence of ZMPSTE24 cause restrictive dermopathy, which is characterized by intrauterine growth retardation, rigid or tight skin associated with abnormal dermal ECM, prominent superficial vessels, defects in bone mineralization, dysplastic clavicles, and early postnatal death (Agarwal et al., 2003; Moulson et al., 2005; Navarro et al., 2005; Shackleton et al., 2005). Hypomorphic *ZMPSTE24* alleles often lead to the accumulation of unprocessed prelamin A in addition to mature lamin A, indicating residual activity of the mutated *ZMPSTE24* protein, with some of these individuals exhibiting progeroid features. In *ZMPSTE24*-null cells, lamin C and prelamin A are expressed, but no mature lamin A is produced (Bergo et al., 2002; Pendas et al., 2002).

3.6. Progeric mouse models

The mutation causing the most common form of progeria is a splicing defect in exon 11 of the *LMNA* gene. This arises due to a *de novo* single-base substitution, a C-to-T transition at nucleotide 1824 of the coding sequence (*LMNA* c.1824C>T, p.G608G). The G608G mutation introduces a cryptic donor splice site resulting in a 150-bp deletion and consequently a 50-amino acid in-frame deletion in prelamin A resulting in the production of a truncated protein called progerin. lamin C expression and translation are unaffected (De Sandre-Giovannoli et al., 2003; Eriksson et al., 2003). Other, individual mutations located throughout the *LMNA* gene, have also been associated in causing progeria (Csoka et al., 2004).

The first gene-targeted *Lmna* mutant with progeroid features was created by the fortuitous introduction of a splicing defect, resulting in the deletion of exon 9 of *Lmna* (*Lmna*$^{\Delta 9/\Delta 9}$). The deletion resulted in the in-frame removal of forty amino acids from the Ig fold domain and produced a truncated laminA protein that remains farnesylated and is expressed at low levels. Mice homozygous for the *Lmna*$^{\Delta 9/\Delta 9}$ mutation develop to term and are overtly normal at birth. Postnatal development is compromised by multiple pathologies resembling HGPS. Loss of subcutaneous fat, decreased bone density, osteoporosis, abnormal dentition, thin hyperkeratotic skin, growth retardation and death by 3–4 weeks were the progeric features characterizing this line (Mounkes, Kozlov, Hernandez, Sullivan, & Stewart, 2003). Subsequent analysis revealed that embryonic fibroblasts homozygous for the mutant allele, despite having abnormally shaped nuclei, were indistinguishable from wild-type fibroblasts in their growth and in gene expression profiles. However fibroblasts isolated from various tissues of postnatal *Lmna*$^{\Delta 9/\Delta 9}$ mice, also had misshapen nuclei, underwent a rapid senescence and death, a characteristic similar to that described for fibroblasts isolated from HGPS patients (Bridger & Kill, 2004; Mounkes et al., 2003). The failure of the postnatal fibroblasts to proliferate was due to the reduced expression of many genes encoding ECM components. By culturing the postnatal *Lmna*$^{\Delta 9/\Delta 9}$ fibroblasts on ECM produced by wild-type cells, it was possible to restore normal rates of proliferation. ECM reduction was associated with defective Wnt signaling with the selective reduction of the Wnt-regulated transcription factor Lef-1, but not other Tcf/Lef factors, a reduction also noted in HGPS patient fibroblasts. Since the levels of the mutant *Lmna*$^{\Delta 9/\Delta 9}$ transcripts in the embryonic and postnatal fibroblasts were indistinguishable, this raises the possibility that embryonic fibroblasts are insensitive to (some) *Lmna* mutations, compared to postnatal cells—a situation that could explain why many *Lmna* mutations do not affect embryogenesis. Two-week-old *Lmna*$^{\Delta 9/\Delta 9}$ mice exhibited extensive vascular smooth muscle apoptosis in the great vessels around the heart that was associated with reduced ECM synthesis and Wnt signaling (Hernandez et al., 2010; Stewart, 2014). These results also raise the issue as to whether the alterations in nuclear shape associated with many of the laminopathies have any significance.

Two successive mouse lines each expressing progerin have since been described. The first, using a "knock-in" strategy replaced the wild-type

Lmna allele with a variant that would only produce progerin ($Lmna^{HG}$) (Yang, Andres, Spielmann, Young, & Fong, 2008). Heterozygous fibroblasts from these mice express large amounts of progerin and have misshapen nuclei. $Lmna^{HG/+}$ mice are normal at birth, but their postweaning growth rate becomes retarded developing many phenotypes common to HGPS, including osteoporosis, alopecia, micrognathia, reduced subcutaneous fat, and osteolysis of the clavicle. These phenotypes are progressive, with the mice dying by 6–7 months of age. Homozygous mice ($Lmna^{HG/HG}$) have severe osteoporosis, spontaneous bone fractures, and die before weaning.

The second HGPS mouse line was generated by introducing the HGPS c.1827C>T;p.Gly609Gly mutation into the mouse *Lmna* gene, making it equivalent to the human HGPS c.1824C>T;p.Gly608Gly mutation in that it introduces a novel splice donor site into the *Lmna* gene (Osorio et al., 2011). Heterozygous mice produce a murine version of progerin, in addition to both lamin A and C proteins. As homozygotes, the mice produced mostly progerin, with very little lamin A, indicating almost a 100% efficiency of the mutated splice donor site at converting lamin A to progerin, with intermediate levels of lamin C being produced. Homozygous $Lmna^{G60G9/G609GG}$ mice are overtly normal to 3 weeks of age. Subsequently, they show a decline in growth, kyphosis, infertility, and a mean longevity of about 3 months, a longer lifespan than that of the $Lmna^{HG/+}$ line, for unknown reasons. Heterozygous $Lmna^{G09G9/+}$ mice are seemingly normal in weight, are fertile, and have a mean lifespan of 8 months, but undergo extensive aortic calcification (Villa-Bellosta et al., 2013). The $Lmna^{G609G/+}$ mice develop many progeric features, including diminished subcutaneous fat, bone density, and a reduction in vascular smooth muscle in the aortic arch. Gene set enrichment analysis revealed the most significant changes in gene expression were those representative of stress responses and DNA damage (Osorio et al., 2011). Overall, the pathology arising from the $Lmna^{G609G}$ mutation is comparatively milder, compared to the HGPS in patients. The mice live to a comparably longer age, the heterozygotes are fertile, and neither genotype develops alopecia. The reasons for such divergence from the human pathology are unclear, although inbred mice are often thought to be more "tolerant" to heterozygous mutations than the human equivalents.

In addition to these lines carrying mutations in the endogenous *Lmna* gene, a few transgenic lines have been derived (Varga et al., 2006). Transgenic mice carrying a human *LMNA* BAC expressing progerin did not manifest any of the early-onset pathologies of progeria. However, within

12 months, they showed a loss of smooth muscle cells in the aorta, a feature reported in a few HGPS autopsies (Stehbens et al., 2001). Other transgenic lines were produced to primarily express progerin in the skin. Postnatal expression of progerin under control of an inducible Keratin-5-promoter resulted in abnormalities in the skin and teeth, loss of dermal fat, hair follicles, sweat glands, abnormal incisors, and a failure of stem regeneration (Sagelius et al., 2008). Subsequent experiments using the same line of mice where expression was induced late in embryogenesis, at the time skin is developing, resulted in the newborns possessing a highly inflamed hyperkeratotic, dry scaly skin, with aberrant (persistent) Lbr expression in the epidermal nuclei (McKenna et al., 2013). To some extent, these mice resembled newborns with restrictive dermopathy (RD) as they died within 5 days of birth, possibly due to the breakdown in the barrier function caused by extensive inflammation that is associated with activation of the NF-kβ pathway (Klement et al., 1996).

3.7. Zmpste24-deficient mice

Humans, with complete loss of *ZMPSTE24*, develop RD, or tight skin and die at birth. In contrast, mice lacking *Zmpste24* do not (Bergo et al., 2002; Navarro et al., 2005; Pendas et al., 2002). *Zmpste24*$^{-/-}$ mice are seemingly normal at birth, but postnatal growth then becomes retarded. Growth retardation is associated with reduced incisor development, kyphosis, hair loss, and an arthritic gait with muscle weakness. Bone abnormalities with osteolytic lesions in the ribs at the costovertebral junction, leading to fractured ribs by ~20 weeks of age, are also present. Bone density is reduced in 3-month-old *Zmpste24*$^{-/-}$ mice although osteoclast numbers are not altered, suggesting a defect in osteogenesis. This phenotype is associated with vacuolated osteoblasts lacking typical stacks of rough ER (Young, Fong, & Michaelis, 2005). Death usually occurs at ~6 months. As with the progerin mutations, the mice have a longer equivalent lifespan than humans lacking *ZMPSTE24* or with progeria. Despite these differences, the mice are frequently regarded as being a model for progeria.

Zmpste24$^{-/-}$ mice exhibit many physiological changes including reduced circulatory IGF-1 levels that maybe associated with reduced postnatal growth, muscle development, bone mineral density, and alopecia. In addition, there is chronic inflammation associated with activation of the NF-kβ pathway (Osorio et al., 2012). Surprisingly, *Zmpste24*$^{-/-}$ mice show increased levels of autophagy, rather than a decline that is associated with normal ageing and in the *Lmna*$^{Sul/Sul}$ and H222P mutations (Marino et al., 2008).

Cultured *Zmpste24*$^{-/-}$ cells show accelerated rates of senescence and increased DNA damage (Liu et al., 2005; Pendas et al., 2002). Accelerated senescence is partly ameliorated by deleting p53, and *in vivo* deletion of p53 improved the postnatal growth and viability (Varela et al., 2005). Surprisingly, *Zmpste24* deficiency was completely ameliorated by a reduction in lamin A levels by deletion of one *Lmna* allele by intercrossing with the *Lmna*$^{Sul/Sul}$ mice (Fong et al., 2004; Varga et al., 2006). This result indicated that the accumulation of farnesylated prelamin A protein is the prime cause of the pathologies. Furthermore, it is probably the retention of the farnesyl group by prelamin A that underlies most of the pathologies, as mice engineered to solely express lamin A, so bypassing prelamin A processing, are seemingly normal and have a longer lifespan. In contrast, non-farnesylated prelamin A only mice, although overtly normal, with none of the major pathological changes associated with *Zmpste24* deficiency, eventually developed cardiomyopathy, indicating retention of the C-terminal 18 amino acid (non-farnesylated) peptide has pathological consequences (Coffinier et al., 2010).

In HGPS, the progerin variant remains farnesylated, as the 50-amino acid deletion removes the second ZMSTE24 endoproteolytic site used to cleave the farnesylated and methylated cysteine. Much evidence has pointed to the retention of the farnesyl group by progerin, as well as other laminA variants, as being the major factor underlying the pathology of HGPS. progerin results in the nuclei becoming grossly misshapen, with thickening of the lamina, alterations in chromatin organization and delays in mitotic progression, with a significant percentage of cells becoming binucleate (Dechat et al., 2007; Liu, Rusinol, Sinensky, Wang, & Zou, 2006; Shumaker et al., 2006). Accompanying these changes are increased levels of DNA damage, though it is not clear if this is due to an increase in the absolute levels of damage or impaired DNA damage repair. HGPS cells show significantly reduced telomere lengths and the expression of exogenous telomerase, to a significant extent, ameliorates DNA damage, resulting in a prolongation, although transient, of HGPS cell proliferation (Benson, Lee, & Aaronson, 2010; Kudlow, Stanfel, Burtner, Johnston, & Kennedy, 2008). It is not entirely clear if all these pathological effects are solely due to the retention of the farnesyl moiety or are compounded by the 50-amino acid deletion, as mice expressing a non-farnesylated form of progerin develop similar though milder pathologies compared to mice expressing the farnesylated form of progerin (Yang, Andres, et al., 2008). Lastly, there are a few cases of progeria that are caused by other mutations in *LMNA*, for example, E145K, where *in vitro* cell proliferation is also retarded (Taimen et al., 2009). How progeria arises from these other progeric mutations remains to be determined, as it unclear if these mutations result in posttranslational processing defects and/or retention of the farnesyl group (Csoka et al., 2004; Plasilova et al., 2004).

3.8. Therapies

Reduction in the levels of the farnesylated progerin has been a major therapeutic goal for progeria. Short-term *in vitro* treatment of HGPS or *ZMPSTE24* null cells with farnesyl transferase inhibitors (FTIs) seemingly rectifies the alterations in nuclear morphology (Glynn & Glover, 2005; Yang et al., 2005). FTI treatment of *Zmpste24* null or $Lmna^{HG}$ mice reduced the rate of weight loss, improved muscle strength, longevity, and reduced the incidence of rib fractures, hallmarks of *Zmpste24* loss (Fong, Frost, et al., 2006). However, the body weights of control wild-type mice were also reduced by the FTIs (Fong, Frost, et al., 2006; Yang et al., 2006; Yang, Qiao, Fong, & Young, 2008).

FTI's alone, do not prevent prelamin A or progerin being geranylgeranylated, which can substitute for, and has the same function as farnesylation (Rusinol & Sinensky, 2006). Consequently, clinical trials have/are using a combination of statins, FTIs, and aminobisphosphonates, which combined, should inhibit all forms of laminA prenylation (Varela et al., 2008). The outcome of the first FTI/Statin combination treatment of HGPS patients was recently reported, and although control studies was not possible, treatment appeared to enhance the children's growth, bone density, and improve vascular function (Gordon et al., 2012). Recently, a new twist to the consequences of defective posttranslational modification to lamin A in progeria was reported implicating the transient carboxy methylation of prelamin A that is mediated by ICMT. Surprisingly, $Zmpste24^{-/-}$ mice that were homozygous for a hypomorphic allele of *Icmt* showed a marked reversal of the progeroid symptoms, with a decrease in bone fractures, and increased body weight and survival time. Cells from these mice also showed elevated activation of the mTOR/AKT pathways that delayed the early senescence associated with *ZMPSTE24* deficiency and HGPS. These fortuitous findings suggest that pharmacological inhibition of ICMT may offer a novel route to the treatment of HGPS (Ibrahim et al., 2013).

Despite considerable effort, there are still lingering issues as to the ultimate effectiveness of FTI/Statin treatments. Treatment of progeria by FTI/ Statins will not reverse it. At best treatment may retard disease progression, requiring a lifetime of drug administration. Furthermore, mice expressing an unfarnesylated form of progerin still develop, albeit a milder form, of progeria, making it uncertain what effects chronic expression of unfarnesylated progerin may have on patients (Yang, Andres, et al., 2008). And, until recently, the effects of FTI treatment were focused on lamin A. Of potential, concern is that FTI treatment also affects B-type lamin dynamics. FTI treatment of human fibroblasts decreased mature laminB1 levels and induced an accumulation of laminB2, with its redistribution to the nucleoplasm (Adam, Butin-Israeli, Cleland, Shimi, & Goldman, 2013). What the consequences of these changes would have *in vivo* are uncertain, as B-type lamins are clearly necessary for development of the nervous system, in particular a farnesylated laminB1, and their requirement by many other tissues (muscle, vasculature?) is uncertain (Jung, Nobumori, et al., 2013). With these reservations, alternative routes to reducing the toxic effects of improperly processed laminA are being considered (Young, Yang, Davies, Jung, & Fong, 2013). One possibility is antisense morpholinos to specifically target the mutant allele in the LmnaG609G mice (Scaffidi & Misteli, 2005). Systemic delivery did have some effect at ameliorating some of the disease phenotypes and increased longevity (Osorio et al., 2011). An even more radical potential route is based on the finding that progerin and lamin A are expressed at very low levels in neuronal cells and in the mouse CNS. This is due to miRNA (mir-9) inhibition of lamin A, but not lamin C, translation in the neuronal cells (Jung et al., 2012). Since laminC only mice are seemingly quite normal, it has been suggested that if it were possible to specifically ablate/inhibit laminA expression in tissues then this may be a more effective, long-lasting form of therapy (Fong, Ng, et al., 2006). Lastly, a few other mutations, mainly localized to the rod domain in *LMNA* have been linked to individuals diagnosed with progeria. It is not at all clear whether these mutations would affect laminA processing/ farnesylation. If they do not, it suggests that other cellular pathways exist that when disrupted lead to progeria. If such pathways exist and are shared with the G608G mutants then they may provide an additional route to seeking potential therapies.

4. LAMINOPATHIES ASSOCIATED WITH MUTATIONS IN THE B-TYPE LAMINS

In contrast to *LMNA*, no autosomal-dominant or recessive mutations within the LMNB1 and LMNB2 genes have been linked to disease, although a polymorphism in LMNB1 may predispose individuals to neural tube defects (De Castro et al., 2012). The lack of any laminB mutants may be because they are expressed at lower levels than the A-type lamins, that only a few cell types solely express the B-type lamins alone, or that either *LMNA* or the other LMNB may compensate for the mutated lamin B gene. However, as described above, there is a requirement for both B-type lamins in development of the CNS (Coffinier et al., 2011; Kim et al., 2011). Whether other tissues, apart from the skin epidermis and liver, have an absolute dependence on the laminBs will have wait for a more extensive analysis on the specific deletion of the two conditional *lmnb* alleles in different tissues. Despite this absence of phenotype, a reduction in lamin B1 levels in cultured cells accelerates their entry into senescence (Dreesen et al., 2013; Freund et al., 2012; Shah et al., 2013; Shimi et al., 2011). In contrast to the loss of lamin B1, duplication of the LMNB1 gene results in adult onset autosomal-dominant leukodystrophy, a progressive and fatal neurological disorder (Padiath et al., 2006). Transgenic mice carrying additional copies of *lmnb1* show cognitive impairment, epilepsy, and motor neuron defects, potentially due to defective myelination (Heng et al., 2013). It is uncertain how duplication of the B1 gene results in such a tissue-specific disease. Two alternatives have been proposed with one suggesting reduced transcription factor (YinYang, YY1) occupancy in the promoter region of the proteolipid protein required for myelin synthesis (Heng et al., 2013). An alternative suggestion was that duplication leads to the generation of autoantibodies causing demyelination (Padiath et al., 2006).

5. DISEASES RESULTING FROM NE PROTEIN MUTATIONS

The very first disease linked to the NE was the X-linked Emery-Dreifuss muscular dystrophy, as a consequence of mutations in the LEM domain protein, EMD, which localizes primarily to the INM (Bione et al., 1994). Many of the symptoms are similar, though milder to the AD-EDMD form caused by *LMNA* mutations.

At least 74 NE integral (NETs) and associated proteins have been identified from proteomic analyses of the NE from nuclei of different tissues (de Las Heras et al., 2013; Korfali et al., 2010). Among the integral NE proteins containing at least one transmembrane domain are EMD, MAN1/LEMD3, LAP1, LAP2, and LBR (Lee & Wilson, 2004). Some of these proteins (LAP2, EMD, and MAN1/LEMD3) share a common 40 amino acid motif (LEM domain), which interacts with the small DNA binding chromatin protein BAF or BANF1 (barrier-to-autointegration factor). In this way, BAF provides a link between LEM domain and DNA (Dorner, Gotzmann, & Foisner, 2007) and mutations within BANF1 were recently linked to cases of atypical progeria (Puente et al., 2011). At least four additional LEM domain proteins LEM2 (NET-25) and LEMs 3–5 may be encoded in the human genome (Brachner, Reipert, Foisner, & Gotzmann, 2005; Ulbert, Antonin, Platani, & Mattaj, 2006). Three of the LEMS, namely EMD, LEM2, and membrane-associated isoforms of LAP2 are localized to the INM in part by their association with the A-type lamins (Dechat et al., 2000). In this way, these proteins are thought to tether chromatin to the nuclear lamina, in particular heterochromatin, in a cell specific and often redundant manner with the other INM protein LBR (Solovei et al., 2013).

5.1. Nuclear envelopathies associated with the LEMs

Heterozygous autosomal-dominant loss-of-function mutations in LEMD3/MAN1 result in osteopoikilosis, Buschke-Ollendorff syndrome, and melorheostosis, familial cutaneous collagenomas, all disorders marked by increased bone density (Hellemans et al., 2004). The N-terminus of the double pass transmembrane MAN1 protein contains a LEM domain, which binds to both lamins A/C and chromatin via BAF (Lin et al., 2000). The C-terminus contains an RRM domain and regulates the TGFβ superfamily regulated SMAD transcription factor function by binding Smads2 and 3. MAN1 competes with the transcription factor Forkhead activin signal transducer 1 for binding to Smad2 and by binding directly to the phosphatase PPM1A, which dephosphorylates Smad2/3, so enhancing Smad transcriptional activity (Bourgeois et al., 2013; Lin, Morrison, Wu, & Worman, 2005).

Embryos homozygous for the gene-trap mutation in LEMD3 (Man1GT/GT) die around E9. Death is associated with defective angiogenesis with disorganized yolk-sac vasculogenesis and vessel formation, phenotypes associated with defective TGFβ signaling pathways in the development of the murine vascular system (Patel-Hett & D'Amore, 2011). *Man1*$^{GT/GT}$ nuclei contain increased levels of phosphorylated Smad2/3, so increasing transcription of Smad regulated genes that results in increased ECM deposition, reduced endothelial cell (EC) proliferation, migration, and angiogenic branching, with a reduced capacity of the ECs to form mature blood vessels (Cohen, Kosti, & Stewart, 2007; Ishimura, Ng, Taira, Young, & Osada, 2006). Interestingly, *Man1*$^{GT/+}$ mice do not exhibit skeletal abnormalities, suggesting that mutations in *Man1/Lemd3* may not cause osteopoikilosis in mice (Cohen et al., 2007; Dheedene et al., 2009).

The lamina-associated polypeptide 2 (LAP2 or Thymopoietin), comprises a family of alternatively spliced polypeptides, α, β, γ, δ, ε, and ζ (Dorner et al., 2007; Wagner & Krohne, 2007). Most of the LAP2 polypeptides are integral NE proteins with the exception of α and ζ. LAP2α is localized exclusively to the nucleoplasm where it interacts with nucleoplasmic lamin A/C. Complexes between LAP2α, lamin A/C, and Rb/E2F may regulate cell cycle exit, by preventing the proteolytic degradation of Rb (Dorner et al., 2007). Proliferating fibroblasts have high levels of LAP2α, and these levels decline once the fibroblasts reach confluence (Dorner et al., 2006). In humans, mutations in LAP2α have been linked to DCM (Taylor et al., 2005). Mice deficient in LAP2α are viable and fertile, although they exhibit erythroid precursor and epidermal hyperproliferation, as well as an increased myofiber-associated stem cell pool with some of these effects due to impaired Rb regulation. The null mice eventually develop impaired systolic function culminating in extensive fibrosis (Gotic, Leschnik, et al., 2010; Gotic, Schmidt, et al., 2010; Naetar et al., 2008).

In contrast to the loss of *Lap2α*, loss of *Emd* had no overt effect *in vivo*, apart from very mild effects on cardiac function, and *in vitro*, a slight delay in myotube differentiation (Melcon et al., 2006; Ozawa et al., 2006). This lack of overt pathology in *Emd* null mice suggested the possibility of redundancy between the various LEM domain proteins. To date, such redundancy or epitasis has not been identified in intercrosses between the *Man1/Lemd3*, *Lap2α* and *Emd* nulls (T. Cohen & C. L. Stewart, unpublished observations). However, EMD binds to the NE protein *Lap1*, and constitutive loss of *Lap1* results in perinatal lethality. Tissue-specific deletion of *Lap1* in skeletal muscle results in postnatal viability, associated with progressive muscular dystrophy. When combined with the loss of EMD, *Lap1/Emd* null mice show greatly exacerbated dystrophy. This suggests that *Lap1* may compensate for the loss of EMD in mouse muscle providing an explanation as to why

$Emd^{-/-}$ mice do not reproduce the pathologies associated with EMD loss in humans (Shin, Spinler, et al., 2013).

In $Lmna^{Sul/Sul}$ myoblasts, Lap2α levels are increased, as are the transcription factors Smads 2 and 3 (Cohen et al., 2013; Melcon et al., 2006). The disruption to the TGFβ signaling pathway in the $Lmna^{Sul/Sul}$ myoblasts maybe of significance since activin mediated increase in Smad activity is the major pathway causing muscle wasting (Chen et al., 2014). Deletion of Lap2α in the $Lmna^{Sul/Sul}$ mice results in improved myoblast fusion, extended viability and increased body weight of the double nulls, although not to the same extent as the loss of Sun1 (Cohen et al., 2013; Solovei et al., 2013). These observations indicate that disruption of the Lap2/Rb and TGFβ pathways contribute to lamin-mediated muscle wasting. They also suggest there is extensive coregulation and integration between different NE proteins and the lamina and it is disruption to this integration that maybe an additional significant contributory factor to the various pathologies resulting from *LMNA* mutations.

5.2. Nuclear envelopathies associated with the lamin B receptor

The lamin B receptor is a multipass INM protein with an N-terminal lamin B and chromatin-interacting Tudor domain. The Tudor domain interacts with multiple chromatin-associated proteins, many enriched in heterochromatin (Makatsori et al., 2004), including the chromatin binding protein HP1 (Ye et al., 1997). During nuclear reassembly after mitosis, the LBR is involved in targeting the NE to chromatin in an importin β-mediated process (Ma et al., 2007). The C-terminal region of LBR contains eight transmembrane domains and exhibits sterol Δ14-reductase activity (Silve, Dupuy, Ferrara, & Loison, 1998).

Heterozygous mutations in the *LBR* gene mostly result in the autosomal-dominant Pelger-Huët anomaly, a disorder marked by decreased nuclear lobulation in granulocytes/neutrophils (Hoffmann et al., 2002). Homozygotes for an *LBR* splicing defect exhibit a more severe blood phenotype and with impaired cognitive development, heart defects, and bone deformities (Hoffmann, Sperling, Olins, & Olins, 2007; Waterham et al., 2003). Additional *LBR* mutations result in autosomal recessive chondrodystrophy, HEM/Greenberg Skeletal Dysplasia, a severe inborn metabolic disorder characterized by hydrops, skeletal abnormalities including shortening, a "moth-eaten appearance," ectopic calcifications of bones, intestinal, lung, and kidney malformations and prenatal lethality (Waterham et al., 2003). Embryonic fibroblasts, homozygous for the LBR mutation, express elevated levels of cholesta-8, 14-dien-3-β-ol, a key metabolite of the sterol-D14-reductase activity of LBR, suggesting that LBR may be essential to proper cholesterol synthesis during development. This has, however, been disputed due to possible redundancy with another cholesterol-metabolizing enzyme Dhcr14 (Wassif et al., 2007).

In mice, missense or nonsense mutations in the *Lbr* gene result in ichthyosis (ic^J) and develop alopecia, postnatal growth retardation, syndactyly, and early lethality (Shultz et al., 2003). Within the hematopoietic compartment of ic^J mice, splenic lymphocytes show heterochromatin clumping. In the peripheral blood, neutrophils and eosinophils are immature and the nuclei are hypolobulated. A gene-trap insertion into the *Lbr* locus ($Lbr^{GT/GT}$) results in the accumulation HP1α in nuclear foci and an increased numbers of micronuclei, suggesting defects in chromosome segregation during mitosis (Ma et al., 2007). $Lbr^{GT/GT}$ and ic^J mice are phenotypically similar, with lymphocytes exhibiting nuclear condensation, with increased numbers of immature granulocyte/neutrophils, with some granulocyte functions (e.g., bacterial killing) not being compromised by the loss of LBR (Cohen et al., 2008; Gaines et al., 2008). More recently, a point mutation within *Lbr* has been linked to lymphopenia, particularly within the T cell compartment, indicating a requirement for LBR in other hematopoietic lineages (Verhagen et al., 2012).

5.3. Nuclear envelopathies involving TorsinA

TorsinA, a member of the AAA-family of ATPases, acts as a chaperone regulating the processing of proteins through the secretory pathway (Hewett et al., 2007). A single deletion of a glutamate residue in the TorsinA protein results in early-onset autosomal-dominant torsion dystonia (DYT1), a disorder characterized by involuntary movements due to hyperactive dopaminergic systems in the striatum (Cookson & Clarimon, 2005). Although TorsinA localizes predominantly to the ER, where it interacts with the luminal protein, LAP1, and pLULL1 (TOR1AIP/NET8) (Goodchild & Dauer, 2005). Despite TorsinA being widely expressed in many different cells types, the mutant form appears to only affect postmigratory neurons indicating these cells have some unique requirement for TorsinA (Granata, Watson, Collinson, Schiavo, & Warner, 2008).

Several mouse lines carrying a mutated torsin have been generated (Goodchild, Kim, & Dauer, 2005). Homozygotes die shortly after birth due to a failure to feed and show abnormal nuclear membrane morphology. Although the heterozygotes are viable and appear normal, subtle behavioral changes are detected (Dang et al., 2005). A more detailed description of mouse Torsin mutants can be found in this volume in Chapter 3 by Dauer and colleagues.

Intriguingly, one of the severe forms of lipodystrophy, Berardinelli-Seip Congenital Lipodystrophy, may also be caused by aberrant relocalization of an ER protein to the NE. Berardinelli-Seip Congenital Lipodystrophy is caused by mutations in BSCL2 or Seipin, a protein normally localized to the ER, where it is required for maintaining white fat (Szymanski et al., 2007). A missense mutation in BSCL2 causes this predominately ER located protein to relocate to the NE where it is apparently nonfunctional (Payne et al., 2008).

5.4. The LINC complexes in development and disease

A significant group of NE proteins, which interact with the lamins, is the LINC complex. The LINC complex consists of two protein families. The first of these, the KASH/Syne/Nesprin family, primarily localize to the ONM. Their conserved C-terminal KASH domains extend into the PNS where they bind to the C-termini of the second family of proteins, the SUN domain family. In mammals, six KASH domain proteins have been identified and all, with the possible exception of KASH6, also known as LRMP or JAW1, interact with different components of the cytoskeleton. KASH/Nesprin/Syne1 and 2, through their N-terminal calponin homology domain, interact with the actin microfilament network. KASH2/Nesprin2/Syne2, in addition, binds the microtubule motor proteins Kinesin1 and cytoplasmic dynein. KASH3/Nesprin3 interacts with the intermediate filament network by binding to the cytoplasmic protein plectin. Nesprin4/KASH4 and KASH5 link with the cytoplasmic microtubular networks through the motor proteins, Kinesin1 and dynein, respectively. As a result, interphase nuclei in many different cell type and tissues directly connect with all three different cytoskeletal networks (and possibly other cytoplasmic organelles), in a cell/tissue-specific manner (Burke & Stewart, 2013).

The Sun family consists of at least six proteins, SUN1, 2, 3, 4 (SPAG4), 5 (SPAG4L), and osteopotentia (Liu et al., 2007; Sohaskey et al., 2010). The two most widespread Sun domain proteins are Sun1 and Sun2 localize to the INM, and are probably expressed in most nucleated cells where they interact with the different KASH proteins. The N-termini of both Sun domain proteins protrude into the nucleoplasm, where they interact to varying extents with the lamins and other nucleoplasmic factors (Crisp et al., 2006; Haque et al., 2006). Consequently, the LINC complexes potentially establish a direct physical connection between the nucleoplasm, via the different cytoskeletal networks, to the cell's membrane and ECM.

To date, a few mutations in the KASH domain genes have been linked to human diseases. No mutations in the SUN domain genes have yet been reported. Some mutations in *SYNE1* and *SYNE2* are associated with an EDMD-like phenotype (Puckelwartz et al., 2010; Zhang, Bethmann, et al., 2007). Whereas other mutations in *SYNE1* are associated with non-muscle pathologies, including cerebellar ataxia and arthrogryposis, the latter of which results in joint contractures (Dupre, Bouchard, Gros-Louis, & Rouleau, 2007). In mice, disrupting the LINC complex by deletion of *Synes1* and *2*, *Suns1* and *2*, and combinations of these deletions, have a range of effects on neuronal migration during CNS development, nuclear positioning in muscle fibers, gametogenesis, hearing and ultimately postnatal viability. Mutation of *Syne1*, by two different groups, produced conflicting results. Deletion of the last exon in *Syne1*, which removes the KASH domain, and was coupled with an abnormal splicing defect that replaced the last 100 amino acids with a novel 61-amino acid peptide, resulted in about 50% of the homozygous mice dying around birth. The remaining survivors developed kyphosis, muscle defects, and cardiomyopathy (Puckelwartz et al., 2009). With the other *Syne1* mutation, where the KASH domain was also eliminated (with no amino acid substitutions being reported), apart from the clustering of nuclei in muscle fibers, and the loss of muscle nuclei under the neuromuscular junctions, the homozygotes were overtly normal (Grady, Starr, Ackerman, Sanes, & Han, 2005; Zhang, Xu, et al., 2007). *Syne2*-deficient mice are also overtly normal, and in contrast to the *Syne1* null mice, they showed normal nuclear distribution in the muscle fibers (Zhang, Xu, et al., 2007). Mice doubly deficient for both the *Syne1* and *2* die perinataly due to respiratory problems that may be associated with abnormal phrenic nerve enervation of the diaphragm (Zhang, Xu, et al., 2007). These results reveal a degree of functional redundancy between *Syne1* and *2*.

A truncated form of the *KASH4/Nesprin4/SYNE4* gene results in a progressive high frequency hearing loss in a few families of middle-eastern origin. Mice deficient in *Kash/Syne4* are overtly normal, though they show defective hearing within 2–3 weeks of birth. The cause of deafness is due to loss of the outer hair cells (Ohcs) in the cochlea, cells that convert sound waves into neuronal signals. Nesprin4/Kash4 localizes to the Ohc nuclei, where it maintains nuclear position at the base of the Ohc, presumably by engaging the nucleus with the interphase cytoplasmic microtubule (MT) network by binding the MT motor protein kinesin-1. In the absence of Nesprin4/Kash4 the nuclei move from a basal to an apical position, disrupting cell polarity. A breakdown in cell polarity may then disrupt Ohc movement in response to sound, culminating in death of the Ohcs. A similar phenotype was also noted in *Sun1* null mice revealing that the Nesprin4/Kash4-Sun1 LINC complex is essential for Ohc function in hearing (Horn, Brownstein, et al., 2013).

Kash5-deficient mice are infertile due to defective gametogenesis in both males and females. In the testes, during meiotic prophase 1, as the chromosomes begin to pair, the telomeres attach to the nuclear periphery, a process that may directly/indirectly require Sun1. The attached telomeres are then moved to cluster under the centrosome, a process known as bouquet formation. Chromosome movement leading to bouquet formation is essential for proper synaptogenesis and chromosomal recombination. Telomere clustering is mediated by Kash5, through its interaction with Sun1, so linking the telomeres to the cytoplasmic MT network via the minus end motor protein cytoplasmic dynein. The function of dynein is to drag telomeres, via their attachment to the Kash5/Sun1 LINC complex, so that they cluster at the nuclear pole proximal to the centrosome. Disruption of the LINC complex by loss of Kash5, where telomeres remain largely tethered to the NE, results in a failure of synapsis. In *Sun1* null mice, which also have defective gametogenesis, telomere tethering is lost and synapsis may still occur, albeit a very low frequency (Ding et al., 2007). In the Kash5 null mice, chromosomal breakage sites are not resolved, and the spermatocytes undergo apoptosis.

With regards to the remaining Kash genes, 3 and 6, Kash3/Nesprin3 null mice are seemingly normal, apart from the disruption to intermediate filament networks in sertoli cells with no obvious effect on gametogenesis (Ketema, Kreft, Secades, Janssen, & Sonnenberg, 2013). Extensive phenotypic analysis of the mice by the German Mouse Clinic failed to identify any physiological, anatomical, or behavioral differences resulting from Kash3/Nesprin3 deficiency (C. L. Stewart & H. Horn, unpublished observations). As for Kash6/LRMP/Jaw1, which is expressed in lymphoid cells and taste receptors (Behrens et al., 1996; Shindo et al., 2010), there are no reports as to what may be the consequences that loss of the protein may have.

Loss of either *Sun1* or *Sun2* alone has no overt effect on the postnatal growth rates or longevity of the null mice, a somewhat surprising result since *Sun1* and *Sun2* are widely expressed. *Sun1* deficiency does however result in defective meiosis in both males and females with both sexes being infertile (Chi et al., 2009; Ding et al., 2007). *Sun1*, but not Sun2, deficient mice are deaf (see above) (Horn, Brownstein, et al., 2013) and in their retinas have a thinner outer nuclear layer, due to reduced rod cell numbers (Yu et al., 2011). Mice that are doubly deficient for *Sun1* and *Sun2* die at birth for reasons that are unclear, though given the requirement of the LINC complex in neuronal migration, defective CNS development is a likely culprit (Lei et al., 2009). These findings suggest, with the exception of a few tissues, there is significant redundancy between *Sun1* and *Sun2* functions in many tissues.

Mouse mutants have revealed that the LINC complex and the B-type lamins have key roles in regulating neuronal migration during development of the CNS. These findings raise the tantalizing possibility that many of the muscular dystrophies, cardiomyopathies, and some ataxias may share a common molecular basis, in that disruption of the different components of the LINC complex affect cytoskeletal organization and function, as well as affecting nucleoplasmic/chromatin organization and gene expression. As with the lamins, conditional alleles of the *Sun* and *Syne* genes will be instrumental in uncovering their respective roles in other postnatal tissues.

5.5. Conclusions

In recent years, our understanding of the function of the cell's nucleus has undergone a renaissance. Rather than considering the NE simply as a porous bag that contains the genome, we now view the Lamina/NE as having a key role in coordinating many cellular processes that include chromatin organization, transcriptional regulation, signaling pathways and organization of the cytoskeleton in regulating nuclear position, cell migration, and mechanical stability. Mouse lines, carrying mutations in many of the NE proteins and lamins, have been instrumental in expanding our insights into these new functions of the NE/lamina. They have provided us with models for many of the diseases resulting from Lamin/NE mutations and are a valuable resource in the search for therapies to treat the diseases caused by the mutations. Although some of the mouse mutants may not faithfully reproduce the phenotypes seen in human disease that they are modeling, by their genetics, they have provided many novel insights into unexpected epistatic interactions between the lamins and other NE proteins. As an example, it is apparent that mutations in one protein alter the expression levels of other NE proteins, and this change in levels in turn contributes to the pathology. Such changes suggest that NE proteins/lamins function as a highly integrated network.

Many of the NE/lamina proteins we know most about, have no enzymatic activity and appear to primarily serve as scaffolds for other proteins with enzymatic activity. Consequently, one pressing need is to identify which are the other proteins that functionally depend on their location to the NE lamina, so having key roles in many of the biological processes regulated by the NE/lamina. Given that the lamins are highly insoluble, identification of other proteins with which they interact has been problematical. This is likely to change, as the BioID technique (proximity-dependent biotin identification), using proximity-based *in vivo* biotinylation, to identify other potential interacting proteins with the lamins (or potentially any other NE protein), offers a realistic and physiologically relevant route to such identification (Roux et al., 2012). It is conceivable that the interactome of any NE/lamina protein can be mapped, not only in cultured cells but also *in vivo* in engineered mice, in different tissues, stages of development, and under different disease conditions. By employing such techniques, we may be eventually able to obtain a real understanding how the architecture of the nucleus is important to its function and that when disrupted leads to disease.

ACKNOWLEDGMENTS

B. B. and C. L. S. are supported by the Singapore Biomedical Research Council and the Singapore Agency for Science Technology and Research, A*STAR.

REFERENCES

Adam, S. A., Butin-Israeli, V., Cleland, M. M., Shimi, T., & Goldman, R. D. (2013). Disruption of lamin B1 and lamin B2 processing and localization by farnesyltransferase inhibitors. *Nucleus*, *4*, 142–150.

Aebi, U., Cohn, J., Buhle, L., & Gerace, L. (1986). The nuclear lamina is a meshwork of intermediate-type filaments. *Nature*, *323*, 560–564.

Agarwal, A. K., Fryns, J. P., Auchus, R. J., & Garg, A. (2003). Zinc metalloproteinase, ZMPSTE24, is mutated in mandibuloacral dysplasia. *Human Molecular Genetics*, *12*, 1995–2001.

Arimura, T., Helbling-Leclerc, A., Massart, C., Varnous, S., Niel, F., Lacene, E., et al. (2005). Mouse model carrying H222P-Lmna mutation develops muscular dystrophy and dilated cardiomyopathy similar to human striated muscle laminopathies. *Human Molecular Genetics*, *14*, 155–169.

Beck, L. A., Hosick, T. J., & Sinensky, M. (1988). Incorporation of a product of mevalonic acid metabolism into proteins of Chinese hamster ovary cell nuclei. *Journal of Cell Biology*, *107*, 1307–1316.

Behrens, T. W., Kearns, G. M., Rivard, J. J., Bernstein, H. D., Yewdell, J. W., & Staudt, L. M. (1996). Carboxyl-terminal targeting and novel post-translational processing of JAW1, a lymphoid protein of the endoplasmic reticulum. *Journal of Biological Chemistry*, *271*, 23528–23534.

Benson, E. K., Lee, S. W., & Aaronson, S. A. (2010). Role of progerin-induced telomere dysfunction in HGPS premature cellular senescence. *Journal of Cell Science*, *123*, 2605–2612.

Bergo, M. O., Gavino, B., Ross, J., Schmidt, W. K., Hong, C., Kendall, L. V., et al. (2002). Zmpste24 deficiency in mice causes spontaneous bone fractures, muscle weakness, and a prelamin A processing defect. *Proceedings of the National Academy of Sciences of the United States of America*, *99*, 13049–13054.

Bertrand, A. T., Renou, L., Papadopoulos, A., Beuvin, M., Lacene, E., Massart, C., et al. (2012). DelK32-lamin A/C has abnormal location and induces incomplete tissue maturation and severe metabolic defects leading to premature death. *Human Molecular Genetics*, *21*, 1037–1048.

Bickmore, W. A. (2013). The spatial organization of the human genome. *Annual Review of Genomics and Human Genetics*, *14*, 67–84.

Bione, S., Maestrini, E., Rivella, S., Mancini, M., Regis, S., Romeo, G., et al. (1994). Identification of a novel X-linked gene responsible for Emery-Dreifuss muscular dystrophy. *Nature Genetics*, *8*, 323–327.

Bonne, G., Di Barletta, M. R., Varnous, S., Becane, H. M., Hammouda, E. H., Merlini, L., et al. (1999). Mutations in the gene encoding lamin A/C cause autosomal dominant Emery-Dreifuss muscular dystrophy. *Nature Genetics*, *21*, 285–288.

Bonne, G., & Levy, N. (2003). LMNA mutations in atypical Werner's syndrome. *Lancet*, *362*, 1585–1586, author reply 1586.

Bourgeois, B., Gilquin, B., Tellier-Lebegue, C., Ostlund, C., Wu, W., Perez, J., et al. (2013). Inhibition of TGF-beta signaling at the nuclear envelope: Characterization of interactions between MAN1, Smad2 and Smad3, and PPM1A. *Science Signaling*, *6*, ra49.

Brachner, A., Reipert, S., Foisner, R., & Gotzmann, J. (2005). LEM2 is a novel MAN1-related inner nuclear membrane protein associated with A-type lamins. *Journal of Cell Science*, *118*, 5797–5810.

Bridger, J. M., & Kill, I. R. (2004). Aging of Hutchinson-Gilford progeria syndrome fibroblasts is characterised by hyperproliferation and increased apoptosis. *Experimental Gerontology*, *39*, 717–724.

Brodsky, G. L., Muntoni, F., Miocic, S., Sinagra, G., Sewry, C., & Mestroni, L. (2000). Lamin A/C gene mutation associated with dilated cardiomyopathy with variable skeletal muscle involvement. *Circulation*, *101*, 473–476.

Burke, B., & Stewart, C. L. (2013). The nuclear lamins: Flexibility in function. *Nature Reviews Molecular Cell Biology*, *14*, 13–24.

Cao, H., & Hegele, R. A. (2000). Nuclear lamin A/C R482Q mutation in Canadian kindreds with Dunnigan-type familial partial lipodystrophy. *Human Molecular Genetics*, *9*, 109–112.

Chaouch, M., Allal, Y., De Sandre-Giovannoli, A., Vallat, J. M., Amer-el-Khedoud, A., Kassouri, N., et al. (2003). The phenotypic manifestations of autosomal recessive axonal Charcot-Marie-Tooth due to a mutation in Lamin A/C gene. *Neuromuscular Disorders*, *13*, 60–67.

Chen, C. Y., Chi, Y. H., Mutalif, R. A., Starost, M. F., Myers, T. G., Anderson, S. A., et al. (2012). Accumulation of the inner nuclear envelope protein Sun1 is pathogenic in progeric and dystrophic laminopathies. *Cell*, *149*, 565–577.

Chen, L., & Oshima, J. (2002). Werner syndrome. *Journal of Biomedicine and Biotechnology*, *2*, 46–54.

Chen, J. L., Walton, K. L., Winbanks, C. E., Murphy, K. T., Thomson, R. E., Makanji, Y., et al. (2014). Elevated expression of activins promotes muscle wasting and cachexia. *FASEB Journal, 4*, 1711–1723.

Chi, Y. H., Cheng, L. I., Myers, T., Ward, J. M., Williams, E., Su, Q., et al. (2009). Requirement for Sun1 in the expression of meiotic reproductive genes and piRNA. *Development, 136*, 965–973.

Choi, J. C., Muchir, A., Wu, W., Iwata, S., Homma, S., Morrow, J. P., et al. (2012). Temsirolimus activates autophagy and ameliorates cardiomyopathy caused by lamin A/C gene mutation. *Science Translational Medicine, 4*, 144ra102.

Coffinier, C., Jung, H. J., Li, Z., Nobumori, C., Yun, U. J., Farber, E. A., et al. (2010). Direct synthesis of lamin A, bypassing prelamin a processing, causes misshapen nuclei in fibroblasts but no detectable pathology in mice. *Journal of Biological Chemistry, 285*, 20818–20826.

Coffinier, C., Jung, H. J., Nobumori, C., Chang, S., Tu, Y., Barnes, R. H., 2nd, et al. (2011). Deficiencies in lamin B1 and lamin B2 cause neurodevelopmental defects and distinct nuclear shape abnormalities in neurons. *Molecular Biology of the Cell, 22*, 4683–4693.

Cohen, T. V., Gnocchi, V. F., Cohen, J. E., Phadke, A., Liu, H., Ellis, J. A., et al. (2013). Defective skeletal muscle growth in lamin A/C-deficient mice is rescued by loss of Lap2alpha. *Human Molecular Genetics, 22*, 2852–2869.

Cohen, T. V., Klarmann, K. D., Sakchaisri, K., Cooper, J. P., Kuhns, D., Anver, M., et al. (2008). The lamin B receptor under transcriptional control of C/EBPepsilon is required for morphological but not functional maturation of neutrophils. *Human Molecular Genetics, 17*, 2921–2933.

Cohen, T. V., Kosti, O., & Stewart, C. L. (2007). The nuclear envelope protein MAN1 regulates TGFbeta signaling and vasculogenesis in the embryonic yolk sac. *Development, 134*, 1385–1395.

Columbaro, M., Mattioli, E., Maraldi, N. M., Ortolani, M., Gasparini, L., D'Apice, M. R., et al. (2013). Oct-1 recruitment to the nuclear envelope in adult-onset autosomal dominant leukodystrophy. *Biochimica et Biophysica Acta, 1832*, 411–420.

Cookson, M. R., & Clarimon, J. (2005). Dystonia and the nuclear envelope. *Neuron, 48*, 875–877.

Corrigan, D. P., Kuszczak, D., Rusinol, A. E., Thewke, D. P., Hrycyna, C. A., Michaelis, S., et al. (2005). Prelamin A endoproteolytic processing in vitro by recombinant Zmpste24. *Biochemical Journal, 387*, 129–138.

Cremer, T., & Cremer, M. (2010). Chromosome territories. *Cold Spring Harbor Perspectives in Biology, 2*, a003889.

Crisp, M., Liu, Q., Roux, K., Rattner, J. B., Shanahan, C., Burke, B., et al. (2006). Coupling of the nucleus and cytoplasm: Role of the LINC complex. *Journal of Cell Biology, 172*, 41–53.

Csoka, A. B., Cao, H., Sammak, P. J., Constantinescu, D., Schatten, G. P., & Hegele, R. A. (2004). Novel lamin A/C gene (LMNA) mutations in atypical progeroid syndromes. *Journal of Medical Genetics, 41*, 304–308.

Cutler, D. A., Sullivan, T., Marcus-Samuels, B., Stewart, C. L., & Reitman, M. L. (2002). Characterization of adiposity and metabolism in Lmna-deficient mice. *Biochemical and Biophysical Research Communications, 291*, 522–527.

Dang, M. T., Yokoi, F., McNaught, K. S., Jengelley, T. A., Jackson, T., Li, J., et al. (2005). Generation and characterization of Dyt1 DeltaGAG knock-in mouse as a model for early-onset dystonia. *Experimental Neurology, 196*, 452–463.

Dauer, W. T., & Worman, H. J. (2009). The nuclear envelope as a signaling node in development and disease. *Developmental Cell, 17*, 626–638.

Davies, B. S., Fong, L. G., Yang, S. H., Coffinier, C., & Young, S. G. (2009). The posttranslational processing of prelamin A and disease. *Annual Review of Genomics and Human Genetics*, *10*, 153–174.
De Castro, S. C., Malhas, A., Leung, K. Y., Gustavsson, P., Vaux, D. J., Copp, A. J., et al. (2012). Lamin b1 polymorphism influences morphology of the nuclear envelope, cell cycle progression, and risk of neural tube defects in mice. *PLoS Genetics*, *8*, e1003059.
de Las Heras, J. I., Meinke, P., Batrakou, D. G., Srsen, V., Zuleger, N., Kerr, A. R., et al. (2013). Tissue specificity in the nuclear envelope supports its functional complexity. *Nucleus*, *4*, 460–477.
De Sandre-Giovannoli, A., Bernard, R., Cau, P., Navarro, C., Amiel, J., Boccaccio, I., et al. (2003). Lamin a truncation in Hutchinson-Gilford progeria. *Science*, *300*, 2055.
De Sandre-Giovannoli, A., Chaouch, M., Kozlov, S., Vallat, J. M., Tazir, M., Kassouri, N., et al. (2002). Homozygous defects in LMNA, encoding lamin A/C nuclear-envelope proteins, cause autosomal recessive axonal neuropathy in human (Charcot-Marie-Tooth disorder type 2) and mouse. *American Journal of Human Genetics*, *70*, 726–736.
Dechat, T., Korbei, B., Vaughan, O. A., Vlcek, S., Hutchison, C. J., & Foisner, R. (2000). Lamina-associated polypeptide 2alpha binds intranuclear A-type lamins. *Journal of Cell Science*, *113*(Pt 19), 3473–3484.
Dechat, T., Shimi, T., Adam, S. A., Rusinol, A. E., Andres, D. A., Spielmann, H. P., et al. (2007). Alterations in mitosis and cell cycle progression caused by a mutant lamin A known to accelerate human aging. *Proceedings of the National Academy of Sciences of the United States of America*, *104*, 4955–4960.
Dedeic, Z., Cetera, M., Cohen, T. V., & Holaska, J. M. (2011). Emerin inhibits Lmo7 binding to the Pax3 and MyoD promoters and expression of myoblast proliferation genes. *Journal of Cell Science*, *124*, 1691–1702.
Dheedene, A., Deleye, S., Hellemans, J., Staelens, S., Vandenberghe, S., & Mortier, G. (2009). The heterozygous Lemd3 +/GT mouse is not a murine model for osteopoikilosis in humans. *Calcified Tissue International*, *85*, 546–551.
Dhe-Paganon, S., Werner, E. D., Chi, Y. I., & Shoelson, S. E. (2002). Structure of the globular tail of nuclear lamin. *Journal of Biological Chemistry*, *277*, 17381–17384.
Ding, X., Xu, R., Yu, J., Xu, T., Zhuang, Y., & Han, M. (2007). SUN1 is required for telomere attachment to nuclear envelope and gametogenesis in mice. *Developmental Cell*, *12*, 863–872.
Dorner, D., Gotzmann, J., & Foisner, R. (2007). Nucleoplasmic lamins and their interaction partners, LAP2alpha, Rb, and BAF, in transcriptional regulation. *FEBS Journal*, *274*, 1362–1373.
Dorner, D., Vlcek, S., Foeger, N., Gajewski, A., Makolm, C., Gotzmann, J., et al. (2006). Lamina-associated polypeptide 2alpha regulates cell cycle progression and differentiation via the retinoblastoma-E2F pathway. *Journal of Cell Biology*, *173*, 83–93.
Dreesen, O., Chojnowski, A., Ong, P. F., Zhao, T. Y., Common, J. E., Lunny, D., et al. (2013). Lamin B1 fluctuations have differential effects on cellular proliferation and senescence. *Journal of Cell Biology*, *200*, 605–617.
Dupre, N., Bouchard, J. P., Gros-Louis, F., & Rouleau, G. A. (2007). Mutations in SYNE-1 lead to a newly discovered form of autosomal recessive cerebellar ataxia. *Médecine Sciences*, *23*, 261–262.
Dwyer, N., & Blobel, G. (1976). A modified procedure for the isolation of a pore complex-lamina fraction from rat liver nuclei. *Journal of Cell Biology*, *70*, 581–591.
Eriksson, M., Brown, W. T., Gordon, L. B., Glynn, M. W., Singer, J., Scott, L., et al. (2003). Recurrent de novo point mutations in lamin A cause Hutchinson-Gilford progeria syndrome. *Nature*, *423*, 293–298.
Ershler, W. B., Ferrucci, L., & Longo, D. L. (2008). Hutchinson-Gilford progeria syndrome. *New England Journal of Medicine*, *358*, 2409–2410, author reply 2410–2401.

Fatkin, D., MacRae, C., Sasaki, T., Wolff, M. R., Porcu, M., Frenneaux, M., et al. (1999). Missense mutations in the rod domain of the lamin A/C gene as causes of dilated cardiomyopathy and conduction-system disease. *New England Journal of Medicine, 341*, 1715–1724.

Finlan, L. E., Sproul, D., Thomson, I., Boyle, S., Kerr, E., Perry, P., et al. (2008). Recruitment to the nuclear periphery can alter expression of genes in human cells. *PLoS Genetics, 4*, e1000039.

Fisher, D. Z., Chaudhary, N., & Blobel, G. (1986). cDNA sequencing of nuclear lamins A and C reveals primary and secondary structural homology to intermediate filament proteins. *Proceedings of the National Academy of Sciences of the United States of America, 83*, 6450–6454.

Fong, L. G., Frost, D., Meta, M., Qiao, X., Yang, S. H., Coffinier, C., et al. (2006). A protein farnesyltransferase inhibitor ameliorates disease in a mouse model of progeria. *Science, 311*, 1621–1623.

Fong, L. G., Ng, J. K., Lammerding, J., Vickers, T. A., Meta, M., Cote, N., et al. (2006). Prelamin A and lamin A appear to be dispensable in the nuclear lamina. *Journal of Clinical Investigation, 116*, 743–752.

Fong, L. G., Ng, J. K., Meta, M., Cote, N., Yang, S. H., Stewart, C. L., et al. (2004). Heterozygosity for Lmna deficiency eliminates the progeria-like phenotypes in Zmpste24-deficient mice. *Proceedings of the National Academy of Sciences of the United States of America, 101*, 18111–18116.

Freund, A., Laberge, R. M., Demaria, M., & Campisi, J. (2012). Lamin B1 loss is a senescence-associated biomarker. *Molecular Biology of the Cell, 23*, 2066–2075.

Fruhbeck, G. (2008). Overview of adipose tissue and its role in obesity and metabolic disorders. *Methods in Molecular Biology, 456*, 1–22.

Fry, M. (2002). The Werner syndrome helicase-nuclease—One protein, many mysteries. *Science of Aging Knowledge Environment, 2002*, re2.

Furukawa, K., & Hotta, Y. (1993). cDNA cloning of a germ cell specific lamin B3 from mouse spermatocytes and analysis of its function by ectopic expression in somatic cells. *EMBO Journal, 12*, 97–106.

Furukawa, K., Inagaki, H., & Hotta, Y. (1994). Identification and cloning of an mRNA coding for a germ cell-specific A-type lamin in mice. *Experimental Cell Research, 212*, 426–430.

Gaines, P., Tien, C. W., Olins, A. L., Olins, D. E., Shultz, L. D., Carney, L., et al. (2008). Mouse neutrophils lacking lamin B-receptor expression exhibit aberrant development and lack critical functional responses. *Experimental Hematology, 36*, 965–976.

Gerace, L., Comeau, C., & Benson, M. (1984). Organization and modulation of nuclear lamina structure. *Journal of Cell Science. Supplement, 1*, 137–160.

Gerace, L., & Huber, M. D. (2012). Nuclear lamina at the crossroads of the cytoplasm and nucleus. *Journal of Structural Biology, 177*, 24–31.

Glynn, M. W., & Glover, T. W. (2005). Incomplete processing of mutant lamin A in Hutchinson-Gilford progeria leads to nuclear abnormalities, which are reversed by farnesyltransferase inhibition. *Human Molecular Genetics, 14*, 2959–2969.

Goldberg, M. W., Huttenlauch, I., Hutchison, C. J., & Stick, R. (2008). Filaments made from A- and B-type lamins differ in structure and organization. *Journal of Cell Science, 121*, 215–225.

Goodchild, R. E., & Dauer, W. T. (2005). The AAA+ protein torsinA interacts with a conserved domain present in LAP1 and a novel ER protein. *Journal of Cell Biology, 168*, 855–862.

Goodchild, R. E., Kim, C. E., & Dauer, W. T. (2005). Loss of the dystonia-associated protein torsinA selectively disrupts the neuronal nuclear envelope. *Neuron, 48*, 923–932.

Gordon, L. B., Harten, I. A., Patti, M. E., & Lichtenstein, A. H. (2005). Reduced adiponectin and HDL cholesterol without elevated C-reactive protein: Clues to the biology of premature atherosclerosis in Hutchinson-Gilford Progeria Syndrome. *Journal of Pediatrics*, *146*, 336–341.

Gordon, L. B., Kleinman, M. E., Miller, D. T., Neuberg, D. S., Giobbie-Hurder, A., Gerhard-Herman, M., et al. (2012). Clinical trial of a farnesyltransferase inhibitor in children with Hutchinson-Gilford progeria syndrome. *Proceedings of the National Academy of Sciences of the United States of America*, *109*, 16666–16671.

Gordon, L. B., McCarten, K. M., Giobbie-Hurder, A., Machan, J. T., Campbell, S. E., Berns, S. D., et al. (2007). Disease progression in Hutchinson-Gilford progeria syndrome: Impact on growth and development. *Pediatrics*, *120*, 824–833.

Gotic, I., Leschnik, M., Kolm, U., Markovic, M., Haubner, B. J., Biadasiewicz, K., et al. (2010). Lamina-associated polypeptide 2alpha loss impairs heart function and stress response in mice. *Circulation Research*, *106*, 346–353.

Gotic, I., Schmidt, W. M., Biadasiewicz, K., Leschnik, M., Spilka, R., Braun, J., et al. (2010). Loss of LAP2 alpha delays satellite cell differentiation and affects postnatal fiber-type determination. *Stem Cells*, *28*, 480–488.

Grady, R. M., Starr, D. A., Ackerman, G. L., Sanes, J. R., & Han, M. (2005). Syne proteins anchor muscle nuclei at the neuromuscular junction. *Proceedings of the National Academy of Sciences of the United States of America*, *102*, 4359–4364.

Granata, A., Watson, R., Collinson, L. M., Schiavo, G., & Warner, T. T. (2008). The dystonia-associated protein torsinA modulates synaptic vesicle recycling. *Journal of Biological Chemistry*, *283*, 7568–7579.

Granger, B., Gueneau, L., Drouin-Garraud, V., Pedergnana, V., Gagnon, F., Ben Yaou, R., et al. (2011). Modifier locus of the skeletal muscle involvement in Emery-Dreifuss muscular dystrophy. *Human Genetics*, *129*, 149–159.

Grossman, E., Medalia, O., & Zwerger, M. (2012). Functional architecture of the nuclear pore complex. *Annual Review of Biophysics*, *41*, 557–584.

Gruenbaum, Y., Margalit, A., Goldman, R. D., Shumaker, D. K., & Wilson, K. L. (2005). The nuclear lamina comes of age. *Nature Reviews Molecular Cell Biology*, *6*, 21–31.

Guelen, L., Pagie, L., Brasset, E., Meuleman, W., Faza, M. B., Talhout, W., et al. (2008). Domain organization of human chromosomes revealed by mapping of nuclear lamina interactions. *Nature*, *453*, 948–951.

Gueneau, L., Bertrand, A. T., Jais, J. P., Salih, M. A., Stojkovic, T., Wehnert, M., et al. (2009). Mutations of the FHL1 gene cause Emery-Dreifuss muscular dystrophy. *American Journal of Human Genetics*, *85*, 338–353.

Haque, F., Lloyd, D. J., Smallwood, D. T., Dent, C. L., Shanahan, C. M., Fry, A. M., et al. (2006). SUN1 interacts with nuclear lamin A and cytoplasmic nesprins to provide a physical connection between the nuclear lamina and the cytoskeleton. *Molecular and Cellular Biology*, *26*, 3738–3751.

Harrison, D. E., Strong, R., Sharp, Z. D., Nelson, J. F., Astle, C. M., Flurkey, K., et al. (2009). Rapamycin fed late in life extends lifespan in genetically heterogeneous mice. *Nature*, *460*, 392–395.

Hellemans, J., Preobrazhenska, O., Willaert, A., Debeer, P., Verdonk, P. C., Costa, T., et al. (2004). Loss-of-function mutations in LEMD3 result in osteopoikilosis, Buschke-Ollendorff syndrome and melorheostosis. *Nature Genetics*, *36*, 1213–1218.

Heng, M. Y., Lin, S. T., Verret, L., Huang, Y., Kamiya, S., Padiath, Q. S., et al. (2013). Lamin B1 mediates cell-autonomous neuropathology in a leukodystrophy mouse model. *Journal of Clinical Investigation*, *123*, 2719–2729.

Hernandez, L., Roux, K. J., Wong, E. S., Mounkes, L. C., Mutalif, R., Navasankari, R., et al. (2010). Functional coupling between the extracellular matrix and nuclear lamina by Wnt signaling in progeria. *Developmental Cell*, *19*, 413–425.

Hewett, J. W., Tannous, B., Niland, B. P., Nery, F. C., Zeng, J., Li, Y., et al. (2007). Mutant torsinA interferes with protein processing through the secretory pathway in DYT1 dystonia cells. *Proceedings of the National Academy of Sciences of the United States of America, 104*, 7271–7276.

Hickson, I. D. (2003). RecQ helicases: Caretakers of the genome. *Nature Reviews Cancer, 3*, 169–178.

Ho, C. Y., Jaalouk, D. E., Vartiainen, M. K., & Lammerding, J. (2013). lamin A/C and emerin regulate MKL1-SRF activity by modulating actin dynamics. *Nature, 497*, 507–511.

Hoffmann, K., Dreger, C. K., Olins, A. L., Olins, D. E., Shultz, L. D., Lucke, B., et al. (2002). Mutations in the gene encoding the lamin B receptor produce an altered nuclear morphology in granulocytes (Pelger-Huet anomaly). *Nature Genetics, 31*, 410–414.

Hoffmann, K., Sperling, K., Olins, A. L., & Olins, D. E. (2007). The granulocyte nucleus and lamin B receptor: Avoiding the ovoid. *Chromosoma, 116*, 227–235.

Hoger, T. H., Grund, C., Franke, W. W., & Krohne, G. (1991). Immunolocalization of lamins in the thick nuclear lamina of human synovial cells. *European Journal of Cell Biology, 54*, 150–156.

Holtz, D., Tanaka, R. A., Hartwig, J., & McKeon, F. (1989). The CaaX motif of lamin A functions in conjunction with the nuclear localization signal to target assembly to the nuclear envelope. *Cell, 59*, 969–977.

Horn, H. F., Brownstein, Z., Lenz, D. R., Shivatzki, S., Dror, A. A., Dagan-Rosenfeld, O., et al. (2013). The LINC complex is essential for hearing. *Journal of Clinical Investigation, 123*, 740–750.

Horn, H. F., Kim, D. I., Wright, G. D., Wong, E. S., Stewart, C. L., Burke, B., et al. (2013). A mammalian KASH domain protein coupling meiotic chromosomes to the cytoskeleton. *Journal of Cell Biology, 202*, 1023–1039.

Ibrahim, M. X., Sayin, V. I., Akula, M. K., Liu, M., Fong, L. G., Young, S. G., et al. (2013). Targeting isoprenylcysteine methylation ameliorates disease in a mouse model of progeria. *Science, 340*, 1330–1333.

Ishimura, A., Ng, J. K., Taira, M., Young, S. G., & Osada, S. (2006). Man1, an inner nuclear membrane protein, regulates vascular remodeling by modulating transforming growth factor beta signaling. *Development, 133*, 3919–3928.

Ivorra, C., Kubicek, M., Gonzalez, J. M., Sanz-Gonzalez, S. M., Alvarez-Barrientos, A., O'Connor, J. E., et al. (2006). A mechanism of AP-1 suppression through interaction of c-Fos with lamin A/C. *Genes & Development, 20*, 307–320.

Jahn, D., Schramm, S., Schnolzer, M., Heilmann, C. J., de Koster, C. G., Schutz, W., et al. (2012). A truncated lamin A in the Lmna -/- mouse line: Implications for the understanding of laminopathies. *Nucleus, 3*, 463–474.

Jung, H. J., Coffinier, C., Choe, Y., Beigneux, A. P., Davies, B. S., Yang, S. H., et al. (2012). Regulation of prelamin A but not lamin C by miR-9, a brain-specific microRNA. *Proceedings of the National Academy of Sciences of the United States of America, 109*, E423–E431.

Jung, H. J., Nobumori, C., Goulbourne, C. N., Tu, Y., Lee, J. M., Tatar, A., et al. (2013). Farnesylation of lamin B1 is important for retention of nuclear chromatin during neuronal migration. *Proceedings of the National Academy of Sciences of the United States of America, 110*, E1923–E1932.

Jung, H. J., Tu, Y., Yang, S. H., Tatar, A., Nobumori, C., Wu, D., et al. (2014). New Lmna knock-in mice provide a molecular mechanism for the 'segmental aging' in Hutchinson-Gilford progeria syndrome. *Human Molecular Genetics, 23*, 1506–1515.

Ketema, M., Kreft, M., Secades, P., Janssen, H., & Sonnenberg, A. (2013). Nesprin-3 connects plectin and vimentin to the nuclear envelope of Sertoli cells but is not required for Sertoli cell function in spermatogenesis. *Molecular Biology of the Cell, 24*, 2454–2466.

Kim, Y., Sharov, A. A., McDole, K., Cheng, M., Hao, H., Fan, C. M., et al. (2011). Mouse B-type lamins are required for proper organogenesis but not by embryonic stem cells. *Science, 334*, 1706–1710.

Kim, Y., & Zheng, Y. (2013). Generation and characterization of a conditional deletion allele for Lmna in mice. *Biochemical and Biophysical Research Communications, 440*, 8–13.

Kim, Y., Zheng, X., & Zheng, Y. (2013). Proliferation and differentiation of mouse embryonic stem cells lacking all lamins. *Cell Research, 23*, 1420–1423.

Kitten, G. T., & Nigg, E. A. (1991). The CaaX motif is required for isoprenylation, carboxyl methylation, and nuclear membrane association of lamin B2. *Journal of Cell Biology, 113*, 13–23.

Klement, J. F., Rice, N. R., Car, B. D., Abbondanzo, S. J., Powers, G. D., Bhatt, P. H., et al. (1996). IkappaBalpha deficiency results in a sustained NF-kappaB response and severe widespread dermatitis in mice. *Molecular and Cellular Biology, 16*, 2341–2349.

Korfali, N., Wilkie, G. S., Swanson, S. K., Srsen, V., Batrakou, D. G., Fairley, E. A., et al. (2010). The leukocyte nuclear envelope proteome varies with cell activation and contains novel transmembrane proteins that affect genome architecture. *Molecular & Cellular Proteomics, 9*, 2571–2585.

Korfali, N., Wilkie, G. S., Swanson, S. K., Srsen, V., de Las Heras, J., Batrakou, D. G., et al. (2012). The nuclear envelope proteome differs notably between tissues. *Nucleus, 3*, 552–564.

Krimm, I., Ostlund, C., Gilquin, B., Couprie, J., Hossenlopp, P., Mornon, J. P., et al. (2002). The Ig-like structure of the C-terminal domain of lamin A/C, mutated in muscular dystrophies, cardiomyopathy, and partial lipodystrophy. *Structure (Camb), 10*, 811–823.

Kubben, N., Voncken, J. W., Konings, G., van Weeghel, M., van den Hoogenhof, M. M., Gijbels, M., et al. (2011). Post-natal myogenic and adipogenic developmental: Defects and metabolic impairment upon loss of A-type lamins. *Nucleus, 2*, 195–207.

Kudlow, B. A., Stanfel, M. N., Burtner, C. R., Johnston, E. D., & Kennedy, B. K. (2008). Suppression of proliferative defects associated with processing-defective lamin A mutants by hTERT or inactivation of p53. *Molecular Biology of the Cell, 19*, 5238–5248.

Kumaran, R. I., Thakar, R., & Spector, D. L. (2008). Chromatin dynamics and gene positioning. *Cell, 132*, 929–934.

Lammerding, J., Fong, L. G., Ji, J. Y., Reue, K., Stewart, C. L., Young, S. G., et al. (2006). Lamins A and C but not lamin B1 regulate nuclear mechanics. *Journal of Biological Chemistry, 281*, 25768–25780.

Lammerding, J., & Lee, R. T. (2005). The nuclear membrane and mechanotransduction: Impaired nuclear mechanics and mechanotransduction in lamin A/C deficient cells. *Novartis Foundation Symposium, 264*, 264–273, discussion 273–268.

Lammerding, J., Schulze, P. C., Takahashi, T., Kozlov, S., Sullivan, T., Kamm, R. D., et al. (2004). Lamin A/C deficiency causes defective nuclear mechanics and mechanotransduction. *Journal of Clinical Investigation, 113*, 370–378.

Lee, J. S., Hale, C. M., Panorchan, P., Khatau, S. B., George, J. P., Tseng, Y., et al. (2007). Nuclear lamin A/C deficiency induces defects in cell mechanics, polarization, and migration. *Biophysical Journal, 93*, 2542–2552.

Lee, K. K., & Wilson, K. L. (2004). All in the family: Evidence for four new LEM-domain proteins Lem2 (NET-25), Lem3, Lem4 and Lem5 in the human genome. *Symposia of the Society for Experimental Biology, 56*, 329–339.

Lei, K., Zhang, X., Ding, X., Guo, X., Chen, M., Zhu, B., et al. (2009). SUN1 and SUN2 play critical but partially redundant roles in anchoring nuclei in skeletal muscle cells in mice. *Proceedings of the National Academy of Sciences of the United States of America, 106*, 10207–10212.

Lelliott, C. J., Logie, L., Sewter, C. P., Berger, D., Jani, P., Blows, F., et al. (2002). Lamin expression in human adipose cells in relation to anatomical site and differentiation state. *Journal of Clinical Endocrinology and Metabolism, 87*, 728–734.

Li, W., Yeo, L. S., Vidal, C., McCorquodale, T., Herrmann, M., Fatkin, D., et al. (2011). Decreased bone formation and osteopenia in lamin a/c-deficient mice. *PLoS One*, *6*, e19313.

Lin, F., Blake, D. L., Callebaut, I., Skerjanc, I. S., Holmer, L., McBurney, M. W., et al. (2000). MAN1, an inner nuclear membrane protein that shares the LEM domain with lamina-associated polypeptide 2 and emerin. *Journal of Biological Chemistry*, *275*, 4840–4847.

Lin, F., Morrison, J. M., Wu, W., & Worman, H. J. (2005). MAN1, an integral protein of the inner nuclear membrane, binds Smad2 and Smad3 and antagonizes transforming growth factor-beta signaling. *Human Molecular Genetics*, *14*, 437–445.

Lin, F., & Worman, H. J. (1993). Structural organization of the human gene encoding nuclear lamin A and nuclear lamin C. *Journal of Biological Chemistry*, *268*, 16321–16326.

Lin, F., & Worman, H. J. (1995). Structural organization of the human gene (LMNB1) encoding nuclear lamin B1. *Genomics*, *27*, 230–236.

Liu, Q., Pante, N., Misteli, T., Elsagga, M., Crisp, M., Hodzic, D., et al. (2007). Functional association of Sun1 with nuclear pore complexes. *Journal of Cell Biology*, *178*, 785–798.

Liu, Y., Rusinol, A., Sinensky, M., Wang, Y., & Zou, Y. (2006). DNA damage responses in progeroid syndromes arise from defective maturation of prelamin A. *Journal of Cell Science*, *119*, 4644–4649.

Liu, B., Wang, J., Chan, K. M., Tjia, W. M., Deng, W., Guan, X., et al. (2005). Genomic instability in laminopathy-based premature aging. *Nature Medicine*, *11*, 780–785.

Ma, Y., Cai, S., Lv, Q., Jiang, Q., Zhang, Q., Sodmergen, et al. (2007). Lamin B receptor plays a role in stimulating nuclear envelope production and targeting membrane vesicles to chromatin during nuclear envelope assembly through direct interaction with importin beta. *Journal of Cell Science*, *120*, 520–530.

Machiels, B. M., Zorenc, A. H., Endert, J. M., Kuijpers, H. J., van Eys, G. J., Ramaekers, F. C., et al. (1996). An alternative splicing product of the lamin A/C gene lacks exon 10. *Journal of Biological Chemistry*, *271*, 9249–9253.

Makatsori, D., Kourmouli, N., Polioudaki, H., Shultz, L. D., McLean, K., Theodoropoulos, P. A., et al. (2004). The inner nuclear membrane protein lamin B receptor forms distinct microdomains and links epigenetically marked chromatin to the nuclear envelope. *Journal of Biological Chemistry*, *279*, 25567–25573.

Maniotis, A. J., Chen, C. S., & Ingber, D. E. (1997). Demonstration of mechanical connections between integrins, cytoskeletal filaments, and nucleoplasm that stabilize nuclear structure. *Proceedings of the National Academy of Sciences of the United States of America*, *94*, 849–854.

Mao, Y. S., Zhang, B., & Spector, D. L. (2011). Biogenesis and function of nuclear bodies. *Trends in Genetics*, *27*, 295–306.

Margalit, A., Brachner, A., Gotzmann, J., Foisner, R., & Gruenbaum, Y. (2007). Barrier-to-autointegration factor—A BAFfling little protein. *Trends in Cell Biology*, *17*, 202–208.

Marino, G., Ugalde, A. P., Salvador-Montoliu, N., Varela, I., Quiros, P. M., Cadinanos, J., et al. (2008). Premature aging in mice activates a systemic metabolic response involving autophagy induction. *Human Molecular Genetics*, *17*, 2196–2211.

Martin, G. M. (1989). Genetic modulation of the senescent phenotype in *Homo sapiens*. *Genome*, *31*, 390–397.

Mattout, A., Pike, B. L., Towbin, B. D., Bank, E. M., Gonzalez-Sandoval, A., Stadler, M. B., et al. (2011). An EDMD mutation in *C. elegans* lamin blocks muscle-specific gene relocation and compromises muscle integrity. *Current Biology*, *21*, 1603–1614.

McKenna, T., Rosengardten, Y., Viceconte, N., Baek, J. H., Grochova, D., & Eriksson, M. (2013). Embryonic expression of the common progeroid lamin A splice mutation arrests postnatal skin development. *Aging Cell*.

McKeon, F. D., Kirschner, M. W., & Caput, D. (1986). Homologies in both primary and secondary structure between nuclear envelope and intermediate filament proteins. *Nature, 319*, 463–468.
Mekhail, K., & Moazed, D. (2010). The nuclear envelope in genome organization, expression and stability. *Nature Reviews Molecular Cell Biology, 11*, 317–328.
Melcon, G., Kozlov, S., Cutler, D. A., Sullivan, T., Hernandez, L., Zhao, P., et al. (2006). Loss of emerin at the nuclear envelope disrupts the Rb1/E2F and MyoD pathways during muscle regeneration. *Human Molecular Genetics, 15*, 637–651.
Merideth, M. A., Gordon, L. B., Clauss, S., Sachdev, V., Smith, A. C., Perry, M. B., et al. (2008). Phenotype and course of Hutchinson-Gilford progeria syndrome. *New England Journal of Medicine, 358*, 592–604.
Morris, G. E. (2001). The role of the nuclear envelope in Emery-Dreifuss muscular dystrophy. *Trends in Molecular Medicine, 7*, 572–577.
Moulson, C. L., Go, G., Gardner, J. M., van der Wal, A. C., Smitt, J. H., van Hagen, J. M., et al. (2005). Homozygous and compound heterozygous mutations in ZMPSTE24 cause the laminopathy restrictive dermopathy. *Journal of Investigative Dermatology, 125*, 913–919.
Mounkes, L. C., Kozlov, S., Hernandez, L., Sullivan, T., & Stewart, C. L. (2003). A progeroid syndrome in mice is caused by defects in A-type lamins. *Nature, 423*, 298–301.
Mounkes, L. C., Kozlov, S. V., Rottman, J. N., & Stewart, C. L. (2005). Expression of an LMNA-N195K variant of A-type lamins results in cardiac conduction defects and death in mice. *Human Molecular Genetics, 14*, 2167–2180.
Muchir, A., Bonne, G., van der Kooi, A. J., van Meegen, M., Baas, F., Bolhuis, P. A., et al. (2000). Identification of mutations in the gene encoding lamins A/C in autosomal dominant limb girdle muscular dystrophy with atrioventricular conduction disturbances (LGMD1B). *Human Molecular Genetics, 9*, 1453–1459.
Muchir, A., Pavlidis, P., Bonne, G., Hayashi, Y. K., & Worman, H. J. (2007). Activation of MAPK in hearts of EMD null mice: Similarities between mouse models of X-linked and autosomal dominant Emery Dreifuss muscular dystrophy. *Human Molecular Genetics, 16*, 1884–1895.
Muchir, A., Pavlidis, P., Decostre, V., Herron, A. J., Arimura, T., Bonne, G., et al. (2007). Activation of MAPK pathways links LMNA mutations to cardiomyopathy in Emery-Dreifuss muscular dystrophy. *Journal of Clinical Investigation, 117*, 1282–1293.
Muchir, A., Shan, J., Bonne, G., Lehnart, S. E., & Worman, H. J. (2009). Inhibition of extracellular signal-regulated kinase signaling to prevent cardiomyopathy caused by mutation in the gene encoding A-type lamins. *Human Molecular Genetics, 18*, 241–247.
Naetar, N., Korbei, B., Kozlov, S., Kerenyi, M. A., Dorner, D., Kral, R., et al. (2008). Loss of nucleoplasmic LAP2alpha-lamin A complexes causes erythroid and epidermal progenitor hyperproliferation. *Nature Cell Biology, 10*, 1341–1348.
Navarro, C. L., Cadinanos, J., De Sandre-Giovannoli, A., Bernard, R., Courrier, S., Boccaccio, I., et al. (2005). Loss of ZMPSTE24 (FACE-1) causes autosomal recessive restrictive dermopathy and accumulation of Lamin A precursors. *Human Molecular Genetics, 14*, 1503–1513.
Nikolova, V., Leimena, C., McMahon, A. C., Tan, J. C., Chandar, S., Jogia, D., et al. (2004). Defects in nuclear structure and function promote dilated cardiomyopathy in lamin A/C-deficient mice. *Journal of Clinical Investigation, 113*, 357–369.
Novelli, G., Muchir, A., Sangiuolo, F., Helbling-Leclerc, A., D'Apice, M. R., Massart, C., et al. (2002). Mandibuloacral dysplasia is caused by a mutation in LMNA-encoding lamin A/C. *American Journal of Human Genetics, 71*, 426–431.
Odgren, P. R., Pratt, C. H., Mackay, C. A., Mason-Savas, A., Curtain, M., Shopland, L., et al. (2010). Disheveled hair and ear (Dhe), a spontaneous mouse Lmna mutation modeling human laminopathies. *PLoS One, 5*, e9959.

Olins, A. L., Zwerger, M., Herrmann, H., Zentgraf, H., Simon, A. J., Monestier, M., et al. (2008). The human granulocyte nucleus: Unusual nuclear envelope and heterochromatin composition. *European Journal of Cell Biology, 87*, 279–290.
Olive, M., Harten, I., Mitchell, R., Beers, J. K., Djabali, K., Cao, K., et al. (2010). Cardiovascular pathology in Hutchinson-Gilford progeria: Correlation with the vascular pathology of aging. *Arteriosclerosis, Thrombosis, and Vascular Biology, 30*, 2301–2309.
Oshima, J., Garg, A., Martin, G. M., & Kennedy, B. K. (2003). LMNA mutations in atypical Werner's syndrome. *Lancet, 362*, 1586.
Osorio, F. G., Barcena, C., Soria-Valles, C., Ramsay, A. J., de Carlos, F., Cobo, J., et al. (2012). Nuclear lamina defects cause ATM-dependent NF-kappaB activation and link accelerated aging to a systemic inflammatory response. *Genes & Development, 26*, 2311–2324.
Osorio, F. G., Navarro, C. L., Cadinanos, J., Lopez-Mejia, I. C., Quiros, P. M., Bartoli, C., et al. (2011). Splicing-directed therapy in a new mouse model of human accelerated aging. *Science Translational Medicine, 3*, 106ra107.
Oza, P., Jaspersen, S. L., Miele, A., Dekker, J., & Peterson, C. L. (2009). Mechanisms that regulate localization of a DNA double-strand break to the nuclear periphery. *Genes & Development, 23*, 912–927.
Ozawa, R., Hayashi, Y. K., Ogawa, M., Kurokawa, R., Matsumoto, H., Noguchi, S., et al. (2006). Emerin-lacking mice show minimal motor and cardiac dysfunctions with nuclear-associated vacuoles. *American Journal of Pathology, 168*, 907–917.
Padiath, Q. S., Saigoh, K., Schiffmann, R., Asahara, H., Yamada, T., Koeppen, A., et al. (2006). Lamin B1 duplications cause autosomal dominant leukodystrophy. *Nature Genetics, 38*, 1114–1123.
Patel-Hett, S., & D'Amore, P. A. (2011). Signal transduction in vasculogenesis and developmental angiogenesis. *International Journal of Developmental Biology, 55*, 353–363.
Payne, V. A., Grimsey, N., Tuthill, A., Virtue, S., Gray, S. L., Dalla Nora, E., et al. (2008). The human lipodystrophy gene BSCL2/seipin may be essential for normal adipocyte differentiation. *Diabetes, 57*, 2055–2060.
Pendas, A. M., Zhou, Z., Cadinanos, J., Freije, J. M., Wang, J., Hultenby, K., et al. (2002). Defective prelamin A processing and muscular and adipocyte alterations in Zmpste24 metalloproteinase-deficient mice. *Nature Genetics, 31*, 94–99.
Peter, M., Kitten, G. T., Lehner, C. F., Vorburger, K., Bailer, S. M., Maridor, G., et al. (1989). Cloning and sequencing of cDNA clones encoding chicken lamins A and B1 and comparison of the primary structures of vertebrate A- and B-type lamins. *Journal of Molecular Biology, 208*, 393–404.
Pilat, U., Dechat, T., Bertrand, A. T., Woisetschlager, N., Gotic, I., Spilka, R., et al. (2013). The muscle dystrophy-causing DeltaK32 lamin A/C mutant does not impair the functions of the nucleoplasmic lamin-A/C-LAP2alpha complex in mice. *Journal of Cell Science, 126*, 1753–1762.
Plasilova, M., Chattopadhyay, C., Pal, P., Schaub, N. A., Buechner, S. A., Mueller, H., et al. (2004). Homozygous missense mutation in the lamin A/C gene causes autosomal recessive Hutchinson-Gilford progeria syndrome. *Journal of Medical Genetics, 41*, 609–614.
Poitelon, Y., Kozlov, S., Devaux, J., Vallat, J. M., Jamon, M., Roubertoux, P., et al. (2012). Behavioral and molecular exploration of the AR-CMT2A mouse model Lmna (R298C/R298C). *Neuromolecular Medicine, 14*, 40–52.
Powell, L., & Burke, B. (1990). Internuclear exchange of an inner nuclear membrane protein (p55) in heterokaryons: In vivo evidence for the interaction of p55 with the nuclear lamina. *Journal of Cell Biology, 111*, 2225–2234.
Puckelwartz, M. J., Kessler, E. J., Kim, G., Dewitt, M. M., Zhang, Y., Earley, J. U., et al. (2010). Nesprin-1 mutations in human and murine cardiomyopathy. *Journal of Molecular and Cellular Cardiology, 48*, 600–608.

Puckelwartz, M. J., Kessler, E., Zhang, Y., Hodzic, D., Randles, K. N., Morris, G., et al. (2009). Disruption of nesprin-1 produces an Emery Dreifuss muscular dystrophy-like phenotype in mice. *Human Molecular Genetics*, *18*, 607–620.

Puente, X. S., Quesada, V., Osorio, F. G., Cabanillas, R., Cadinanos, J., Fraile, J. M., et al. (2011). Exome sequencing and functional analysis identifies BANF1 mutation as the cause of a hereditary progeroid syndrome. *American Journal of Human Genetics*, *88*, 650–656.

Raharjo, W. H., Enarson, P., Sullivan, T., Stewart, C. L., & Burke, B. (2001). Nuclear envelope defects associated with LMNA mutations cause dilated cardiomyopathy and Emery-Dreifuss muscular dystrophy. *Journal of Cell Science*, *114*, 4447–4457.

Ramos, F. J., Chen, S. C., Garelick, M. G., Dai, D. F., Liao, C. Y., Schreiber, K. H., et al. (2012). Rapamycin reverses elevated mTORC1 signaling in lamin A/C-deficient mice, rescues cardiac and skeletal muscle function, and extends survival. *Science Translational Medicine*, *4*, 144ra103.

Reddy, K. L., Zullo, J. M., Bertolino, E., & Singh, H. (2008). Transcriptional repression mediated by repositioning of genes to the nuclear lamina. *Nature*, *452*, 243–247.

Rober, R. A., Sauter, H., Weber, K., & Osborn, M. (1990). Cells of the cellular immune and hemopoietic system of the mouse lack lamins A/C: Distinction versus other somatic cells. *Journal of Cell Science*, *95*(Pt 4), 587–598.

Rober, R. A., Weber, K., & Osborn, M. (1989). Differential timing of nuclear lamin A/C expression in the various organs of the mouse embryo and the young animal: A developmental study. *Development*, *105*, 365–378.

Roux, K. J., Kim, D. I., Raida, M., & Burke, B. (2012). A promiscuous biotin ligase fusion protein identifies proximal and interacting proteins in mammalian cells. *Journal of Cell Biology*, *196*, 801–810.

Rowat, A. C., Jaalouk, D. E., Zwerger, M., Ung, W. L., Eydelnant, I. A., Olins, D. E., et al. (2013). Nuclear envelope composition determines the ability of neutrophil-type cells to passage through micron-scale constrictions. *Journal of Biological Chemistry*, *288*, 8610–8618.

Rusinol, A. E., & Sinensky, M. S. (2006). Farnesylated lamins, progeroid syndromes and farnesyl transferase inhibitors. *Journal of Cell Science*, *119*, 3265–3272.

Sagelius, H., Rosengardten, Y., Hanif, M., Erdos, M. R., Rozell, B., Collins, F. S., et al. (2008). Targeted transgenic expression of the mutation causing Hutchinson-Gilford progeria syndrome leads to proliferative and degenerative epidermal disease. *Journal of Cell Science*, *121*, 969–978.

Scaffidi, P., & Misteli, T. (2005). Reversal of the cellular phenotype in the premature aging disease Hutchinson-Gilford progeria syndrome. *Nature Medicine*, *11*, 440–445.

Schirmer, E. C., Florens, L., Guan, T., Yates, J. R., 3rd., & Gerace, L. (2003). Nuclear membrane proteins with potential disease links found by subtractive proteomics. *Science*, *301*, 1380–1382.

Shackleton, S., Lloyd, D. J., Jackson, S. N., Evans, R., Niermeijer, M. F., Singh, B. M., et al. (2000). LMNA, encoding lamin A/C, is mutated in partial lipodystrophy. *Nature Genetics*, *24*, 153–156.

Shackleton, S., Smallwood, D. T., Clayton, P., Wilson, L. C., Agarwal, A. K., Garg, A., et al. (2005). Compound heterozygous ZMPSTE24 mutations reduce prelamin A processing and result in a severe progeroid phenotype. *Journal of Medical Genetics*, *42*, e36.

Shah, P. P., Donahue, G., Otte, G. L., Capell, B. C., Nelson, D. M., Cao, K., et al. (2013). Lamin B1 depletion in senescent cells triggers large-scale changes in gene expression and the chromatin landscape. *Genes & Development*, *27*, 1787–1799.

Shimi, T., Butin-Israeli, V., Adam, S. A., Hamanaka, R. B., Goldman, A. E., Lucas, C. A., et al. (2011). The role of nuclear lamin B1 in cell proliferation and senescence. *Genes & Development*, *25*, 2579–2593.

Shimi, T., Pfleghaar, K., Kojima, S., Pack, C. G., Solovei, I., Goldman, A. E., et al. (2008). The A- and B-type nuclear lamin networks: Microdomains involved in chromatin organization and transcription. *Genes & Development, 22*, 3409–3421.
Shin, J. Y., Mendez-Lopez, I., Wang, Y., Hays, A. P., Tanji, K., Lefkowitch, J. H., et al. (2013). Lamina-associated polypeptide-1 interacts with the muscular dystrophy protein emerin and is essential for skeletal muscle maintenance. *Developmental Cell, 26*, 591–603.
Shin, J. W., Spinler, K. R., Swift, J., Chasis, J. A., Mohandas, N., & Discher, D. E. (2013). Lamins regulate cell trafficking and lineage maturation of adult human hematopoietic cells. *Proceedings of the National Academy of Sciences of the United States of America, 110*, 18892–18897.
Shindo, Y., Kim, M. R., Miura, H., Yuuki, T., Kanda, T., Hino, A., et al. (2010). Lrmp/Jaw1 is expressed in sweet, bitter, and umami receptor-expressing cells. *Chemical Senses, 35*, 171–177.
Shnyrova, A., Frolov, V. A., & Zimmerberg, J. (2008). ER biogenesis: Self-assembly of tubular topology by protein hairpins. *Current Biology, 18*, R474–R476.
Shultz, L. D., Lyons, B. L., Burzenski, L. M., Gott, B., Samuels, R., Schweitzer, P. A., et al. (2003). Mutations at the mouse ichthyosis locus are within the lamin B receptor gene: A single gene model for human Pelger-Huet anomaly. *Human Molecular Genetics, 12*, 61–69.
Shumaker, D. K., Dechat, T., Kohlmaier, A., Adam, S. A., Bozovsky, M. R., Erdos, M. R., et al. (2006). Mutant nuclear lamin A leads to progressive alterations of epigenetic control in premature aging. *Proceedings of the National Academy of Sciences of the United States of America, 103*, 8703–8708.
Silve, S., Dupuy, P. H., Ferrara, P., & Loison, G. (1998). Human lamin B receptor exhibits sterol C14-reductase activity in Saccharomyces cerevisiae. *Biochimica et Biophysica Acta, 1392*, 233–244.
Simha, V., Agarwal, A. K., Oral, E. A., Fryns, J. P., & Garg, A. (2003). Genetic and phenotypic heterogeneity in patients with mandibuloacral dysplasia-associated lipodystrophy. *Journal of Clinical Endocrinology and Metabolism, 88*, 2821–2824.
Sinensky, M., Fantle, K., Trujillo, M., McLain, T., Kupfer, A., & Dalton, M. (1994). The processing pathway of prelamin A. *Journal of Cell Science, 107*(Pt 1), 61–67.
Sohaskey, M. L., Jiang, Y., Zhao, J. J., Mohr, A., Roemer, F., & Harland, R. M. (2010). Osteopotentia regulates osteoblast maturation, bone formation, and skeletal integrity in mice. *Journal of Cell Biology, 189*, 511–525.
Solovei, I., Wang, A. S., Thanisch, K., Schmidt, C. S., Krebs, S., Zwerger, M., et al. (2013). LBR and lamin A/C sequentially tether peripheral heterochromatin and inversely regulate differentiation. *Cell, 152*, 584–598.
Soullam, B., & Worman, H. J. (1995). Signals and structural features involved in integral membrane protein targeting to the inner nuclear membrane. *Journal of Cell Biology, 130*, 15–27.
Speckman, R. A., Garg, A., Du, F., Bennett, L., Veile, R., Arioglu, E., et al. (2000). Mutational and haplotype analyses of families with familial partial lipodystrophy (Dunnigan variety) reveal recurrent missense mutations in the globular C-terminal domain of lamin A/C. *American Journal of Human Genetics, 66*, 1192–1198.
Steglich, B., Sazer, S., & Ekwall, K. (2013). Transcriptional regulation at the yeast nuclear envelope. *Nucleus, 4*, 379–389.
Stehbens, W. E., Delahunt, B., Shozawa, T., & Gilbert-Barness, E. (2001). Smooth muscle cell depletion and collagen types in progeric arteries. *Cardiovascular Pathology, 10*, 133–136.
Stewart, C. L. (2014). Unpublished observations.
Stewart, C., & Burke, B. (1987). Teratocarcinoma stem cells and early mouse embryos contain only a single major lamin polypeptide closely resembling lamin B. *Cell, 51*, 383–392.

Stewart, C. L., Roux, K. J., & Burke, B. (2007). Blurring the boundary: The nuclear envelope extends its reach. *Science*, *318*, 1408–1412.
Sullivan, T., Escalante-Alcalde, D., Bhatt, H., Anver, M., Bhat, N., Nagashima, K., et al. (1999). Loss of A-type lamin expression compromises nuclear envelope integrity leading to muscular dystrophy. *Journal of Cell Biology*, *147*, 913–920.
Swift, J., Ivanovska, I. L., Buxboim, A., Harada, T., Dingal, P. C., Pinter, J., et al. (2013). Nuclear lamin-A scales with tissue stiffness and enhances matrix-directed differentiation. *Science*, *341*, 1240104.
Szymanski, K. M., Binns, D., Bartz, R., Grishin, N. V., Li, W. P., Agarwal, A. K., et al. (2007). The lipodystrophy protein seipin is found at endoplasmic reticulum lipid droplet junctions and is important for droplet morphology. *Proceedings of the National Academy of Sciences of the United States of America*, *104*, 20890–20895.
Taimen, P., Pfleghaar, K., Shimi, T., Moller, D., Ben-Harush, K., Erdos, M. R., et al. (2009). A progeria mutation reveals functions for lamin A in nuclear assembly, architecture, and chromosome organization. *Proceedings of the National Academy of Sciences of the United States of America*, *106*, 20788–20793.
Taylor, M. R., Slavov, D., Gajewski, A., Vlcek, S., Ku, L., Fain, P. R., et al. (2005). Thymopoietin (lamina-associated polypeptide 2) gene mutation associated with dilated cardiomyopathy. *Human Mutation*, *26*, 566–574.
Ulbert, S., Antonin, W., Platani, M., & Mattaj, I. W. (2006). The inner nuclear membrane protein Lem2 is critical for normal nuclear envelope morphology. *FEBS Letters*, *580*, 6435–6441.
van Engelen, B. G., Muchir, A., Hutchison, C. J., van der Kooi, A. J., Bonne, G., & Lammens, M. (2005). The lethal phenotype of a homozygous nonsense mutation in the lamin A/C gene. *Neurology*, *64*, 374–376.
van Tintelen, J. P., Hofstra, R. M., Katerberg, H., Rossenbacker, T., Wiesfeld, A. C., du Marchie Sarvaas, G. J., et al. (2007). High yield of LMNA mutations in patients with dilated cardiomyopathy and/or conduction disease referred to cardiogenetics outpatient clinics. *American Heart Journal*, *154*, 1130–1139.
Varela, I., Cadinanos, J., Pendas, A. M., Gutierrez-Fernandez, A., Folgueras, A. R., Sanchez, L. M., et al. (2005). Accelerated ageing in mice deficient in Zmpste24 protease is linked to p53 signalling activation. *Nature*, *437*, 564–568.
Varela, I., Pereira, S., Ugalde, A. P., Navarro, C. L., Suarez, M. F., Cau, P., et al. (2008). Combined treatment with statins and aminobisphosphonates extends longevity in a mouse model of human premature aging. *Nature Medicine*, *14*, 767–772.
Varga, R., Eriksson, M., Erdos, M. R., Olive, M., Harten, I., Kolodgie, F., et al. (2006). Progressive vascular smooth muscle cell defects in a mouse model of Hutchinson-Gilford progeria syndrome. *Proceedings of the National Academy of Sciences of the United States of America*, *103*, 3250–3255.
Venables, R. S., McLean, S., Luny, D., Moteleb, E., Morley, S., Quinlan, R. A., et al. (2001). Expression of individual lamins in basal cell carcinomas of the skin. *British Journal of Cancer*, *84*, 512–519.
Vergnes, L., Peterfy, M., Bergo, M. O., Young, S. G., & Reue, K. (2004). Lamin B1 is required for mouse development and nuclear integrity. *Proceedings of the National Academy of Sciences of the United States of America*, *101*, 10428–10433.
Verhagen, A. M., de Graaf, C. A., Baldwin, T. M., Goradia, A., Collinge, J. E., Kile, B. T., et al. (2012). Reduced lymphocyte longevity and homeostatic proliferation in lamin B receptor-deficient mice results in profound and progressive lymphopenia. *Journal of Immunology*, *188*, 122–134.
Vigouroux, C., Magre, J., Vantyghem, M. C., Bourut, C., Lascols, O., Shackleton, S., et al. (2000). Lamin A/C gene: Sex-determined expression of mutations in Dunnigan-type

familial partial lipodystrophy and absence of coding mutations in congenital and acquired generalized lipoatrophy. *Diabetes*, *49*, 1958–1962.

Villa-Bellosta, R., Rivera-Torres, J., Osorio, F. G., Acin-Perez, R., Enriquez, J. A., Lopez-Otin, C., et al. (2013). Defective extracellular pyrophosphate metabolism promotes vascular calcification in a mouse model of Hutchinson-Gilford progeria syndrome that is ameliorated on pyrophosphate treatment. *Circulation*, *127*, 2442–2451.

Vorburger, K., Lehner, C. F., Kitten, G. T., Eppenberger, H. M., & Nigg, E. A. (1989). A second higher vertebrate B-type lamin. cDNA sequence determination and in vitro processing of chicken lamin B2. *Journal of Molecular Biology*, *208*, 405–415.

Wagner, N., & Krohne, G. (2007). LEM-Domain proteins: New insights into lamin-interacting proteins. *International Review of Cytology*, *261*, 1–46.

Wassif, C. A., Brownson, K. E., Sterner, A. L., Forlino, A., Zerfas, P. M., Wilson, W. K., et al. (2007). HEM dysplasia and ichthyosis are likely laminopathies and not due to 3beta-hydroxysterol Delta14-reductase deficiency. *Human Molecular Genetics*, *16*, 1176–1187.

Waterham, H. R., Koster, J., Mooyer, P., Noort Gv, G., Kelley, R. I., Wilcox, W. R., et al. (2003). Autosomal recessive HEM/Greenberg skeletal dysplasia is caused by 3 beta-hydroxysterol delta 14-reductase deficiency due to mutations in the lamin B receptor gene. *American Journal of Human Genetics*, *72*, 1013–1017.

Weber, K., Plessmann, U., & Traub, P. (1989). Maturation of nuclear lamin A involves a specific carboxy-terminal trimming, which removes the polyisoprenylation site from the precursor; implications for the structure of the nuclear lamina. *FEBS Letters*, *257*, 411–414.

Wojtanik, K. M., Edgemon, K., Viswanadha, S., Lindsey, B., Haulzik, M., Chen, W., et al. (2009). A search for the role of LMNA in adipose: A novel mouse model of lipodystrophy based on the Dunnigan's partial lipodystrophy mutation. *Journal of Lipid Research*, *50*, 1068–1079.

Wolda, S. L., & Glomset, J. A. (1988). Evidence for modification of lamin B by a product of mevalonic acid. *Journal of Biological Chemistry*, *263*, 5997–6000.

Wolf, C. M., Wang, L., Alcalai, R., Pizard, A., Burgon, P. G., Ahmad, F., et al. (2008). Lamin A/C haploinsufficiency causes dilated cardiomyopathy and apoptosis-triggered cardiac conduction system disease. *Journal of Molecular and Cellular Cardiology*, *44*, 293–303.

Worman, H. J., & Bonne, G. (2007). "Laminopathies": A wide spectrum of human diseases. *Experimental Cell Research*, *313*, 2121–2133.

Wu, W., Iwata, S., Homma, S., Worman, H. J., & Muchir, A. (2014). Depletion of extracellular signal-regulated kinase 1 in mice with cardiomyopathy caused by lamin A/C gene mutation partially prevents pathology before isoenzyme activation. *Human Molecular Genetics*, *23*, 1–11.

Yang, S. H., Andres, D. A., Spielmann, H. P., Young, S. G., & Fong, L. G. (2008). Progerin elicits disease phenotypes of progeria in mice whether or not it is farnesylated. *Journal of Clinical Investigation*, *118*, 3291–3300.

Yang, S. H., Bergo, M. O., Toth, J. I., Qiao, X., Hu, Y., Sandoval, S., et al. (2005). Blocking protein farnesyltransferase improves nuclear blebbing in mouse fibroblasts with a targeted Hutchinson-Gilford progeria syndrome mutation. *Proceedings of the National Academy of Sciences of the United States of America*, *102*, 10291–10296.

Yang, S. H., Chang, S. Y., Yin, L., Tu, Y., Hu, Y., Yoshinaga, Y., et al. (2011). An absence of both lamin B1 and lamin B2 in keratinocytes has no effect on cell proliferation or the development of skin and hair. *Human Molecular Genetics*, *20*, 3537–3544.

Yang, S. H., Meta, M., Qiao, X., Frost, D., Bauch, J., Coffinier, C., et al. (2006). A farnesyltransferase inhibitor improves disease phenotypes in mice with a Hutchinson-Gilford progeria syndrome mutation. *Journal of Clinical Investigation*, *116*, 2115–2121.

Yang, S. H., Qiao, X., Fong, L. G., & Young, S. G. (2008). Treatment with a farnesyltransferase inhibitor improves survival in mice with a Hutchinson-Gilford progeria syndrome mutation. *Biochimica et Biophysica Acta*, *1781*, 36–39.

Ye, Q., Callebaut, I., Pezhman, A., Courvalin, J. C., & Worman, H. J. (1997). Domain-specific interactions of human HP1-type chromodomain proteins and inner nuclear membrane protein LBR. *Journal of Biological Chemistry*, *272*, 14983–14989.

Young, S. G., Fong, L. G., & Michaelis, S. (2005). Prelamin A, Zmpste24, misshapen cell nuclei, and progeria—New evidence suggesting that protein farnesylation could be important for disease pathogenesis. *Journal of Lipid Research*, *46*, 2531–2558.

Young, S. G., Jung, H. J., Coffinier, C., & Fong, L. G. (2012). Understanding the roles of nuclear A- and B-type lamins in brain development. *Journal of Biological Chemistry*, *287*, 16103–16110.

Young, S. G., Yang, S. H., Davies, B. S., Jung, H. J., & Fong, L. G. (2013). Targeting protein prenylation in progeria. *Science Translational Medicine*, *5*, 171ps173.

Yu, J., Lei, K., Zhou, M., Craft, C. M., Xu, G., Xu, T., et al. (2011). KASH protein Syne-2/Nesprin-2 and SUN proteins SUN1/2 mediate nuclear migration during mammalian retinal development. *Human Molecular Genetics*, *20*, 1061–1073.

Zhang, Q., Bethmann, C., Worth, N. F., Davies, J. D., Wasner, C., Feuer, A., et al. (2007). Nesprin-1 and −2 are involved in the pathogenesis of Emery Dreifuss muscular dystrophy and are critical for nuclear envelope integrity. *Human Molecular Genetics*, *16*, 2816–2833.

Zhang, J., Lian, Q., Zhu, G., Zhou, F., Sui, L., Tan, C., et al. (2011). A human iPSC model of Hutchinson Gilford progeria reveals vascular smooth muscle and mesenchymal stem cell defects. *Cell Stem Cell*, *8*, 31–45.

Zhang, X., Xu, R., Zhu, B., Yang, X., Ding, X., Duan, S., et al. (2007). Syne-1 and Syne-2 play crucial roles in myonuclear anchorage and motor neuron innervation. *Development*, *134*, 901–908.

CHAPTER TWO

Mouse Models of Lipodystrophy and Their Significance in Understanding Fat Regulation

Justin J. Rochford[1]
Rowett Institute of Nutrition and Health, University of Aberdeen, Foresterhill, Aberdeen, UK
[1]Corresponding author: e-mail address: j.rochford@abdn.ac.uk

Contents

Abstract

Adipose tissue plays a critical role in human metabolic health. This is most dramatically illustrated by the severe metabolic disease that occurs in syndromes of lipodystrophy where individuals fail to develop or maintain appropriate adipose tissue mass. The most severe form of this disorder is congenital generalized lipodystrophy (CGL). Individuals

Current Topics in Developmental Biology, Volume 109
ISSN 0070-2153
http://dx.doi.org/10.1016/B978-0-12-397920-9.00005-6

with CGL have a striking paucity of adipose tissue and typically display severe metabolic disease with insulin resistance and dyslipidemia. Understanding of the metabolic consequences of lipodystrophies and their underlying molecular mechanisms will provide new information regarding the development and function of human adipose tissue. Mouse models of these conditions offer key resources to investigate this *in vivo*. Adipocyte dysfunction is believed to underlie the development of metabolic disease in obesity. Hence, understanding how one might beneficially manipulate adipose tissue by studying genes whose disruption causes lipodystrophy is likely to suggest novel means to improve metabolic health in common obesity.

1. INTRODUCTION

1.1. Lipodystrophies: Rare disorders with relevance for common obesity

Obesity is now a major, worldwide health problem as excessive weight gain is strongly associated with a raft of conditions such as type 2 diabetes, cardiovascular disease, dyslipidemia, and several forms of cancer(Ferguson, Gallagher, Scheinman, Damouni, & LeRoith, 2013). Identifying novel therapies to reduce obesity and/or treat its consequences represent a key healthcare challenge. A widely accepted model of how obesity causes metabolic disease proposes that adipose tissue in any given individual has the capacity to expand only to a given point (Bluher, 2013; Sun, Kusminski, & Scherer, 2011; Unger, Clark, Scherer, & Orci, 2010; Unger & Scherer, 2010). In an initial phase of adipose tissue expansion, adipocyte precursors proliferate and so the fat tissue grows by a combination of both hyperplasia and hypertrophy as new adipocytes form and existing ones fill with additional lipid. However, as the maximum expansion capacity is reached, further lipid is initially stored mostly through hypertrophy leading to larger adipocytes with impaired function. As these hypertrophied adipocytes struggle to accommodate further nutrients, particularly circulating lipids, these nutrients spill over into nonadipose tissues (Bluher, 2013; Sun et al., 2011; Unger et al., 2010; Unger & Scherer, 2010; Virtue & Vidal-Puig, 2010). This lipid accumulates ectopically in the liver, skeletal muscle, pancreas, and cardiovascular tissues and impairs the function of these tissues causing metabolic disease. In this way, while adipose mass in obesity is abundant, it is insufficient in terms of its buffering capacity in the face of continued nutrient excess. This mechanism is believed to explain why individuals with syndromes of lipodystrophy, in which there is a paucity and/or dysfunction of adipose tissue, display some similar metabolic abnormalities to obese individuals (Garg, 2011; Vantyghem et al., 2012; Vatier et al., 2013; Vigouroux, Caron-Debarle, Le Dour, Magre, & Capeau, 2011). In both cases, the adipose tissue is unable to perform its appropriate task of safely storing lipids (Fig. 2.1). Therefore, defining the pathophysiological mechanisms via which lipodystrophy causes insulin resistance and other metabolic

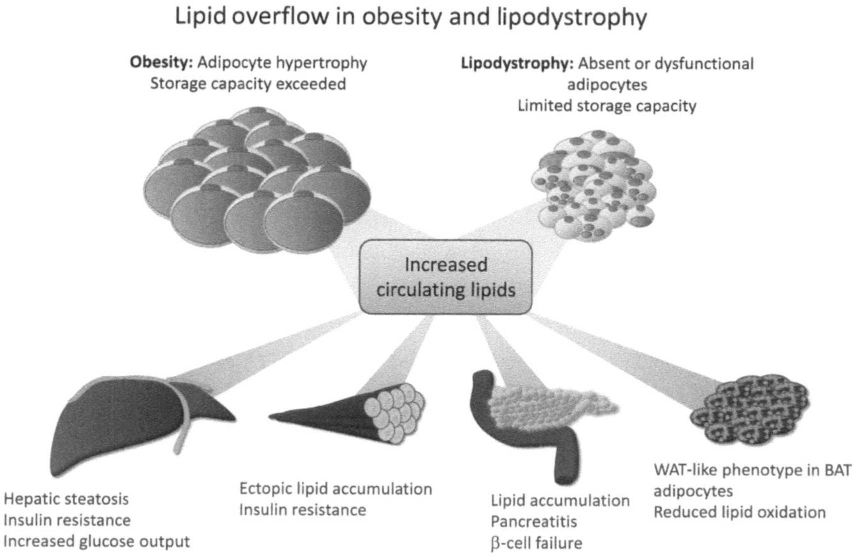

Figure 2.1 Lipid overflow and lipotoxicity may be a common driver of metabolic disease in both obesity and lipodystrophy. Excessive nutrient storage in the adipocytes of obese individuals leads the adipose tissue to reach a limit of expansion. Further nutrients, particularly lipids, can then no longer be stored safely in adipocytes. Hence, while fat tissue is abundant, there is a state of relative adipose tissue insufficiency. This is similar to the case in lipodystrophy where there is a primary defect of adipose tissue development and/or function. In both cases, the lipids spill over into the circulation and subsequently accumulate ectopically in other tissues where their effects are harmful. Accumulation in the liver and skeletal muscle may cause insulin resistance and so impaired glucose metabolism. Lipid accumulation in the pancreas can cause β-cell dysfunction, particularly when allied to increased demand for insulin secretion from this organ as the body attempts to overcome peripheral insulin resistance. In rodents, brown adipose tissue (BAT) can also become less able to oxidize lipids and accumulate more and larger lipid droplets, adopting an appearance more like white adipose tissue. The relevance of this in humans is not yet clear although significant BAT depots can be detected in healthy individuals. Overall, the lipotoxic effects deriving from an inability to store lipids safely in adipose tissue may explain the similarities in metabolic disease observed in obese and lipodystrophic individuals despite the disparity in adiposity. (See the color plate.)

consequences are not only relevant to these rare disorders. Such studies may also provide new insights regarding the development of metabolic disease in more common obesity and highlight novel pathways that may be targeted therapeutically.

In addition to their relevance to our understanding of human metabolic diseases, lipodystrophy genes also offer starting points to gain new insights regarding the developmental pathways by which human adipose tissue develops. In many cases, we know relatively little regarding why these genes are important for appropriate human fat development and the precise molecular roles of the proteins they encode. There are many unanswered questions regarding the developmental stages of the adipocyte life cycle at which these proteins are important and why some of them appear important for the development or function of some adipose depots but not others (Garg, 2011; Vantyghem et al., 2012). Overall there is much to learn. Many of these questions can only be answered using sophisticated genetic techniques in rodents where different developmental pathways can be systematically manipulated. The currently available models of these conditions have only begun to scratch the surface of these questions. This chapter aims to summarize the major advances so far and highlight some of the key points for future investigation.

1.2. The metabolic importance of adipose tissue as an energy store and endocrine organ

Adipose tissue normally provides a critical safe store for dietary lipids and is a highly evolved, specialized tissue. Adipocytes within the adipose tissue are remarkable cells. These cells are typically around 100 micrometers in diameter with over 90% of the cell volume consisting of a single large lipid droplet. Despite this, adipocytes are far from just inert stores of lipid. In addition to acutely regulating the supply of lipid to and from the droplet, adipocytes perform important endocrine secretory roles (Rosen & Spiegelman, 2014). Of particular note, adipose tissue is the source of leptin, a key regulator of appetite, and adiponectin, which is secreted in large quantities mostly, if not exclusively, from adipocytes (Cottrell & Mercer, 2012; Dalamaga et al., 2013; Gautron & Elmquist, 2011; Groeneveld, Huang-Doran, & Semple, 2012; Konner & Bruning, 2012; Mantzoros et al., 2011). Systemic leptin levels provide the body with a readout of adiposity such that greater fat stores lead to increased circulating leptin levels. In this way, leptin reports levels of adiposity to key feeding centers of the brain and influences satiety to alter eating behavior. Adiponectin is a circulating adipokine relevant to insulin sensitivity. The precise functional effects of adiponectin on the insulin sensitivity of different tissues continue to be a major area of interest with much left to discover (Groeneveld et al., 2012). However, there is a strong correlation between adiponectin levels and insulin sensitivity. Thus, "healthy" adipocytes secrete abundant adiponectin with positive systemic effects on insulin sensitivity. In obesity, adiponectin levels fall despite increased adiposity and insulin sensitivity declines. In lipodystrophy, particularly in congenital generalized lipodystrophy (CGL), the lack of fat leads to low levels of leptin (so hunger) and reduced adiponectin (so insulin resistance). In obesity, increased adiposity leads to higher leptin but without additional suppressive effect on appetite. Conversely adiponectin levels fall as adipocytes become dysfunctional.

Leptin acts via receptors in the arcuate nucleus of the hypothalamus in the brain to suppress appetite (Dalamaga et al., 2013; Gautron & Elmquist, 2011; Konner & Bruning, 2012; Mantzoros et al., 2011). It appears that leptin is unlikely to acutely regulate appetite in a meaningful way in states of normal to high adiposity. Rather, the fall of leptin levels in states of significantly reduced adiposity provides an increased drive to eat. However, while leptin may not provide a useful tool to reduce food intake in common obesity, it can be therapeutically beneficial in patients with lipodystrophy (Chong, Lupsa, Cochran, & Gorden, 2010; Oral et al., 2002; Safar Zadeh et al., 2013). Moreover, the severe obesity and hyperphagia in humans with disruptions of leptin, the leptin receptor, or key components of neuronal pathways downstream of leptin demonstrates the critical role of leptin in metabolic homeostasis. Mouse models of altered leptin levels, signaling or downstream neuronal pathways have provided many critical insights regarding appetite regulation most of which translate relatively faithfully to the situation in humans (Dalamaga et al., 2013; Dubern & Clement, 2012; Gautron & Elmquist, 2011; Mantzoros et al., 2011; Yeo & Heisler, 2012). As such, these murine models of obesity have proven invaluable research tools. Unfortunately, this has not proven to be so strongly the case for murine models of lipodystrophy, as detailed below. Despite this, mouse models of human lipodystrophy have already provided some extremely important and relevant new information regarding human adipose tissue development and function.

2. LIPODYSTROPHY SYNDROMES

Lipodystrophies are a range of rare human disorders characterized by altered patterns of adiposity. These may be either acquired or inherited

(Capeau et al., 2010; Garg, 2011; Rochford, 2010). Acquired lipodystrophies have been reviewed elsewhere and the underlying mechanisms remain poorly defined in most cases (Capeau et al., 2010; Galescu, Bhangoo, & Ten, 2013; Garg, 2011; Mallewa et al., 2008). This chapter will focus on inherited lipodystrophies in which the causative genes have been identified.

2.1. Congenital generalized lipodystrophies

At the most severe end of the spectrum are syndromes of CGL, also referred to as Berardinelli–Seip congenital lipodystrophy (BSCL), in which affected individuals have very little functional adipose tissue in almost all depots. These are recessively inherited disorders and only four genes have been identified whose disruption causes CGL, namely, *AGPAT2*, *BSCL2*, *CAV1*, and *PTRF* (Capeau et al., 2010; Garg, 2011; Mallewa et al., 2008; Rochford, 2010; Vantyghem et al., 2012). While *AGPAT2*, *BSCL2*, and *CAV1* disruption affect adipose tissue relatively selectively, *PTRF* mutations cause a more complex phenotype as will be described later. Around 95% of CGL cases can be explained by mutations in either *AGPAT2* or *BSCL2*. A small number of "unsolved" cases of CGL remain, suggesting that other genetic causes have yet to be discovered. Mice with targeted disruption of *Agpat2*, *Bscl2*, *Cav1*, and *Ptrf* have all been reported and these will be discussed in more detail below.

2.2. Familial partial lipodystrophies

Familial partial lipodystrophies (FPLD) are a phenotypically diverse range of conditions in which, as the name suggests, not all adipose depots are affected. Mutations in the genes *LMNA*, *PPARG*, *ZMPSTE24*, *AKT2*, *CIDEC*, and *PLIN1* have all been reported to cause forms of FPLD (Capeau et al., 2010; Gandotra, Le Dour, et al., 2011; Garg, 2011; Rochford, 2010; Vantyghem et al., 2012). Heterozygous mutations in *LMNA* and *PPARG* account for more than half of the cases of FPLD. Compound heterozygous or homozygous mutations in *LMNA* have also been reported, as is the case for *ZMPSTE24*. So far only one family has been reported with a heterozygous mutation in *AKT2* and one individual with homozygous disruption of *CIDEC* (George et al., 2004; Rubio-Cabezas et al., 2009). Three different pedigrees have been reported in which individuals with FPLD are heterozygous for either of two frameshift mutations in PLIN1 (Gandotra, Le Dour, et al., 2011). With the possible exception of *CIDEC*, the mutations causing

FPLD do not cause a simple loss of function of the encoded proteins. Hence, in most cases, knockout mice cannot be expected to model the human disease, as discussed below.

2.3. Other syndromes with lipodystrophy

A syndrome featuring lipodystrophy has also been reported due to mutations in PSMB8 (Agarwal et al., 2010; Kitamura et al., 2011). It appears that in this case lipodystrophy is secondary to inflammatory disease rather than altered adipocyte development or function *per se*. However, mice in which PSMB8 has been disrupted do exist and offer the opportunity for future more detailed study of this aspect of the phenotype (Fehling et al., 1994). Recently, a mutation POLD1, the gene encoding the catalytic subunit of DNA polymerase delta, has been shown to cause a multisystem disease including subcutaneous lipodystrophy but also mandibular hypoplasia, deafness, and hypogonadism in males (Weedon et al., 2013). The causative, heterozygous single-codon deletion abolishes DNA polymerase activity and mildly impairs 3′- to 5′-exonuclease activity. Evidently, the pathogenic effects of this mutation, causing loss of very selective aspects of function in such a critical protein is only likely to be modeled by specific knockin of the disease allele in mice.

2.4. Metabolic consequences of inappropriate adiposity

FPLD or CGL patients typically accumulate lipid ectopically in the liver, muscle, pancreas, and kidneys due to the inability of their adipose depots to appropriately buffer postprandial lipid levels (Capeau et al., 2010; Gandotra, Le Dour, et al., 2011; Garg, 2011; Rochford, 2010; Vantyghem et al., 2012). This ectopic lipid accumulation is generally held to cause lipotoxic effects in these tissues leading to insulin resistance, cardiovascular disease, and pancreatitis, depending on its severity. However, the precise mechanisms by which lipid accumulation leads to these effects remains incompletely understood. One notable feature is that the severity of metabolic disease observed between CGL and FPLD patients, and within syndromes of FPLD, does not correlate simply with the absolute decrease in adipose mass. Indeed, in some FPLD patients, the absolute reduction in adipose mass is minor but the metabolic phenotype may be relatively severe. This is likely to result from a combination of effects including differences in the relative importance of different adipose depots for metabolic health and the precise functions of the molecules involved. For example, some genes such as *PPARG* and *AKT2* may exert effects both on the development of adipocytes *per se* and also on the insulin sensitivity of adipocytes (Pearce, Komander, & Alessi, 2010; Rodriguez-Cuenca et al., 2012; Rosen & Spiegelman, 2014). Hence, the overall effect becomes a complex combination of altered adipogenesis and the metabolic function of the adipocytes that form in these individuals.

2.5. Mouse models recapitulating inherited lipodystrophies

All cases of CGL in humans display a recessive or compound heterozygous pattern of inheritance and also include examples with very significant rearrangements of the affected genes, with large deletions or nonsense mutations very early in the coding sequence (Capeau et al., 2010; Garg, 2011). Together, this indicates that simple loss of function of these genes can cause CGL. For this reason, modeling CGL in mice can, broadly speaking, be achieved by generating knockout models. Indeed, knockout mice with disruption of each of the four genes whose disruption causes CGL have been reported and have provided valuable new information regarding this condition.

The situation for FPLD is much more complex as these are often conditions with autosomal dominant inheritance, and in several cases, it appears that the causative mutation confers some dominant negative functions on the mutant protein (Capeau et al., 2010; Garg, 2011). To appropriately model this in mice requires the development of knockin mice such that heterozygous mice can be bred in which the mutation of the human gene is mimicked by mutation of corresponding or homologous residue(s) in the mouse gene. For this reason, examples of this are far fewer and this type of knockin model of FPLD has only been reported for *Pparg* so far (Gray, Nora, et al., 2006; Tsai et al., 2004). However, mutations in Zmpste24, CIDEC, and perilipin that cause FPLD appear to do so by disrupting at least some significant functions of the proteins (Agarwal, Fryns, Auchus, & Garg, 2003; Gandotra, Le Dour, et al., 2011; Rubio-Cabezas et al., 2009). As such, the knockout mouse models that exist for these proteins are also capable of providing relevant insights regarding the corresponding forms of FPLD in humans.

An additional complexity when attempting to model FPLD in mice is that the major effect may be to alter the distribution rather than absolute, overall levels of adipose tissue. Indeed, the alteration of adiposity in affected FLPD patients may be difficult to discern without an expert clinical eye.

Despite an apparently subtle effect on total adiposity, the patients suffer disproportionately severe metabolic disease implying a significant metabolic role of the specific adipose depots affected, even though these depots may be quantitatively small. The differences between human and murine fat distribution and the remaining uncertainties regarding which depots correlate with which between these two species are key limitations. Along with the differences in lipid metabolism between humans and mice, this is probably a major reason why mouse models of FPLD struggle to easily recapitulate the metabolic disease observed in humans with FPLD.

3. GENES MUTATED IN CONGENITAL GENERALIZED LIPODYSTROPHY

3.1. *AGPAT2* (CGL1)

The *AGPAT2* gene encodes the enzyme 1-acylglycerol-3-phosphate *O*-acyltransferase 2 (AGPAT2). This was the first locus at which mutations were identified causing CGL, hence the designation CGL1 for CGL caused by *AGPAT2* mutations (Garg et al., 1999). However, the causative gene was only identified subsequently, shortly after the identification of *BSCL2*/seipin (see below) (Agarwal et al., 2002). Patients with CGL1 almost entirely lack adipose tissue in all depots although some so-called mechanical adipose tissue remains. Affected individuals suffer hypertriglyceridemia, hepatic steatosis, and severe insulin resistance (Van Maldergem et al., 2002). The majority of the phenotype appears explicable by the inability to develop appropriate adipose stores, although secondary effects in other tissues may exacerbate the situation.

3.1.1 AGPAT2 in adipocyte development

AGPAT2 is highly expressed in adipose tissue and catalyzes the generation of phosphatidic acid (PA) from lysophosphatidic acid (LPA) by addition of a fatty acyl moiety at the *sn*-2 position of the glycerol backbone. Initially, this implied a clear molecular mechanism whereby AGPAT2 mutations might cause lipodystrophy as PA is a key intermediate in the generation of triacylglycerol (TG) (Fig. 2.2). The loss of this important lipid biosynthetic pathway would leave developing adipocytes unable to synthesize TG leading to a lack of functional adipocytes. However, subsequent cellular studies have revealed that AGPAT2 plays a key, cell autonomous role in the process of adipogenesis, almost certainly independent of any role in TG generation (Gale et al., 2006; Ramanathan et al., 2013). Preadipocytes lacking AGPAT2

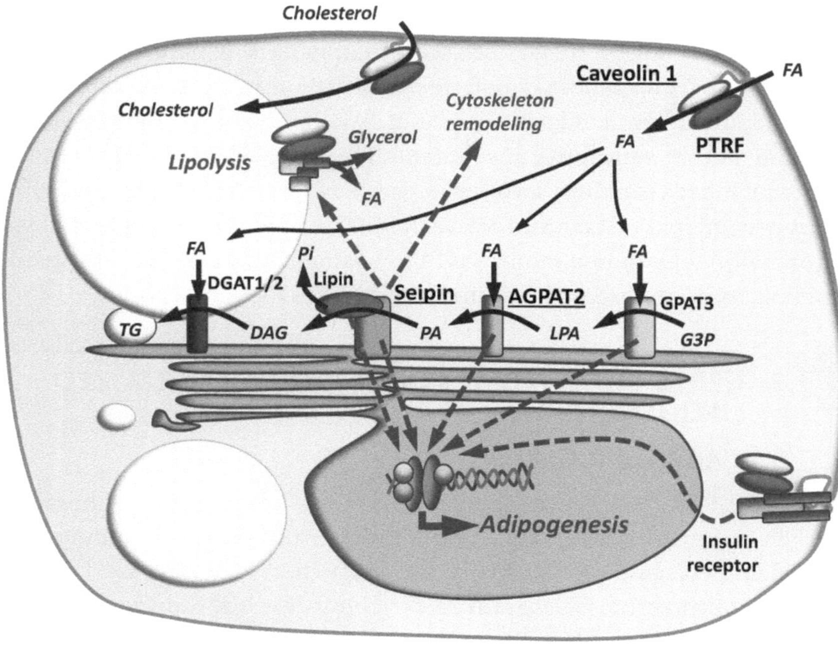

Figure 2.2 Molecular mechanisms via which disruption of CGL genes may cause generalized lipodystrophy. The proteins encoded by genes whose disruption causes CGL (whose names are underlined) have multiple roles in adipocyte development and function. AGPAT2 converts lysophosphatidic acid (LPA) to phosphatidic acid (PA) by the addition of a fatty acid (FA) moiety. This is an important intermediate in the generation of triacylglycerol (TG). Initially, glycerol 3-phosphate (G3P) is converted to LPA by glycerol 3-phosphate acyltransferase 3 (GPAT3), then LPA to PA by AGPAT2. PA is dephosphorylated by lipin enzymes, which are PA phosphatases, to generate diacylglycerol (DAG). DAG is then converted to TG by diacylglycerol acyltransferases (DGATs). Seipin, the product of the *BSCL2* gene, can recruit lipin to the ER membrane, potentiating its function and thereby affecting the levels of PA. However, seipin loss also increases lipolysis in differentiating cells and, in addition, can affect lipid droplet budding and/or morphology. A role for seipin in regulating the cytoskeleton during adipogenesis has also been proposed. The loss of GPAT3, AGPAT2, seipin, or lipin 1 all prevent the normal induction of adipogenic gene expression in cultured models of adipocyte differentiation (red broken arrows). It is not clear how this is mediated but loss of DGAT does not affect adipogenic gene expression, only the synthesis of TG, implying that a lipid intermediate in this pathway or its derivative influences adipogenic gene expression. Both caveolin 1 and PTRF/cavin proteins are essential for the formation or maintenance of caveolae in adipocytes. These proteins are important for FA uptake, cholesterol uptake, cholesterol and TG accumulation in the lipid droplet, the regulation of lipolysis, and signaling by cell-surface growth factor receptors including the insulin receptor. Insulin signaling is important for controlling glucose uptake and lipolysis in the mature adipocyte. However, it may also provide important proadipogenic signals and several insulin-responsive intracellular signaling intermediates play critical roles in adipogenesis. (See the color plate.)

are unable to induce the necessary program of gene expression such that the adipogenic transcriptional program stalls in these cells, before the accumulation of significant TG is normally observed. The molecular mechanisms underlying this phenomenon are not fully understood. However, it has been proposed recently that the loss of AGPAT2 reduces the levels of PA which, as a precursor in the production of phosphatidylinositol 3-phosphate (PIP_3), leads to reduced PIP_3 levels (Subauste et al., 2012). PIP_3 is a critical signaling lipid in the PI3-kinase/Akt intracellular signaling pathways which is known to play important roles in adipogenesis (Fischer-Posovszky, Tews, Horenburg, Debatin, & Wabitsch, 2012; Rosen & MacDougald, 2006). Thus, reduced Akt activity and growth factor responsiveness in AGPAT2-deficient cells would lead to impaired proadipogenic signals. The same study also suggested that an additional pathway may actively suppress adipogenesis in AGPAT2-deficient preadipocytes. By raising the levels of LPA, the generation of cyclic PA (cPA) may be increased. cPA has been shown to inhibit the activity of the key adipogenic transcription factor PPARγ (Tsukahara et al., 2010). Hence, by this parallel pathway, AGPAT2 loss might also suppress this central transcriptional controller of adipogenesis. Further work is needed to show whether rescue of either or both of these adipogenic pathways can rescue adipogenesis in AGPAT2-deficient cells and it remains possible that other pathways controling adipogenesis are also lost with AGPAT2 disruption. Analysis of the functional deficiencies of pathogenic mutants of AGPAT2 occurring in CGL patients may offer some insights. An initial study indicated that loss of the catalytic activity of AGPAT2 was likely to play a key role in the molecular pathogenesis of several such mutations. This supported the notion that loss of the acyltransferase activity of AGPAT2 is key to the disease (Haque, Garg, & Agarwal, 2005). However, our own study showed that at least two pathogenic mutations of AGPAT2 cause destabilization of the protein such that it is not expressed in developing preadipocytes (Ramanathan et al., 2013). A more comprehensive analysis of pathogenic mutants of AGPAT2, particularly those causing single amino acid substitutions, is likely to prove more insightful in future. However, overall it seems likely that the cell autonomous regulation of key signaling lipids underlies the critical contribution of AGPAT2 in adipogenesis and is the reason that disruption of this gene causes CGL.

3.1.2 AGPAT2 knockout mice

Mice lacking AGPAT2 develop a generalized lipodystrophy evident from birth and so appear to model CGL1 well (Cortes et al., 2009; Vogel et al., 2011). Given that no significant adipose tissue develops in these mice and the precursors of adipocytes are not easily defined, it is not trivial to use them to investigate molecular mechanisms underlying their failure to develop adipose tissue. The first report of these mice therefore focussed on examining the pathophysiology of this form of CGL, particularly focusing on the significant hepatic steatosis in the mice, consistent with that observed in CGL1 patients. The authors observed that lipogenic genes and the rate of *de novo* fatty acid biosynthesis were significantly upregulated in the livers of AGPAT2 null mice (Cortes et al., 2009). The expression of monoacylglycerol acyltransferase 1 (MGAT1) was also increased, as were some lysophosphatidic acid phosphatases (LPPs). This provided evidence for a compensatory increase in a pathway whereby diacylglycerol (DAG) may be synthesized for TG synthesis by LPPs generating monoacylglycerol (MG) from LPA then MGAT1 converting MG to DAG. This would circumvent the need for AGPAT2 to generate this pool of DAG and subsequently TG. The authors of this study also used the model to examine the effect of low-fat diet feeding and observed a significant reduction in hepatic steatosis in AGPAT2 null mice on this diet. A subsequent study by the same group used viral transduction to express or reexpress AGPAT1 or AGPAT2, respectively, in the livers of AGPAT2 null mice to determine whether the loss of AGPAT activity played a significant role in the development of hepatic steatosis in this model (Agarwal et al., 2011). Notwithstanding the caveat that this was a relatively acute reexpression in a background of chronic deficiency, there was no evidence of reduced liver lipid levels. Hence, it appears that the steatosis observed with AGPAT2 deficiency is secondary to the failure to store lipids appropriately in adipose tissue. Most recently, these investigators have examined the effect of leptin treatment of AGPAT2 null mice (Cortes et al., 2014). Leptin treatment for 28 days normalized hepatic steatosis and reduced plasma triglyceride, insulin, and glucose levels. At the molecular level, the expression of carbohydrate-responsive element-binding protein was reduced with leptin treatment implying that the beneficial effects of leptin may be partly mediated through this transcription factor. Importantly, deletion of leptin receptors specifically in hepatocytes did not prevent the beneficial effect of leptin treatment in AGPAT2 null mice (Cortes et al., 2014). This provided new evidence that the capacity of leptin to improve hepatic steatosis and metabolic disease in lipodystrophic patients is not due to direct effects in the liver. Together, these studies demonstrate the potential value of mouse models of CGL to test hypotheses regarding the molecular pathophysiology of generalized lipodystrophies. Importantly, they also provide the capacity to determine the effects of dietary or hormonal interventions and their potential therapeutic benefit.

A second group independently generated a model of AGPAT2 deficiency and included more detailed study of the mice at early developmental stages. They observed that there were small depots of both white and brown adipose tissue containing amoeboid adipocytes in the first few days after birth (Vogel et al., 2011). However, necrosis was observed in these tissues as the mice aged leading to the generalized lack of fat expected from the human condition. Overall, the study suggests that a failure of adipogenesis may not fully explain the deficiency of adipose tissue when AGPAT2 is absent. Rather, there may be a combination of impaired development and subsequent loss of these residual depots. Future studies using time and cell-type defined ablation of AGPAT2 are likely to be required to resolve this question.

3.2. *BSCL2* (CGL2)

The *BSCL2* gene (for BSCL type 2) encodes the protein seipin. CGL patients with mutations in *BSCL2* suffer the most severe lipodystrophy described thus far with almost no detectable adipose tissue in any depots (Magre et al., 2001). Unlike AGPAT2 loss, loss of seipin also appears to prevent the development of mechanical adipose tissue such as that beneath the scalp, retroorbitally and on the palms, and soles of the feet (Agarwal et al., 2002; Agarwal, Simha, et al., 2003; Garg et al., 1999; Magre et al., 2001). As with CGL1, these patients suffer severe metabolic disease as a result (Magre et al., 2001; Van Maldergem et al., 2002).

The severity of the phenotype in these individuals allied to the lack of dramatic additional developmental defects implies a selective and essential role for seipin in human adipose tissue development. Despite this, it was not until several years after the identification of seipin as the protein disrupted in CGL2, that a cell autonomous role for this protein in adipogenesis was demonstrated (Chen et al., 2009; Payne et al., 2008). Seipin is an endoplasmic reticulum (ER) membrane resident protein, with two transmembrane domains such that its N and C termini are located in the cell cytoplasm (Lundin et al., 2006). While loss of seipin causes severe generalized lipodystrophy, two specific mutations abrogating an N glycosylation site in the protein cause an autosomal dominant form of distal hereditary motor neuropathy (dHMN) (Windpassinger et al., 2004). Seipin is highly expressed in the brain with a broad but defined pattern (Garfield et al.,

2012; Magre et al., 2001) and the mutations in seipin causing dHMN lead to aggregation of the mutant protein (Windpassinger et al., 2004). There is no evidence for lipodystrophy in these dHMN patients and so the relevance, if any, of this to the role of seipin in adipose tissue development is unclear. Initial case reports of CGL2 patients included intellectual disability as a feature of CGL2 (Van Maldergem et al., 2002). However, this does not seem to be a consistent feature of affected individuals and so it is not clear whether this results from a specific neuronal function of seipin that is lost when *BSCL2* is disrupted.

3.2.1 Regulation of lipid droplets by seipin

Despite significant recent interest in this area, the precise molecular mechanism via which seipin disruption leads to a failure of adipogenesis remains unclear. Several possible pathways have been identified, some or all of which may contribute to the failure of adipogenesis *in vivo* and additional roles of seipin may yet be uncovered. A variety of studies that have identified or characterized the yeast orthologue of seipin, Fld1p, have reported that when yeast lacking Fld1p were loaded with oleate they accumulate disordered and/or larger lipid droplets than wild-type yeast (Fei et al., 2008; Fei, Shui, et al., 2011; Szymanski et al., 2007; Wolinski, Kolb, Hermann, Koning, & Kohlwein, 2011). While this implied a role in lipid droplet formation that could be relevant to the development of lipodystrophy total TG levels are increased in this model which appears paradoxical for a gene whose loss causes lipodystrophy in humans. This phenotype of lipid droplet morphology has been described in multiple cell types and model organisms including human fibroblasts and *Drosophila* but the molecular mechanisms behind it remain undefined (Cartwright & Goodman, 2012; Fei, Du, & Yang, 2011; Fei et al., 2008; Fei, Shui, et al., 2011; Szymanski et al., 2007; Tian et al., 2011; Wolinski et al., 2011; Yang et al., 2013). Moreover, a recent study has suggested that the evolutionarily conserved core sequence present in the yeast seipin mediates this lipid droplet effect but the adipogenic capacity of seipin additionally requires the cytoplasmic C-terminus of the mammalian protein not present in Fld1p (Yang et al., 2013). This suggests that the functions of seipin in lipid droplet size and adipogenesis may be distinct. Adding to the complexity, lipid loaded lymphoblastoid cells derived from BSCL2 patients were found to have smaller rather than larger lipid droplets than those in control cells and evidence of reduced steroyl-CoA desaturase 1 activity (Boutet et al., 2009). This may yet prove a more relevant finding for the role of seipin in CGL2.

3.2.2 Role(s) of seipin in adipogenesis

Initial studies using mouse embryonic fibroblasts (MEFs) isolated from seipin null mice reported increased lipolysis during adipogenesis in these cells (Chen et al., 2012; Prieur et al., 2013) (Fig. 2.2). Chen and colleagues found that the initial phases of adipogenesis appeared to proceed relatively normally but that unrestrained lipolysis downstream of increased cyclic AMP-activated protein kinase (PKA) activity appeared to cause a loss of adipocyte development (Chen et al., 2012). The molecular mechanism behind this phenomenon remains unclear. In our lab, we identified a direct role for seipin in recruiting the PA phosphatase lipin 1 to the ER membrane (Sim, Dennis, et al., 2013). Lipin 1 lies immediately downstream of AGPAT2 in the same lipid biosynthetic pathway and is known to play a critical role in adipogenesis (Fig. 2.2). Seipin loss or overexpression caused corresponding increases or decreases in PA levels, respectively, indicating that the regulation of lipin 1 by seipin could be functionally important (Sim, Dennis, et al., 2013). Moreover, several pathogenic mutants of seipin fail to appropriately bind lipin 1. However, it remains to be determined whether the loss of lipin 1 targeting is responsible for the failure of adipogenesis in seipin-deficient cells. In addition to these mechanisms, Yang et al. have recently reported that seipin may interact with 14-3-3β during adipogenesis (Yang et al., 2014). They propose that because 14-3-3β also interacts with the actin-severing protein cofilin-1, in this way seipin indirectly influences cytoskeleton remodeling, a process known to be important for efficient adipogenesis. However, the dependence of adipogenesis on the 14-3-3β/seipin interaction remains to be demonstrated. Hence in summary, while three plausible mechanisms exist by which seipin may inhibit adipogenesis, the importance of each in the pathogenesis of CGL2 remains uncertain. It should also be noted that these hypotheses need not be mutually exclusive, particularly as we have recently shown that human seipin forms stable dodecamers, while a similar nine-subunit oligomer has been shown for the yeast Fld1p (Binns, Lee, Hilton, Jiang, & Goodman, 2010; Sim, Talukder, et al., 2013). These large complexes would be well placed to bring together multiple binding partners leading to diverse effects when seipin expression is lost. We also observed that some pathogenic single amino acid substitution mutations in seipin prevent the formation of these dodecamers, which may be important in the failure of these mutants to function appropriately (Sim, Talukder, et al., 2013). Overall, while there have been recent advances in our knowledge of seipin, its structure and potential functions, it remains unclear which of these may underlie its importance in adipogenesis and it seems plausible that other functions may remain to be identified.

3.2.3 Seipin-deficient mouse models

Three independent groups have reported the generation and characterization of mice with targeted disruption of the *Bscl2* gene (Chen et al., 2012; Cui et al., 2011; Prieur et al., 2013). The similarities and inconsistencies between these models and their relevance to CGL2 is the subject of an excellent recent review (Dollet, Magre, Cariou, & Prieur, 2014). The main points will be summarized here but readers are directed to this comprehensive review for a more detailed analysis.

Overall, all three models of seipin deficiency displayed classic characteristics of CGL2 in humans including a major reduction in adipose tissue mass, severe hepatic steatosis, and dysregulated glucose metabolism (Chen et al., 2012; Cui et al., 2011; Prieur et al., 2013). Surprisingly, seipin-deficient mice did not have high plasma triglyceride or fatty acid levels as are seen in CGL2 patients. As described for AGPAT2 above, the lack of adipose tissue makes it difficult to conclude a great deal regarding the molecular mechanisms underlying the lipodystrophy in these initial models. Two of the groups generated MEFs from these mice which have provided some useful insights regarding potential pathogenic mechanisms and act as a valuable resource for cellular studies as described above (Chen et al., 2012; Prieur et al., 2013). However, there were subtle differences in the stage at which adipogenesis failed in these two reports. Prieur et al. observed an early defect with reduced C/EBPβ and subsequent impaired induction of PPARγ and C/EBPα expression within the first 48 h of differentiation (Prieur et al., 2013). In contrast, Chen and colleagues found that these initial stages proceeded normally but that the adipocyte phenotype was not maintained subsequently (Chen et al., 2012).

Careful examination of seipin-deficient mice revealed some residual adipose tissue containing immature adipocytes (Chen et al., 2012; Cui et al., 2011; Prieur et al., 2013). BAT mass was also reduced and remaining cells did not have the characteristic multilocular, brown adipocyte phenotype (Chen et al., 2012). Prieur and colleagues observed that treatment of seipin-deficient MEFs with thiazolidinediones (TZDs) increased their capacity to form adipocytes *in vitro*, an effect similar to that previously reported in 3T3-L1 preadipocytes in which seipin expression had been knocked down with siRNA (Chen et al., 2009; Prieur et al., 2013). Extending these observations *in vivo* they showed that pioglitazone treatment of seipin knockout mice could modestly but significantly rescue both white and brown adipose depots (Prieur et al., 2013). This had significant metabolic benefits, almost entirely normalizing hepatic TG levels, fasting insulin and glucose levels, and insulin tolerance and also correcting triglyceride rich lipoprotein clearance. Leptin and adiponectin levels were also improved. Significant further work will be required to determine whether TZD-type drugs could be beneficial in seipin deficiency. However, there is encouraging evidence from a single existing case report of rosiglitazone treatment improving insulin sensitivity and leptin levels in a CGL2 patient (Victoria et al., 2010). Overall, this work again illustrates the potential of mouse models of CGL to investigate potential novel therapeutic interventions.

In all three seipin knockout models, investigators have made some analyses of the residual adipose tissue (Chen et al., 2012; Cui et al., 2011; Prieur et al., 2013). This should probably be interpreted with caution as it may be anomalous by definition, given that these depots are not typically apparent in CGL2 patients. Moreover, it will have been subject to a severely altered metabolic milieu and so may display secondary changes not a direct consequence of the intrinsic absence of seipin. The most recent of these studies reported altered lipid metabolism with increased oxidation and some BAT-like, oxidative characteristics in residual epididymal adipose tissue (Chen et al., 2013). In addition, the fatty acid desaturation ratio was increased, at odds with previous findings in lymphoblastoid cells from CGL2 patients where this was decreased (Boutet et al., 2009; Chen et al., 2013). Overall, the relevance of these data is unclear but may prove valuable as these models of lipodystrophy are studied further.

3.3. *CAV1* (CGL3) and *PTRF* (CGL4)

The progressive generalized lipodystrophy observed in mice with targeted disruption of caveolin 1 led investigators to screen the corresponding gene, *CAV1*, in several patients with CGL that was not due to *AGPAT2* or *BSCL2* mutations (Kim et al., 2008; Razani et al., 2002). Hence, this has been designated CGL3. This study identified one patient homozygous for a premature stop mutation in *CAV1* which led to undetectable levels of the caveolin 1 protein and an apparent inability to support the formation of caveolae in cells (Kim et al., 2008). The affected patient displayed classic features of CGL including lipoatrophy with near complete loss of subcutaneous and visceral adipose depots although vestigial adipose tissue was observed in dorsal subcutaneous regions. Severe insulin resistance, hypertriglyceridemia, and muscular hypertrophy were also evident. However, some unusual features were also observed including the absence of increased linear growth in childhood seen in other forms of CGL and abnormal calcium homeostasis was also reported.

3.3.1 Caveolin 1 is a key protein of caveolae

Caveolin 1 is a principle protein of caveolae in adipocytes and caveolin 1 and the caveolae themselves play critical roles in lipid uptake and supply to the lipid droplet in these cells (Fig. 2.2) (Le Lay, Blouin, Hajduch, & Dugail, 2009; Le Lay, Hajduch, et al., 2009). In addition, caveolae provide key environments for some hormone and growth factor signaling, including that by insulin. Caveolae are believed to be a focal point for the activation of the insulin receptor (IR), binding of IR substrates, localized generation of PIP_3 and so activation of downstream intracellular signaling by the IR (Le Lay, Blouin, et al., 2009; Parton & del Pozo, 2013). IR internalization has also been proposed to occur through caveolae and these organelles may also represent the major site of accumulation of the insulin-responsive glucose transporter GLUT-4 at the cell surface following insulin stimulation (Karlsson, Thorn, Parpal, Stralfors, & Gustavsson, 2002).

Caveolin 1 has also been shown to have roles in cellular trafficking both of proteins and lipids and can be detected associated to lipid droplets in mature adipocytes (Le Lay, Blouin, et al., 2009; Parton & del Pozo, 2013). Caveolin 1 plays a key role in the stabilization of caveolae in cells where it constitutes the major caveolin isoform such as adipose tissue. Hence, the loss of caveolin 1 has the effect of preventing caveolae formation in these cells (Razani et al., 2002). As adipocytes have probably the highest density of caveolae in the body, it is perhaps not surprising that this tissue is so significantly affected by disruption of caveolin 1. Caveolin 1 appears important for lipid absorption from the gut and so nonadipose effects, such as altered nutrient supply from the diet, may also contribute to the overall phenotype of caveolin 1 disruption (Le Lay, Blouin, et al., 2009; Parton & del Pozo, 2013). However, no alteration of dietary fat absorption was observed in initial studies of caveolin 1 null mice (Razani et al., 2002).

3.3.2 Caveolin 1 null mice

Caveolin 1 deficiency in mice has clear effects on lipid transport in adipocytes (Le Lay, Blouin, et al., 2009; Parton & del Pozo, 2013; Pilch & Liu, 2011). Caveolin is a cholesterol-binding protein and may be involved in the trafficking of cholesterol from the plasma membrane to the lipid droplet (Le Lay, Blouin, et al., 2009; Parton & del Pozo, 2013). Cholesterol is a significant component of the TG-rich lipid droplet in mature adipocytes and caveolin 1 is also found associated with lipid droplets in these cells. Moreover, the levels of cholesterol in lipid droplets are substantially reduced in adipocytes from caveolin 1 null mice (Le Lay et al., 2006). Fibroblasts from these mice display reduced fatty acid uptake, while several fatty acid-binding proteins are found in caveolae (Le Lay, Blouin, et al., 2009; Parton & del Pozo, 2013). Adipocyte lipolysis is also reduced in caveolin null mice. This appears to involve increased activity of PKA although the mechanisms underlying this are unclear (Cohen et al., 2004). Caveolin has even been proposed to bind the lipid droplet coat protein perilipin in caveolae at the plasma membrane (Aboulaich, Vener, & Stralfors, 2006). Similarly, a role for caveolin 1 has been proposed in the synthesis of TG at the plasma membrane, in addition to the ER, where this process is commonly held to occur (Ortegren, Aboulaich, Ost, & Stralfors, 2007). More recently, Martin et al. have reported that caveolin deficiency in mice leads to instability of adipocytes with increased cell death and inflammation resulting in fibrosis (Martin et al., 2012). Evidently, this provides a plausible explanation for the progressive lipodystrophy observed in these animals.

Given their previous findings that *CAV1* polymorphisms might be linked to insulin resistance in hypertensive individuals, Chuengsamarn et al. examined whether direct renin inhibitors might ameliorate these features in caveolin 1 null mice (Chuengsamarn et al., 2013). They found that one such drug, aliskiren, reduced systolic blood pressure, triglyceride levels, and insulin resistance as measured by HOMA-IR in cav1 KO mice.

Caveolin knockout mice have also been used to study several other aspects of caveolin biology including endothelial function, retinal epithelium function, tumorigenesis, pulmonary fibrosis, blood brain barrier permeability, immune function, scleroderma, cardiac function, and neurodegeneration (Parton & del Pozo, 2013). The relevance of these to CGL3 is not clear. Indeed reexpression of caveolin 1 specifically in endothelial cells rescued the vascular and cardiac phenotypes and macrophage infiltration but not the lipoatrophic phenotype demonstrating that these consequences of caveolin 1 loss can be functionally separated (Briand, Le Lay, Sessa, Ferre, & Dugail, 2011; Murata et al., 2007). These wider roles of caveolin 1 are not discussed in detail here, other than to highlight the multiple roles of the protein *in vivo*.

It is clear that the roles of caveolin 1 in adipocytes are somewhat pleiotropic, regulating insulin sensitivity due to its roles in IR signaling from caveolae, influencing cholesterol and fatty acid uptake, and altering the storage and hydrolysis of TG at the lipid droplet (Le Lay, Blouin, et al., 2009; Parton & del Pozo, 2013). There does not appear to be a significant effect of caveolin 1 loss on adipogenesis *per se*, although this has not been extensively examined in detail (Cohen et al., 2004; Le Lay et al., 2006). Hence, lipodystrophy due to CAV1 mutations appears likely to be mechanistically distinct from that due to AGPAT2 and BSCL2 mutations.

3.3.3 PTRF is required for caveolae formation

Polymerase 1 and transcript release factor (PTRF), also known as cavin, is another key component of caveolae and mutations in the *PTRF* gene were found in several individuals causing a syndrome of generalized lipodystrophy with muscular dystrophy (CGL4) (Hayashi et al., 2009; Shastry et al., 2010). PTRF plays a critical role in the stabilization of caveolar protein complexes and thereby the caveolae themselves (Chadda & Mayor, 2008; Hill et al., 2008) (Fig. 2.2). Pathogenic mutations in *PTRF* appear to interfere with the interaction of PTRF with caveolin proteins and thereby disrupt the caveolae. For this reason, it is likely that the molecular mechanisms underlying lipodystrophy due to *PTRF* mutations share many features with that due to *CAV1* mutations with defects in insulin signaling and lipid storage and metabolism. However, as PTRF is important for caveolae formation in cell types where caveolin 1 is not the major caveolin and this is likely to explain the additional more complex phenotype of these patients, particularly the muscular dystrophy. As its name suggests, PTRF has a role in releasing the polymerase complex from the transcript during ribosomal RNA transcription (Jansa, Mason, Hoffmann-Rohrer, & Grummt, 1998). Evidently, this may further expand the repertoire of pathways whereby PTRF disruption can alter adipocyte development and function, beyond those caused by lack of caveolin 1.

3.3.4 PTRF null mice

PTRF disruption in mice causes a global loss of caveolae and a syndrome featuring reduced adipose tissue mass, hyperinsulinemia, glucose intolerance, and raised circulating triglycerides (Liu et al., 2008). The phenotype was more severe than that in caveolin 1 knockout mice; however, the authors did not describe muscular dystrophy in this model. As with caveolin 1 disruption, the phenotype of the mouse model led to the identification of *PTRF* mutations as a cause of lipodystrophy in humans, immediately illustrating the value of this model. This also suggests that, as with other mice in which CGL genes have been disrupted, PTRF knockout mice may be a valuable model for learning more regarding the corresponding human condition and investigating novel potential therapies.

4. GENES DISRUPTED IN FAMILIAL PARTIAL LIPODYSTROPHIES

4.1. *LMNA* and *ZMPSTE24*

The *LMNA* gene encodes the nuclear lamina proteins lamin A and lamin C. Mutations in lamin A/C can cause several syndromes collectively described as laminopathies which include lipodystrophy but also muscular dystrophies, cardiomyopathies, neuropathies, and progeroid syndromes (Worman & Bonne, 2007). At least 11 different diseases have been described resulting from more than 300 different *LMNA* mutations. In some cases, *LMNA* mutations cause complex phenotypes featuring lipodystrophy with progeria (mandibuloacral dysplasia-associated lipodystrophy, type 1), while others selectively result in FPLD (Dunnigan variety) (Cao & Hegele, 2000; Hegele, Anderson, Wang, Jones, & Cao, 2000; Hegele, Cao, Anderson, & Hramiak, 2000; Shackleton et al., 2000). Lamins are synthesized as prelamin proteins which are then posttranslationally processed including farnesylation and carboxymethylation of a CAAX motif on their C-terminus (Dechat et al., 2008). In lamin A, the protease Zmpste24 subsequently cleaves this motif to generate the mature protein. For this reason, mutations in the gene encoding Zmpste24, *ZMPSTE24*, can cause another form of FPLD with progeria called mandibuloacral dysplasia type B (MAD-B) (Agarwal, Fryns, Auchus, & Garg, 2003). Interestingly, the residual protease activity of pathogenic mutant forms of Zmpste24 has been correlated to the phenotype of affected individuals whereby retention of partial activity leads to MAD-B and complete loss of activity to the very severe, rare and lethal laminopathy restrictive dermopathy (Barrowman, Wiley, Hudon-Miller, Hrycyna, & Michaelis, 2012). Disruption of Zmpste24 in mice causes a similar phenotype and accumulation of prelamin A (Pendas et al., 2002).

4.1.1 Mechanisms via which LMNA *mutations may cause FPLD*

Despite the many studies investigating lamin A/C function, the reason why mutations in the same gene cause such diverse phenotypes and how mutation of such a widely expressed nuclear lamina protein could affect specific tissues remains poorly understood. Lamins have diverse roles in the nucleus including maintaining the structural integrity of the nuclear membrane, influencing nuclear pore complexes and binding to chromatin and multiple transcription factors (Dechat et al., 2008). One mechanism via which lamin

A/C mutation is proposed to influence adipose development is via interaction with the transcription factor SREBP1c (Lloyd, Trembath, & Shackleton, 2002). SREBP1c plays a key role in adipocyte differentiation, inducing the expression of many genes of the mature adipocyte (Kim & Spiegelman, 1996). In addition SREBP1c is proposed to be in producing an endogenous ligand that activates PPARγ in developing adipocytes (Kim, Wright, Wright, & Spiegelman, 1998).

SREBP1c is produced as a precursor protein inserted in the ER membrane that is activated following transport to the Golgi and proteolytic cleavage. This generates an N-terminal fragment that translocates to the nucleus and acts as a transcription factor (Bengoechea-Alonso & Ericsson, 2007). While several studies have suggested a link between lamin A/C and SREBP1c, there is little consensus about the precise mechanism involved. Mutant lamins have been proposed variously to impair binding to SREBP1c (Lloyd et al., 2002) or to sequester SREBP1c away from its nuclear targets (Capanni et al., 2003, 2005). Further studies show that expression of the R482W and R482Q mutants of lamin A/C that cause FPLD can inhibit adipogenesis in cultured 3T3-L1 preadipocytes, however, overexpression of the wild-type protein also has a similar effect making the relevance of this uncertain (Boguslavsky, Stewart, & Worman, 2006). While impaired SREBP1c function could contribute to the development of lipodystrophy in individuals with *LMNA* or *ZMPSTE24* mutations, consistent mechanistic insights remain elusive. Given the multiple roles of lamin A/C, it seems unlikely that only SREBP1c mediated mechanisms are involved. Multiple other transcription factors have been shown to interact with, or be regulated by, lamins, and some of these such as pRb can also influence adipogenesis (Classon, Kennedy, Mulloy, & Harlow, 2000). Lamins are involved in chromatin organization, RNA polymerase II-driven transcription, mitotic nuclear membrane disassembly and reassembly, the formation of the nuclear pore complexes, the structural integrity of the nuclear envelope and its communication with the cytoplasm and cytoskeleton (Dechat et al., 2008). In most cases, the importance of these functions in adipose tissue development has not been investigated and several might be expected to link lamin A/C function to adipocyte differentiation and/or function.

4.1.2 Mouse models of LMNA disruption

Such is the complexity of the different phenotypes resulting from different mutations in *LMNA*; this is evidently a case where a knockin model of mice specifically bearing mutations causing lipodystrophy in humans will be required to gain a fuller understanding the molecular pathogenesis of this condition. Existing lamin models have not proven particularly insightful in the study of FPLD. Lamin null mice are viable and develop to term without major abnormalities (Sullivan et al., 1999). They subsequently show severely retarded postnatal growth and develop muscular dystrophy, consistent with the capacity of some *LMNA* mutations to cause Emery–Dreifus muscular dystrophy (EDMD).

Adipocyte-selective transgenic overexpression of the FPLD mutant R482Q form of lamin A in mice using the aP2 promoter did result in a lipodystrophic phenotype (Wojtanik et al., 2009). These mice display a progressively decreased ability to develop fat with age, impaired glucose tolerance, and hepatic steatosis. They appear to have a decreased ability to renew adipose tissue as they age, rather than any defect of adipose tissue lipolysis. Evidently, this mouse may be modeling some aspects of FPLD but it is subject to several caveats including the disproportionate expression of the disease allele and its tissue selective expression which do not recapitulate the situation in affected patients.

Knockin models have been reported for *LMNA* mutations causing EDMD or progeria which provide useful models of these conditions. Similar models for FPLD mutations are likely to be informative in the defining how LMNA mutations can cause partial lipodystrophy. *Zmpste24* null mice do display lipodystrophy, along with cardiomyopathy and muscular dystrophy (Bergo et al., 2002; Pendas et al., 2002). The adipose tissue aspect of the phenotype has not been extensively studied and, given that the effects are probably mediated via lamin processing, lamin mutant mice are likely to provide more tractable models to specifically investigate the molecular pathogenesis of these forms of FPLD without the complexity of other disease phenotypes.

4.2. PPARG

PPARγ is a critical transcription factor that plays a central role in the regulation of adipogenesis controling the expression of a plethora of genes characteristic of insulin sensitive, metabolically active, mature adipocytes (Tontonoz & Spiegelman, 2008). Hence, it may be unsurprising that impaired PPARγ function can lead to the development of lipodystrophy.

4.2.1 PPARγ is a key regulator of adipocyte development and function

PPARγ null mice die *in utero* due to defects in placental development (Barak et al., 1999). However, chimeric rescue experiments in mice to circumvent embryonic lethality have demonstrated that cells lacking PPARγ cannot develop into mature adipocytes *in vivo* (Barak et al., 1999; Rosen et al., 1999). Thus, a certain level of PPARγ activity appears essential for survival *in utero* implying that the mutant forms observed in humans must retain some residual activity. Multiple different PPARγ mutations have been identified which affect the function of the protein in different ways, as determined by molecular and functional studies. These have been reviewed in detail elsewhere (Jeninga, Gurnell, & Kalkhoven, 2009). The *PPARG* gene encodes two isoforms, PPARγ_1 and PPARγ_2, the latter containing an additional 28 amino acids at the N-terminus due to the use of an alternative promoter (Tontonoz & Spiegelman, 2008). Both PPARγ_1 and PPARγ_2 have highest expression in adipose tissue, while PPARγ_2 is almost exclusively expressed in adipocytes. Genome wide analyses of PPARγ binding to target DNA sequences in adipocytes have confirmed its critical involvement regulating many key genes of the adipogenic and lipogenic programs, and also demonstrated that it cooperates with the C/CAAT enhancer-binding protein C/EBPα, another critical transcription factor in adipogenesis, on many of these targets (Lefterova et al., 2008; Nielsen et al., 2008). While endogenous ligands for PPARγ are believed to exist in adipocytes, their identity remains unclear (Tontonoz & Spiegelman, 2008). In contrast, the TZD class of antidiabetic drugs act as potent ligands for PPARγ and have been widely used clinically in the past and extensively studied both *in vitro* and *in vivo* (Semple, Chatterjee, & O'Rahilly, 2006; Tontonoz & Spiegelman, 2008).

The mechanism via which impaired PPARγ function causes lipodystrophy and metabolic disease is likely to principally involve effects on lipids. Many of the genes regulated by PPARγ in adipocytes increase the uptake of lipid substrates or their storage and metabolism (Nielsen et al., 2008; Tontonoz & Spiegelman, 2008). PPARγ activation also appears to induce a futile cycle of hydrolysis and reesterification of TG (Guan et al., 2002). Overall, this tends to trap lipids in adipose tissue reducing plasma fatty acids and decreasing lipid accumulation in nonadipose tissues. As the ectopic storage of lipid in liver, skeletal muscle, and the pancreas has deleterious effects on insulin sensitivity and secretion, PPARγ dysfunction may drive systemic insulin resistance via this mechanism (Medina-Gomez, Gray, & Vidal-Puig, 2007). Normal PPARγ function is also likely to benefit insulin sensitivity within target cells by inducing the expression of proteins such as the insulin-sensitive glucose transporter GLUT-4 (Tontonoz & Spiegelman, 2008). PPARγ activation also increases expression of the circulating adipokine adiponectin, which may improve insulin sensitivity in skeletal muscle and liver (Yamauchi et al., 2001, 2002). PPARγ activation by TZDs also decreases the expression of proinflammatory cytokines such as TNFα and resistin in adipocytes and macrophages, which can contribute to insulin resistance (Tontonoz & Spiegelman, 2008). PPARγ dysfunction is likely, therefore, to cause partial lipodystrophy and insulin resistance through multiple mechanisms. Significant adipose depots remain in FPLD patients with PPARγ mutations implying there may be only a modest, if any, impairment of adipogenesis *per se*. Rather, pathogenic PPARγ mutations may mainly impair adipocyte function and alter lipid storage and insulin sensitivity in multiple metabolically important tissues.

4.2.2 PPARγ P467L knockin mice

Knockin mice bearing the dominant negative P465L mutation of PPARγ, corresponding to the FPLD-causing P467L mutation in humans, were generated independently by two groups (Gray, Dalla Nora, et al., 2006; Tsai et al., 2004). Tsai et al. reported that the total adipose mass was unchanged in their heterozygous mice. However, the distribution had shifted such that intra-abdominal fat mass was reduced and subcutaneous adipose tissue increased. They also observed reduced interscapular BAT mass (Tsai et al., 2004). The mice had normal insulin sensitivity and glycemia while glucose tolerance was improved. However, following high-fat feeding the displayed mildly elevated plasma insulin and the mice were hypertensive. Gray et al. also observed no significant changes in total adipose tissue mass in heterozygous P465L PPARγ (*P465L/wt*) mice and no altered insulin sensitivity or glucose tolerance on either chow or high-fat diet (Gray, Dalla Nora, et al., 2006). However, the *P465L/wt* mice had impaired postprandial lipid metabolism and elevated hepatic TG levels when fed a high-fat diet. Analysis of BAT in these mice revealed reduced brown adipocyte recruitment and an impaired thermogenic capacity with BAT adipocytes containing larger lipid droplets (Gray, Dalla Nora, et al., 2006). The investigators also observed a significantly reduced capacity of stromal cells isolated from *P465L/wt* mice to differentiate into adipocytes in culture (Gray, Dalla Nora, et al., 2006). They subsequently crossed their mice onto a background of the leptin-deficient, obese and hyperphagic *ob/ob* mice to generate mice with homozygous leptin disruption and heterozygous for the P465L allele of PPARγ (*P465L/ob* mice). This additional metabolic stress was able to unmask a phenotype relevant to the partial lipodystrophy observed in the FPLD patients. The *P465L/ob* mice appear restricted in their capacity to expand their adipose tissue and so gain less weight than their *ob/ob* littermates. The loss of this protective weight gain in the face of extreme nutrient excess leads to severe insulin resistance in the

P465L/ob mice (Gray, Dalla Nora, et al., 2006). In a similar manner, the *P465L/wt* mice were also crossed with *Ins2–Akita* mice, a model with a mutation in the *Ins2* gene causing ER stress-induced β-cell apoptosis, reduced plasma insulin, and so consistently elevated glycemia from early life (Pendse, Johnson, Tsai, & Maeda, 2010). The resulting doubly heterozygous mice (*P465L/Akita* mice) had worsened hyperglycemia, reduced adiposity, and decreased leptin and adiponectin levels compared with their *Ins2–Akita* littermates. This appears to result from impaired adipose tissue insulin sensitivity and lipid storage.

Several subsequent studies have used the *P465L/wt* mice to examine other aspects of PPARγ biology. These include roles in vascular function (Beyer et al., 2008), an interaction with hypertension in the development of cardiac fibrosis (Kis et al., 2009), atherogenesis (Pendse et al., 2012), and lipolysis (Rodriguez-Cuenca et al., 2012). In each case, the subtle partial loss of function has allowed the generation of viable mice in which the involvement of PPARγ in these processes can be investigated. Overall, these studies of mice carrying the P465L mutation in PPARγ serve to illustrate that mice may not immediately and simply model inherited partial lipodystrophies identified in humans, probably due to altered lipid metabolism and oxidative capacity. Rather, a further metabolic stress may be required in order to unveil the relevant phenotype.

4.3. AKT2

AKT2 is a threonine/serine kinase with essential roles in growth factor and insulin signaling and is one of three AKT isoforms, AKT1 (also called PKBα), AKT2 (PKBβ), and AKT3 (PKBγ). AKT is activated in response to insulin and growth factors via phosphoinositide 3-kinase (PI3K) by phosphorylation of two residues, T308 and S473 (numbered for AKT1). Active AKT is capable of phosphorylating a wide variety of substrates with roles in cell differentiation, proliferation, survival, nutrient uptake/storage, and protein synthesis (Pearce et al., 2010). A single family with partial lipodystrophy due to a mutation in the *AKT2* gene was identified in 2004 (George et al., 2004).The mutation of *AKT2* identified in individuals with lipodystrophy causes an amino acid change of R274H in the AKT2 protein. R274 is highly conserved in AKT isoforms across multiple species and lies in the catalytic loop of the enzyme. This amino acid forms critical hydrogen bonds with phosphorylated T309 (equivalent to T308 in AKT1) and appears important for the appropriate positioning of substrates for phosphorylation by AKT2

(Yang et al., 2002). Affected individuals are heterozygous for this mutation which appears to act in a dominant negative manner, inhibiting both wild-type AKT2 and AKT1 (George et al., 2004). Thus, the R274H mutation would be expected to inhibit both AKT1 and AKT2 in cell types where AKT2 is abundantly expressed, but not in cells selectively expressing AKT1.

4.3.1 Potential mechanisms underlying FPLD due to AKT2 *mutations*

Lipodystrophy and insulin resistance in individuals carrying the *AKT2* mutation is likely to involve modestly impaired adipogenesis exacerbated by insulin resistance and poor metabolic function in both adipose and nonadipose tissues. Impaired insulin action in skeletal muscle, where AKT is important for insulin stimulated glycogen synthesis, translocation of GLUT-4, and thereby glucose transport, almost certainly contributes significantly to insulin resistance in these patients (Bouzakri et al., 2006). Likewise AKT also regulates the suppression of hepatic gluconeogenesis by insulin by acting as an upstream regulator of signaling to inhibit the expression of glucose-6-phosphatase and phosphoenolpyruvate carboxykinase mRNA. In adipocytes AKT, particularly AKT2, phosphorylates AS160, a Rab GTPase-activating protein which promotes Glut-4 translocation to the plasma membrane in response to insulin. AKT may also contribute to the inhibition of lipolysis by indirectly reducing PKA activity, so suppressing the activation of adipose triglyceride lipase (ATGL) by CGI-58 and the phosphorylation and activation of hormone-sensitive lipase (HSL) (Giorgino, Laviola, & Eriksson, 2005).

At the transcriptional level, AKT has been proposed to regulate adipogenesis through inhibition of various forkhead transcription factors including Foxo1, Foxc2, and Foxa2, each of which can negatively regulate adipocyte differentiation (Davis, Moldes, & Farmer, 2004; Gerin et al., 2009; Nakae et al., 2003; Wolfrum et al., 2003). Phosphorylation of forkhead factors by AKT causes translocation from the nucleus and so loss of transcriptional activity. Constitutively active Foxo1 increases the expression of CHOP10, an inhibitory C/EBP isoform, and inhibits PPARγ expression and the induction several adipocyte genes such as GLUT-4 (Nakae et al., 2003). Active Foxa2 induces the expression of Pref1, an inhibitor of adipogenesis, while Foxc2 activity inhibits PPARγ expression and activity and the induction of C/EBPα and multiple PPARγ target genes (Davis et al., 2004; Wolfrum et al., 2003). In addition to the regulation of forkhead factors, AKT has been suggested to influence SREBP1c expression, translocation, processing, activation, and degradation (Krycer, Sharpe,

Luu, & Brown, 2010). Although not yet examined in differentiating adipocytes expressing AKT2-R274H, it is possible that disturbance of these pathways by this mutant AKT restricts adipogenesis. AKT also has proliferative/prosurvival effects (Jiang et al., 2005; Levine, Feng, Mak, You, & Jin, 2006). Hence, it is possible that AKT2-R274H may render developing adipocytes prone to reduced proliferation or increased cell death.

Overall, the R274H mutation in AKT2 could therefore affect both insulin sensitivity and the transcriptional program of adipogenesis, causing the phenotype of insulin resistance and partial lipodystrophy. However, this has yet to be experimentally tested in detail at the cellular level.

4.3.2 Mouse models of altered AKT function

Knockin mice bearing the R274H mutation of AKT2 would be required to model this form of partial lipodystrophy most accurately. However, *Akt* knockout mice exist and provide some relevant insights. Mice lacking Akt2 show impaired glucose metabolism and age-dependent decreases in adipose mass (Cho et al., 2001; Garofalo et al., 2003), while loss of Akt1 does not impair glucose metabolism but leads to growth retardation. Knockout of both Akt1 and Akt2 leads to early postnatal death with a complex phenotype featuring defects of skeletal muscle, skin, and bone but also significantly impaired adipose tissue development (Peng et al., 2003). Studies in MEFs from Akt knockout mice have suggested that Akt1 may play a greater role in adipogenesis than Akt2 (Peng et al., 2003; Yun et al., 2008). However, this may be somewhat misleading because Akt1 is the major isoform in MEFs but may be less relevant *in vivo*. Moreover, additional loss of Akt2 in Akt1 null MEFs further inhibits adipogenesis suggesting that Akt2 is also involved in adipocyte development. This is consistent with studies comparing adipose development *in vivo* in Akt1, Akt2, and double-knockout animals (Garofalo et al., 2003; Peng et al., 2003).

4.4. CIDEC

CIDEC is a member of the *c*ell death-*i*nducing *D*NA fragmentation factor-α-like *e*ffector (CIDE) family of proteins, which were initially characterized as potential mitochondrial inducers of apoptosis (Inohara, Koseki, Chen, Wu, & Nunez, 1998). However, it now appears that the true physiological roles of these proteins (CIDEA, CIDEB, and CIDEC) are likely to be in regulating metabolism. Mice lacking either CIDEA or CIDEB are lean and resistant to obesity. Loss of CIDEA causes increased brown adipose tissue mitochondrial uncoupling, while CIDEB disruption leads to increased fatty acid oxidation and decreased lipogenesis in the liver (Li et al., 2007; Zhou et al., 2003). CIDEC is highly expressed in white adipose tissue, leading to its alternative name, Fsp27 (for *f*at *s*pecific *p*rotein of *27* kDa) in mice (Danesch, Hoeck, & Ringold, 1992). Mutation of *CIDEC* was identified as a cause of partial lipodystrophy in humans when an affected individual was shown to be homozygous for a mutation causing a premature stop codon E186X in the CIDEC protein (Rubio-Cabezas et al., 2009).

4.4.1 Cidec-*deficient mice*

Mice with targeted disruption of Fsp27 have been generated and some aspects of the observed phenotype are likely to be relevant to the molecular pathogenesis of FPLD due to mutation of *CIDEC*. However, there are some significant differences regarding the physiological consequences of CIDEC/ Fsp27 mutation in humans and mice. It should also be acknowledged that the pathogenic mutation identified in humans generates a prematurely truncated protein which may not be directly comparable in terms of molecular mechanism to complete loss of the corresponding Fsp27 protein in mice.

Disruption of Fsp27 in mice causes the development of multilocular adipocytes in adipose tissue (Nishino et al., 2008; Toh et al., 2008) and Fsp27 knockdown in differentiating cultured adipocytes reduces lipid droplet size (Keller et al., 2008; Puri et al., 2007). Similarly, the patient homozygous for the E138X mutation in CIDEC also has characteristic multiple lipid droplets in white adipocytes (Rubio-Cabezas et al., 2009). However, Fsp27 loss protects mice from diet-induced obesity and insulin resistance, promoting a more brown fat-like morphology and increasing mitochondrial function (Nishino et al., 2008; Toh et al., 2008). In contrast, human *CIDEC* disruption leads to insulin-resistant diabetes with dyslipidemia, once again highlighting the potential difficulties of extrapolating findings regarding metabolism in mice to humans (Rubio-Cabezas et al., 2009). The high fasting plasma fatty acid levels observed with *CIDEC* disruption in the lipodystrophic patient were not so marked in the Fsp27 null mice. In addition, unlike the mice, there was no evidence of increased mitochondrial number in the adipose tissue of the individual with lipodystrophy (Nishino et al., 2008; Rubio-Cabezas et al., 2009; Toh et al., 2008). Together, these factors may contribute to discrepancy in the phenotypes observed with disruption of *CIDEC* in mice versus humans.

CIDEC is localized to lipid droplets and the ER in adipocytes and appears to translocate to lipid droplets in response to lipolytic stimulation (Nian et al., 2010; Puri et al., 2007; Rubio-Cabezas et al., 2009). Moreover, binding to the lipid droplet appears to stabilize CIDEC, delaying its ubiquitin-dependent degradation which otherwise occurs rapidly (Nian et al., 2010). Critically, the E138X form of CIDEC is truncated early in the functionally important and conserved CIDEC domain and does not associate with lipid droplets (Rubio-Cabezas et al., 2009). Recently, CIDEC has been reported to interact physically with the lipid droplet coat protein perilipin 1 (Grahn et al., 2013). As described below, perilipin 1 is a key component of lipolytic machinery in mature adipocytes. Overexpression of either CIDEC or perilipin 1 in a cultured human adipocyte cell line was shown to increase TG content and decrease lipolysis. However, the effect was not synergistic and expression of both proteins together resulted in no further change in these parameters. However, the combined expression of both CIDEC and perilipin did increase the average size of lipid droplets significantly over those seen with each protein expressed alone in these cells, implying roles of both in lipolysis but a synergistic role in droplet fission/fusion dynamics (Grahn et al., 2013). Thus, CIDEC expression influences the pattern of lipid accumulation in cultured adipocytes and plays an important role in the formation of a single lipid droplet in mature adipocytes *in vivo*. This may involve altering lipid droplet fission and fusion, the synthesis of phospholipids coating the lipid droplet or altering lipogenesis or lipolysis. However, the precise molecular function of CIDEC in adipocytes remains to be fully determined.

4.5. PLIN1

Perilipin 1 is the major lipid droplet coat protein in mature adipocytes and plays a critical role in the regulation of lipolysis, the process via which fatty acids and glycerol are liberated from TG in the lipid droplet (Girousse & Langin, 2012; Zechner et al., 2012). The majority of perilipin is associated with the lipid droplet, although a small but significant proportion has been reported to be bound to the ER membrane where it may be involved in the function of lipids droplets that bud from this organelle (Skinner, Harris, Shew, Abumrad, & Wolins, 2013). In unstimulated cells where the equilibrium favors lipid storage, perilipin binds CGI-58, also called ABHD5, an activator of the major triglyceride lipase ATGL. In this way, CGI-58 is sequestered away from ATGL and lipolysis is suppressed (Girousse & Langin, 2012; Zechner et al., 2012). Following stimulation of the adipocyte by adrenergic agonists, typically via sympathetic innervation *in vivo*, PKA becomes activated and this directly phosphorylates several residues on perilipin which leads to the dissociation of CGI-58. The liberated CGI-58 is able to bind to ATGL which accumulates at the lipid droplet in its active form and hydrolyzes TG to generate DAG with the release of fatty acid. DAG is the major substrate for HSL, which hydrolyzes DAG to MG. HSL is also phosphorylated by PKA which stimulates its activity in lipolytically active cells. Monoacylglycerol lipase then catalyzes the final step to generate glycerol and fatty acid from MG.

4.5.1 Pathogenic mutation in PLIN1, the gene encoding perilipin

Two mutations have been identified in PLIN1 within three pedigrees that cause FPLD (Gandotra, Le Dour, et al., 2011). In each case, the mutations cosegregated with lipodystrophy, dyslipidemia, and insulin resistance in these pedigrees. Detailed histological analysis of the adipose tissue revealed a significant reduction in adipocyte size, macrophage infiltration, and fibrosis (Gandotra, Le Dour, et al., 2011). Both mutations caused frameshifts that led to the substitution of a nonsense protein sequence in place of the wild-type C-terminal domain of perilipin. In each case, the absent sequence included three PKA phosphorylation sites known to be phosphorylated in response to lipolytic stimuli and important for the regulation of lipolysis. Unlike wild-type perilipin, the FPLD mutant forms of the protein were unable to suppress lipolysis and promote TG accumulation into lipid droplets in a well-characterized nonadipocyte cell assay (Gandotra, Le Dour, et al., 2011). Subsequent analysis revealed that neither mutant form of perilipin could bind CGI-58, the activator of ATGL (Gandotra, Lim, et al., 2011). As a result, these mutant forms of perilipin were unable to disrupt the interaction between CGI-58 and ATGL predicted to lead to increased ATGL activity is increased in the basal state *in vivo*.

4.5.2 Mouse models of Plin1 *disruption*

While no mice have been generated bearing either of these pathogenic perilipin mutations identified in FPLD, there are some similarities with perilipin null mice (Martinez-Botas et al., 2000; Tansey et al., 2001). Other mouse models in which regulators of lipolysis are specifically altered can also provide some useful insights but are beyond the scope of this review. However, an excellent recent review of this field is provided by Girouse et al. (Girousse & Langin, 2012). *Plin1* mutant mice have reduced adipose mass, adipocyte size, and adipocyte TG content but normal body weight due to increased lean mass (Martinez-Botas et al., 2000; Tansey et al., 2001). While no change in lipolysis was detected *in vivo*, adipocytes isolated from of *Plin1* null mice have increased basal lipolysis *in vitro* and blunted stimulated lipolysis. These studies preceded the discovery of ATGL and its regulation by CGI-58 and so this was not examined. However, it seems likely that the loss of perilipin increased the activation of ATGL by CGI-58 as proposed in the FPLD patients with perilipin mutations. Histological analysis of perilipin knockout mice showed adipose tissue histology reminiscent of the fibrosis observed in patients with PLIN1 mutations (Gandotra, Le Dour, et al., 2011; Tansey et al., 2001). However, there was no evidence of hepatic steatosis and energy expenditure was increased. Indeed, the knockout mice appeared resistant to obesity on a high-fat diet or when crossed with *db/db* mice (Martinez-Botas et al., 2000; Tansey et al., 2001). Of interest, perilipin null mice were more susceptible to glucose intolerance following high-fat diet feeding and a later study observed glucose intolerance and insulin resistance in these mice as they aged (Saha, Kojima, Martinez-Botas, Sunehag, & Chan, 2004). Overall, this again implies that the metabolic consequences of a genetic change that could lipodystrophy in humans may not lead to any adverse metabolic change in mice without an additional stress, be that genetic, as in the case of PPARγ or with diet or aging as in the case of perilipin.

5. LIMITATIONS OF THE CURRENT MODELS AND PERSPECTIVES FOR THE FUTURE

Studies of knockout mouse models of generalized lipodystrophies have already proven valuable. The first step of demonstrating that disruption specifically of the genes involved can indeed cause CGL is an important one. Indeed, in the case of CGL3, it was the phenotype of the knockout mice that led the investigators to discover that CAV1 disruption can cause CGL in humans and by association, the subsequent identification of PTRF mutations in CGL4. The CGL mouse models have already revealed some valuable insights regarding the likely etiology and progression of lipodystrophy due to these gene disruptions. They have permitted detailed analysis of metabolic parameters in carefully controlled conditions and allowed for the effects of precise dietary or drug manipulations to be investigated. They have also allowed for detailed study of the effects of CGL on other tissues and how these effects change over time. MEF cells from these mice have given some insights regarding the molecular mechanisms by which disruption of CGL genes cause the condition. However, the overall lack of fat in the mice themselves makes them relatively ill-suited to the discovery of the molecular mechanisms by which this tissue fails to develop. Such insights are likely to come in future from more sophisticated inducible and tissue-specific disruptions of CGL genes in which hypotheses generated in cultured cell models can be rigorously tested. Despite this the knockout models will always have great value in investigating potential therapies for CGL, additional studies of the disease progression and as a model in which to examine potential rescue of adipose tissue development by reconstitution of dysfunctional pathways.

As described above, there remains only one true murine model of FPLD reported to date, the *Pparg* P465L knockin mouse (Gray, Nora, et al., 2006; Tsai et al., 2004). Evidently, the autosomal dominant inheritance of several cases of FPLD requires the generation of knockin models of the gene mutations involved for their accurate study to be possible. However, in the case of *CIDEC*, the *Fsp27* knockout mouse is likely to represent an informative model of this condition (Nishino et al., 2008; Rubio-Cabezas et al., 2009; Toh et al., 2008). Perhaps, the greatest challenge in this area is that the initial studies of the *Pparg* P465L mouse failed to show as significant a metabolic phenotype as that observed in humans with no evidence of the insulin resistance that is manifest in the patients (Gray, Nora, et al., 2006; Tsai et al., 2004). This likely underlines the differences between human and mouse metabolism, particularly in the handling and oxidation of lipids, when mice are maintained in standard housing conditions. The work of Gray et al. clearly demonstrates the potential need for an additional metabolic stress to uncover more aspects of the human phenotype in the *Pparg* P465L mouse (Gray, Nora, et al., 2006). Hence, as new models become available it may be that a combination of high-fat diet, thermoneutral housing and/or specific genetic backgrounds are required to unmask an FPLD phenotype and permit the more accurate study of this condition in mice.

Another potentially relevant aspect is highlighted by recent studies using lineage tracing to define populations of adipocyte precursors *in vivo*. It is now evident that there are different populations of adipocyte stem cells that may reside in a variety of niches. These may make quantitatively different contributions to different adipose depots in the body and the adipocytes they form may have different metabolic properties (Rosen & Spiegelman, 2014). As such it is possible that these different adipocyte precursors may be differentially affected by the disruption of FPLD genes. For example, the *Pparg* P465L mouse, like the human patients has been reported to have subtly altered distribution of adipose tissue (Hegele, Cao, Frankowski, Mathews, & Leff, 2002; Savage et al., 2003; Tsai et al., 2004). Difficulties in defining precisely what constitutes the corresponding depots between species and the very modest metabolic effect of this in the mice makes for a complex situation. However, it is possible that FPLD is caused by a near complete loss of adipocytes arising from a specific stem cell population which gives rise only to a subset of adipocytes in selective depots. As the precise identity of these adipocyte stem cells becomes clearer FPLD mouse models coupled to lineage tracing techniques may provide a unique way to investigate this possibility.

6. CONCLUSIONS

In summary, there are evidently significant strengths and weaknesses in the use of mouse models to learn more about the molecular pathogenesis and metabolic consequences of gene disruptions that cause lipodystrophy in humans. So far it appears that observations made in models of CGL are most likely to translate accurately from mice to humans and these have already yielded some valuable initial insights. However, with the emergence of more specific knockin models of FPLD mutant alleles in mice, it is likely that these too will teach us more about the molecular functions of the affected genes. So far, the majority of studies using mouse models of lipodystrophy could be considered as initial, if thorough, characterizations. Future studies should include crossing these models with mice bearing mutations affecting specific developmental pathways or adipocyte functions to probe the pathogenic mechanisms involved. Alternatively, the reexpression of potential downstream effectors by viral or inducible transgene technologies is also likely to prove informative. Finally, the generation of tissue-specific or inducible models for the disruption of these genes will also be valuable. It is these more sophisticated, combined genetic techniques that are likely to push forward the next phase of discovery in this exciting area. This should reveal precisely why these genes are important for adipocyte development and function with the opportunity to suggest new therapies for all metabolic diseases involving adipose tissue dysfunction.

REFERENCES

Aboulaich, N., Vener, A. V., & Stralfors, P. (2006). Hormonal control of reversible translocation of perilipin B to the plasma membrane in primary human adipocytes. *Journal of Biological Chemistry*, *281*, 11446–11449.

Agarwal, A. K., Arioglu, E., De Almeida, S., Akkoc, N., Taylor, S. I., Bowcock, A. M., et al. (2002). AGPAT2 is mutated in congenital generalized lipodystrophy linked to chromosome 9q34. *Nature Genetics*, *31*, 21–23.

Agarwal, A. K., Fryns, J. P., Auchus, R. J., & Garg, A. (2003). Zinc metalloproteinase, ZMPSTE24, is mutated in mandibuloacral dysplasia. *Human Molecular Genetics*, *12*, 1995–2001.

Agarwal, A. K., Simha, V., Oral, E. A., Moran, S. A., Gorden, P., O'Rahilly, S., et al. (2003). Phenotypic and genetic heterogeneity in congenital generalized lipodystrophy. *Journal of Clinical Endocrinology and Metabolism*, *88*, 4840–4847.

Agarwal, A. K., Sukumaran, S., Cortes, V. A., Tunison, K., Mizrachi, D., Sankella, S., et al. (2011). Human 1-acylglycerol-3-phosphate O-acyltransferase isoforms 1 and 2: Biochemical characterization and inability to rescue hepatic steatosis in Agpat2(-/-) gene lipodystrophic mice. *Journal of Biological Chemistry*, *286*, 37676–37691.

Agarwal, A. K., Xing, C., DeMartino, G. N., Mizrachi, D., Hernandez, M. D., Sousa, A. B., et al. (2010). PSMB8 encoding the beta5i proteasome subunit is mutated in joint contractures, muscle atrophy, microcytic anemia, and panniculitis-induced lipodystrophy syndrome. *American Journal of Human Genetics*, *87*, 866–872.

Barak, Y., Nelson, M. C., Ong, E. S., Jones, Y. Z., Ruiz-Lozano, P., Chien, K. R., et al. (1999). PPAR gamma is required for placental, cardiac, and adipose tissue development. *Molecular Cell*, *4*, 585–595.

Barrowman, J., Wiley, P. A., Hudon-Miller, S. E., Hrycyna, C. A., & Michaelis, S. (2012). Human ZMPSTE24 disease mutations: Residual proteolytic activity correlates with disease severity. *Human Molecular Genetics*, *21*, 4084–4093.

Bengoechea-Alonso, M. T., & Ericsson, J. (2007). SREBP in signal transduction: Cholesterol metabolism and beyond. *Current Opinion in Cell Biology*, *19*, 215–222.

Bergo, M. O., Gavino, B., Ross, J., Schmidt, W. K., Hong, C., Kendall, L. V., et al. (2002). Zmpste24 deficiency in mice causes spontaneous bone fractures, muscle weakness, and a prelamin A processing defect. *Proceedings of the National Academy of Sciences of the United States of America*, *99*, 13049–13054.

Beyer, A. M., Baumbach, G. L., Halabi, C. M., Modrick, M. L., Lynch, C. M., Gerhold, T. D., et al. (2008). Interference with PPARgamma signaling causes cerebral vascular dysfunction, hypertrophy, and remodeling. *Hypertension*, *51*, 867–871.

Binns, D., Lee, S., Hilton, C. L., Jiang, Q. X., & Goodman, J. M. (2010). Seipin is a discrete homooligomer. *Biochemistry*, *49*, 10747–10755.

Bluher, M. (2013). Adipose tissue dysfunction contributes to obesity related metabolic diseases. *Best Practice & Research Clinical Endocrinology & Metabolism*, *27*, 163–177.

Boguslavsky, R. L., Stewart, C. L., & Worman, H. J. (2006). Nuclear lamin A inhibits adipocyte differentiation: Implications for Dunnigan-type familial partial lipodystrophy. *Human Molecular Genetics*, *15*, 653–663.

Boutet, E., El Mourabit, H., Prot, M., Nemani, M., Khallouf, E., Colard, O., et al. (2009). Seipin deficiency alters fatty acid Delta9 desaturation and lipid droplet formation in Berardinelli-Seip congenital lipodystrophy. *Biochimie*, *91*, 796–803.

Bouzakri, K., Zachrisson, A., Al-Khalili, L., Zhang, B. B., Koistinen, H. A., Krook, A., et al. (2006). siRNA-based gene silencing reveals specialized roles of IRS-1/Akt2 and IRS-2/Akt1 in glucose and lipid metabolism in human skeletal muscle. *Cell Metabolism*, *4*, 89–96.

Briand, N., Le Lay, S., Sessa, W. C., Ferre, P., & Dugail, I. (2011). Distinct roles of endothelial and adipocyte caveolin-1 in macrophage infiltration and adipose tissue metabolic activity. *Diabetes*, *60*, 448–453.

Cao, H., & Hegele, R. A. (2000). Nuclear lamin A/C R482Q mutation in Canadian kindreds with Dunnigan-type familial partial lipodystrophy. *Human Molecular Genetics*, *9*, 109–112.

Capanni, C., Cenni, V., Mattioli, E., Sabatelli, P., Ognibene, A., Columbaro, M., et al. (2003). Failure of lamin A/C to functionally assemble in R482L mutated familial partial lipodystrophy fibroblasts: Altered intermolecular interaction with emerin and implications for gene transcription. *Experimental Cell Research*, *291*, 122–134.

Capanni, C., Mattioli, E., Columbaro, M., Lucarelli, E., Parnaik, V. K., Novelli, G., et al. (2005). Altered pre-lamin A processing is a common mechanism leading to lipodystrophy. *Human Molecular Genetics, 14*, 1489–1502.
Capeau, J., Magre, J., Caron-Debarle, M., Lagathu, C., Antoine, B., Bereziat, V., et al. (2010). Human lipodystrophies: Genetic and acquired diseases of adipose tissue. *Endocrine Development, 19*, 1–20.
Cartwright, B. R., & Goodman, J. M. (2012). Seipin: From human disease to molecular mechanism. *Journal of Lipid Research, 53*, 1042–1055.
Chadda, R., & Mayor, S. (2008). PTRF triggers a cave in. *Cell, 132*, 23–24.
Chen, W., Chang, B., Saha, P., Hartig, S. M., Li, L., Reddy, V. T., et al. (2012). Berardinelli-Seip congenital lipodystrophy 2/seipin is a cell-autonomous regulator of lipolysis essential for adipocyte differentiation. *Molecular and Cellular Biology, 32*, 1099–1111.
Chen, W., Yechoor, V. K., Chang, B. H., Li, M. V., March, K. L., & Chan, L. (2009). The human lipodystrophy gene product Berardinelli-Seip congenital lipodystrophy 2/seipin plays a key role in adipocyte differentiation. *Endocrinology, 150*, 4552–4561.
Chen, W., Zhou, H., Liu, S., Fhaner, C. J., Gross, B. C., Lydic, T. A., et al. (2013). Altered lipid metabolism in residual white adipose tissues of bscl2 deficient mice. *PLoS One, 8*, e82526.
Cho, H., Mu, J., Kim, J. K., Thorvaldsen, J. L., Chu, Q., Crenshaw, E. B., 3rd., et al. (2001). Insulin resistance and a diabetes mellitus-like syndrome in mice lacking the protein kinase Akt2 (PKB beta). *Science (New York), 292*, 1728–1731.
Chong, A. Y., Lupsa, B. C., Cochran, E. K., & Gorden, P. (2010). Efficacy of leptin therapy in the different forms of human lipodystrophy. *Diabetologia, 53*, 27–35.
Chuengsamarn, S., Garza, A. E., Krug, A. W., Romero, J. R., Adler, G. K., Williams, G. H., et al. (2013). Direct renin inhibition modulates insulin resistance in caveolin-1-deficient mice. *Metabolism, Clinical and Experimental, 62*, 275–281.
Classon, M., Kennedy, B. K., Mulloy, R., & Harlow, E. (2000). Opposing roles of pRB and p107 in adipocyte differentiation. *Proceedings of the National Academy of Sciences of the United States of America, 97*, 10826–10831.
Cohen, A. W., Razani, B., Schubert, W., Williams, T. M., Wang, X. B., Iyengar, P., et al. (2004). Role of caveolin-1 in the modulation of lipolysis and lipid droplet formation. *Diabetes, 53*, 1261–1270.
Cortes, V. A., Cautivo, K. M., Rong, S., Garg, A., Horton, J. D., & Agarwal, A. K. (2014). Leptin ameliorates insulin resistance and hepatic steatosis in Agpat2-/- lipodystrophic mice independent of hepatocyte leptin receptors. *Journal of Lipid Research, 55*, 276–288.
Cortes, V. A., Curtis, D. E., Sukumaran, S., Shao, X., Parameswara, V., Rashid, S., et al. (2009). Molecular mechanisms of hepatic steatosis and insulin resistance in the AGPAT2-deficient mouse model of congenital generalized lipodystrophy. *Cell Metabolism, 9*, 165–176.
Cottrell, E. C., & Mercer, J. G. (2012). Leptin receptors. In Hans-Georg Joost (Ed.), *Handbook of experimental pharmacology* (pp. 3–21).
Cui, X., Wang, Y., Tang, Y., Liu, Y., Zhao, L., Deng, J., et al. (2011). Seipin ablation in mice results in severe generalized lipodystrophy. *Human Molecular Genetics, 20*, 3022–3030.
Dalamaga, M., Chou, S. H., Shields, K., Papageorgiou, P., Polyzos, S. A., & Mantzoros, C. S. (2013). Leptin at the intersection of neuroendocrinology and metabolism: Current evidence and therapeutic perspectives. *Cell Metabolism, 18*, 29–42.
Danesch, U., Hoeck, W., & Ringold, G. M. (1992). Cloning and transcriptional regulation of a novel adipocyte-specific gene, FSP27. CAAT-enhancer-binding protein (C/EBP) and C/EBP-like proteins interact with sequences required for differentiation-dependent expression. *Journal of Biological Chemistry, 267*, 7185–7193.

Davis, K. E., Moldes, M., & Farmer, S. R. (2004). The forkhead transcription factor FoxC2 inhibits white adipocyte differentiation. *Journal of Biological Chemistry*, *279*, 42453–42461.
Dechat, T., Pfleghaar, K., Sengupta, K., Shimi, T., Shumaker, D. K., Solimando, L., et al. (2008). Nuclear lamins: Major factors in the structural organization and function of the nucleus and chromatin. *Genes & Development*, *22*, 832–853.
Dollet, L., Magre, J., Cariou, B., & Prieur, X. (2014). Function of seipin: New insights from Bscl2/seipin knockout mouse models. *Biochimie*, *96*, 166–172.
Dubern, B., & Clement, K. (2012). Leptin and leptin receptor-related monogenic obesity. *Biochimie*, *94*, 2111–2115.
Fehling, H. J., Swat, W., Laplace, C., Kuhn, R., Rajewsky, K., Muller, U., et al. (1994). MHC class I expression in mice lacking the proteasome subunit LMP-7. *Science (New York, NY)*, *265*, 1234–1237.
Fei, W., Du, X., & Yang, H. (2011). Seipin, adipogenesis and lipid droplets. *Trends in Endocrinology and Metabolism*, *22*, 204–210.
Fei, W., Shui, G., Gaeta, B., Du, X., Kuerschner, L., Li, P., et al. (2008). Fld1p, a functional homologue of human seipin, regulates the size of lipid droplets in yeast. *Journal of Cell Biology*, *180*, 473–482.
Fei, W., Shui, G., Zhang, Y., Krahmer, N., Ferguson, C., Kapterian, T. S., et al. (2011). A role for phosphatidic acid in the formation of "supersized" lipid droplets. *PLoS Genetics*, *7*, e1002201.
Ferguson, R. D., Gallagher, E. J., Scheinman, E. J., Damouni, R., & LeRoith, D. (2013). The epidemiology and molecular mechanisms linking obesity, diabetes, and cancer. *Vitamins and Hormones*, *93*, 51–98.
Fischer-Posovszky, P., Tews, D., Horenburg, S., Debatin, K. M., & Wabitsch, M. (2012). Differential function of Akt1 and Akt2 in human adipocytes. *Molecular and Cellular Endocrinology*, *358*, 135–143.
Gale, S. E., Frolov, A., Han, X., Bickel, P. E., Cao, L., Bowcock, A., et al. (2006). A regulatory role for 1-acylglycerol-3-phosphate-O-acyltransferase 2 in adipocyte differentiation. *Journal of Biological Chemistry*, *281*, 11082–11089.
Galescu, O., Bhangoo, A., & Ten, S. (2013). Insulin resistance, lipodystrophy and cardiometabolic syndrome in HIV/AIDS. *Reviews in Endocrine & Metabolic Disorders*, *14*, 133–140.
Gandotra, S., Le Dour, C., Bottomley, W., Cervera, P., Giral, P., Reznik, Y., et al. (2011). Perilipin deficiency and autosomal dominant partial lipodystrophy. *New England Journal of Medicine*, *364*, 740–748.
Gandotra, S., Lim, K., Girousse, A., Saudek, V., O'Rahilly, S., & Savage, D. B. (2011). Human frame shift mutations affecting the carboxyl terminus of perilipin increase lipolysis by failing to sequester the adipose triglyceride lipase (ATGL) coactivator AB-hydrolase-containing 5 (ABHD5). *Journal of Biological Chemistry*, *286*, 34998–35006.
Garfield, A. S., Chan, W. S., Dennis, R. J., Ito, D., Heisler, L. K., & Rochford, J. J. (2012). Neuroanatomical characterisation of the expression of the lipodystrophy and motor-neuropathy gene Bscl2 in adult mouse brain. *PLoS One*, *7*, e45790.
Garg, A. (2011). Clinical review#: Lipodystrophies: Genetic and acquired body fat disorders. *Journal of Clinical Endocrinology and Metabolism*, *96*, 3313–3325.
Garg, A., Wilson, R., Barnes, R., Arioglu, E., Zaidi, Z., Gurakan, F., et al. (1999). A gene for congenital generalized lipodystrophy maps to human chromosome 9q34. *Journal of Clinical Endocrinology and Metabolism*, *84*, 3390–3394.
Garofalo, R. S., Orena, S. J., Rafidi, K., Torchia, A. J., Stock, J. L., Hildebrandt, A. L., et al. (2003). Severe diabetes, age-dependent loss of adipose tissue, and mild growth deficiency in mice lacking Akt2/PKB beta. *Journal of Clinical Investigation*, *112*, 197–208.

Gautron, L., & Elmquist, J. K. (2011). Sixteen years and counting: An update on leptin in energy balance. *Journal of Clinical Investigation, 121*, 2087–2093.
George, S., Rochford, J. J., Wolfrum, C., Gray, S. L., Schinner, S., Wilson, J. C., et al. (2004). A family with severe insulin resistance and diabetes due to a mutation in AKT2. *Science (New York), 304*, 1325–1328.
Gerin, I., Bommer, G. T., Lidell, M. E., Cederberg, A., Enerback, S., & Macdougald, O. A. (2009). On the role of FOX transcription factors in adipocyte differentiation and insulin-stimulated glucose uptake. *Journal of Biological Chemistry, 284*, 10755–10763.
Giorgino, F., Laviola, L., & Eriksson, J. W. (2005). Regional differences of insulin action in adipose tissue: Insights from in vivo and in vitro studies. *Acta Physiologica Scandinavica, 183*, 13–30.
Girousse, A., & Langin, D. (2012). Adipocyte lipases and lipid droplet-associated proteins: Insight from transgenic mouse models. *International Journal of Obesity, 36*, 581–594.
Grahn, T. H., Zhang, Y., Lee, M. J., Sommer, A. G., Mostoslavsky, G., Fried, S. K., et al. (2013). FSP27 and PLIN1 interaction promotes the formation of large lipid droplets in human adipocytes. *Biochemical and Biophysical Research Communications, 432*, 296–301.
Gray, S. L., Dalla Nora, E., Backlund, E. C., Manieri, M., Virtue, S., Noland, R. C., et al. (2006). Decreased brown adipocyte recruitment and thermogenic capacity in mice with impaired peroxisome proliferator-activated receptor (P465L PPARgamma) function. *Endocrinology, 147*, 5708–5714.
Gray, S. L., Nora, E. D., Grosse, J., Manieri, M., Stoeger, T., Medina-Gomez, G., et al. (2006). Leptin deficiency unmasks the deleterious effects of impaired peroxisome proliferator-activated receptor gamma function (P465L PPARgamma) in mice. *Diabetes, 55*, 2669–2677.
Groeneveld, M. P., Huang-Doran, I., & Semple, R. K. (2012). Adiponectin and leptin in human severe insulin resistance—Diagnostic utility and biological insights. *Biochimie, 94*, 2172–2179.
Guan, H. P., Li, Y., Jensen, M. V., Newgard, C. B., Steppan, C. M., & Lazar, M. A. (2002). A futile metabolic cycle activated in adipocytes by antidiabetic agents. *Nature Medicine, 8*, 1122–1128.
Haque, W., Garg, A., & Agarwal, A. K. (2005). Enzymatic activity of naturally occurring 1-acylglycerol-3-phosphate-O-acyltransferase 2 mutants associated with congenital generalized lipodystrophy. *Biochemical and Biophysical Research Communications, 327*, 446–453.
Hayashi, Y. K., Matsuda, C., Ogawa, M., Goto, K., Tominaga, K., Mitsuhashi, S., et al. (2009). Human PTRF mutations cause secondary deficiency of caveolins resulting in muscular dystrophy with generalized lipodystrophy. *Journal of Clinical Investigation, 119*, 2623–2633.
Hegele, R. A., Anderson, C. M., Wang, J., Jones, D. C., & Cao, H. (2000). Association between nuclear lamin A/C R482Q mutation and partial lipodystrophy with hyperinsulinemia, dyslipidemia, hypertension, and diabetes. *Genome Research, 10*, 652–658.
Hegele, R. A., Cao, H., Anderson, C. M., & Hramiak, I. M. (2000). Heterogeneity of nuclear lamin A mutations in Dunnigan-type familial partial lipodystrophy. *Journal of Clinical Endocrinology and Metabolism, 85*, 3431–3435.
Hegele, R. A., Cao, H., Frankowski, C., Mathews, S. T., & Leff, T. (2002). PPARG F388L, a transactivation-deficient mutant, in familial partial lipodystrophy. *Diabetes, 51*, 3586–3590.
Hill, M. M., Bastiani, M., Luetterforst, R., Kirkham, M., Kirkham, A., Nixon, S. J., et al. (2008). PTRF-Cavin, a conserved cytoplasmic protein required for caveola formation and function. *Cell, 132*, 113–124.

Inohara, N., Koseki, T., Chen, S., Wu, X., & Nunez, G. (1998). CIDE, a novel family of cell death activators with homology to the 45 kDa subunit of the DNA fragmentation factor. *EMBO Journal, 17*, 2526–2533.

Jansa, P., Mason, S. W., Hoffmann-Rohrer, U., & Grummt, I. (1998). Cloning and functional characterization of PTRF, a novel protein which induces dissociation of paused ternary transcription complexes. *EMBO Journal, 17*, 2855–2864.

Jeninga, E. H., Gurnell, M., & Kalkhoven, E. (2009). Functional implications of genetic variation in human PPARgamma. *Trends in Endocrinology and metabolism, 20*, 380–387.

Jiang, Z. Y., Zhou, Q. L., Holik, J., Patel, S., Leszyk, J., Coleman, K., et al. (2005). Identification of WNK1 as a substrate of Akt/protein kinase B and a negative regulator of insulin-stimulated mitogenesis in 3T3-L1 cells. *Journal of Biological Chemistry, 280*, 21622–21628.

Karlsson, M., Thorn, H., Parpal, S., Stralfors, P., & Gustavsson, J. (2002). Insulin induces translocation of glucose transporter GLUT4 to plasma membrane caveolae in adipocytes. *FASEB Journal, 16*, 249–251.

Keller, P., Petrie, J. T., De Rose, P., Gerin, I., Wright, W. S., Chiang, S. H., et al. (2008). Fat-specific protein 27 regulates storage of triacylglycerol. *Journal of Biological Chemistry, 283*, 14355–14365.

Kim, C. A., Delepine, M., Boutet, E., El Mourabit, H., Le Lay, S., Meier, M., et al. (2008). Association of a homozygous nonsense caveolin-1 mutation with Berardinelli-Seip congenital lipodystrophy. *Journal of Clinical Endocrinology and Metabolism, 93*, 1129–1134.

Kim, J. B., & Spiegelman, B. M. (1996). ADD1/SREBP1 promotes adipocyte differentiation and gene expression linked to fatty acid metabolism. *Genes & Development, 10*, 1096–1107.

Kim, J. B., Wright, H. M., Wright, M., & Spiegelman, B. M. (1998). ADD1/SREBP1 activates PPARgamma through the production of endogenous ligand. *Proceedings of the National Academy of Sciences of the United States of America, 95*, 4333–4337.

Kis, A., Murdoch, C., Zhang, M., Siva, A., Rodriguez-Cuenca, S., Carobbio, S., et al. (2009). Defective peroxisomal proliferators activated receptor gamma activity due to dominant-negative mutation synergizes with hypertension to accelerate cardiac fibrosis in mice. *European Journal of Heart Failure, 11*, 533–541.

Kitamura, A., Maekawa, Y., Uehara, H., Izumi, K., Kawachi, I., Nishizawa, M., et al. (2011). A mutation in the immunoproteasome subunit PSMB8 causes autoinflammation and lipodystrophy in humans. *Journal of Clinical Investigation, 121*, 4150–4160.

Konner, A. C., & Bruning, J. C. (2012). Selective insulin and leptin resistance in metabolic disorders. *Cell Metabolism, 16*, 144–152.

Krycer, J. R., Sharpe, L. J., Luu, W., & Brown, A. J. (2010). The Akt-SREBP nexus: Cell signaling meets lipid metabolism. *Trends in Endocrinology and Metabolism, 21*, 268–276.

Le Lay, S., Blouin, C. M., Hajduch, E., & Dugail, I. (2009). Filling up adipocytes with lipids. Lessons from caveolin-1 deficiency. *Biochimica et Biophysica Acta, 1791*, 514–518.

Le Lay, S., Hajduch, E., Lindsay, M. R., Le Liepvre, X., Thiele, C., Ferre, P., et al. (2006). Cholesterol-induced caveolin targeting to lipid droplets in adipocytes: A role for caveolar endocytosis. *Traffic (Copenhagen, Denmark), 7*, 549–561.

Le Lay, S., Li, Q., Proschogo, N., Rodriguez, M., Gunaratnam, K., Cartland, S., et al. (2009). Caveolin-1-dependent and -independent membrane domains. *Journal of Lipid Research, 50*, 1609–1620.

Lefterova, M. I., Zhang, Y., Steger, D. J., Schupp, M., Schug, J., Cristancho, A., et al. (2008). PPARgamma and C/EBP factors orchestrate adipocyte biology via adjacent binding on a genome-wide scale. *Genes & Development, 22*, 2941–2952.

Levine, A. J., Feng, Z., Mak, T. W., You, H., & Jin, S. (2006). Coordination and communication between the p53 and IGF-1-AKT-TOR signal transduction pathways. *Genes & Development, 20*, 267–275.

Li, J. Z., Ye, J., Xue, B., Qi, J., Zhang, J., Zhou, Z., et al. (2007). Cideb regulates diet-induced obesity, liver steatosis, and insulin sensitivity by controlling lipogenesis and fatty acid oxidation. *Diabetes, 56*, 2523–2532.

Liu, L., Brown, D., McKee, M., Lebrasseur, N. K., Yang, D., Albrecht, K. H., et al. (2008). Deletion of Cavin/PTRF causes global loss of caveolae, dyslipidemia, and glucose intolerance. *Cell Metabolism, 8*, 310–317.

Lloyd, D. J., Trembath, R. C., & Shackleton, S. (2002). A novel interaction between lamin A and SREBP1: Implications for partial lipodystrophy and other laminopathies. *Human Molecular Genetics, 11*, 769–777.

Lundin, C., Nordstrom, R., Wagner, K., Windpassinger, C., Andersson, H., von Heijne, G., et al. (2006). Membrane topology of the human seipin protein. *FEBS Letters, 580*, 2281–2284.

Magre, J., Delepine, M., Khallouf, E., Gedde-Dahl, T., Jr., Van Maldergem, L., Sobel, E., et al. (2001). Identification of the gene altered in Berardinelli-Seip congenital lipodystrophy on chromosome 11q13. *Nature Genetics, 28*, 365–370.

Mallewa, J. E., Wilkins, E., Vilar, J., Mallewa, M., Doran, D., Back, D., et al. (2008). HIV-associated lipodystrophy: A review of underlying mechanisms and therapeutic options. *Journal of Antimicrobial Chemotherapy, 62*, 648–660.

Mantzoros, C. S., Magkos, F., Brinkoetter, M., Sienkiewicz, E., Dardeno, T. A., Kim, S. Y., et al. (2011). Leptin in human physiology and pathophysiology. *American Journal of Physiology, Endocrinology and Metabolism, 301*, E567–E584.

Martin, S., Fernandez-Rojo, M. A., Stanley, A. C., Bastiani, M., Okano, S., Nixon, S. J., et al. (2012). Caveolin-1 deficiency leads to increased susceptibility to cell death and fibrosis in white adipose tissue: Characterization of a lipodystrophic model. *PLoS One*, 7, e46242.

Martinez-Botas, J., Anderson, J. B., Tessier, D., Lapillonne, A., Chang, B. H., Quast, M. J., et al. (2000). Absence of perilipin results in leanness and reverses obesity in Lepr(db/db) mice. *Nature Genetics, 26*, 474–479.

Medina-Gomez, G., Gray, S., & Vidal-Puig, A. (2007). Adipogenesis and lipotoxicity: Role of peroxisome proliferator-activated receptor gamma (PPARgamma) and PPARgammacoactivator-1 (PGC1). *Public Health Nutrition, 10*, 1132–1137.

Murata, T., Lin, M. I., Huang, Y., Yu, J., Bauer, P. M., Giordano, F. J., et al. (2007). Reexpression of caveolin-1 in endothelium rescues the vascular, cardiac, and pulmonary defects in global caveolin-1 knockout mice. *Journal of Experimental Medicine, 204*, 2373–2382.

Nakae, J., Kitamura, T., Kitamura, Y., Biggs, W. H., 3rd., Arden, K. C., & Accili, D. (2003). The forkhead transcription factor Foxo1 regulates adipocyte differentiation. *Developmental Cell, 4*, 119–129.

Nian, Z., Sun, Z., Yu, L., Toh, S. Y., Sang, J., & Li, P. (2010). Fat-specific protein 27 undergoes ubiquitin-dependent degradation regulated by triacylglycerol synthesis and lipid droplet formation. *Journal of Biological Chemistry, 285*, 9604–9615.

Nielsen, R., Pedersen, T. A., Hagenbeek, D., Moulos, P., Siersbaek, R., Megens, E., et al. (2008). Genome-wide profiling of PPARgamma:RXR and RNA polymerase II occupancy reveals temporal activation of distinct metabolic pathways and changes in RXR dimer composition during adipogenesis. *Genes & Development, 22*, 2953–2967.

Nishino, N., Tamori, Y., Tateya, S., Kawaguchi, T., Shibakusa, T., Mizunoya, W., et al. (2008). FSP27 contributes to efficient energy storage in murine white adipocytes by promoting the formation of unilocular lipid droplets. *Journal of Clinical Investigation, 118*, 2808–2821.

Oral, E. A., Simha, V., Ruiz, E., Andewelt, A., Premkumar, A., Snell, P., et al. (2002). Leptin-replacement therapy for lipodystrophy. *New England Journal of Medicine, 346*, 570–578.

Ortegren, U., Aboulaich, N., Ost, A., & Stralfors, P. (2007). A new role for caveolae as metabolic platforms. *Trends in Endocrinology and Metabolism, 18*, 344–349.
Parton, R. G., & del Pozo, M. A. (2013). Caveolae as plasma membrane sensors, protectors and organizers. *Nature Reviews, 14*, 98–112.
Payne, V. A., Grimsey, N., Tuthill, A., Virtue, S., Gray, S. L., Dalla Nora, E., et al. (2008). The human lipodystrophy gene BSCL2/seipin may be essential for normal adipocyte differentiation. *Diabetes, 57*, 2055–2060.
Pearce, L. R., Komander, D., & Alessi, D. R. (2010). The nuts and bolts of AGC protein kinases. *Nature Reviews, 11*, 9–22.
Pendas, A. M., Zhou, Z., Cadinanos, J., Freije, J. M., Wang, J., Hultenby, K., et al. (2002). Defective prelamin A processing and muscular and adipocyte alterations in Zmpste24 metalloproteinase-deficient mice. *Nature Genetics, 31*, 94–99.
Pendse, A. A., Johnson, L. A., Kim, H. S., McNair, M., Nipp, C. T., Wilhelm, C., et al. (2012). Pro- and antiatherogenic effects of a dominant-negative P465L mutation of peroxisome proliferator-activated receptor-gamma in apolipoprotein E-Null mice. *Arteriosclerosis, Thrombosis, and Vascular Biology, 32*, 1436–1444.
Pendse, A. A., Johnson, L. A., Tsai, Y. S., & Maeda, N. (2010). Pparg-P465L mutation worsens hyperglycemia in Ins2-Akita female mice via adipose-specific insulin resistance and storage dysfunction. *Diabetes, 59*, 2890–2897.
Peng, X. D., Xu, P. Z., Chen, M. L., Hahn-Windgassen, A., Skeen, J., Jacobs, J., et al. (2003). Dwarfism, impaired skin development, skeletal muscle atrophy, delayed bone development, and impeded adipogenesis in mice lacking Akt1 and Akt2. *Genes & Development, 17*, 1352–1365.
Pilch, P. F., & Liu, L. (2011). Fat caves: Caveolae, lipid trafficking and lipid metabolism in adipocytes. *Trends in Endocrinology and Metabolism, 22*, 318–324.
Prieur, X., Dollet, L., Takahashi, M., Nemani, M., Pillot, B., Le May, C., et al. (2013). Thiazolidinediones partially reverse the metabolic disturbances observed in Bscl2/seipin-deficient mice. *Diabetologia, 56*, 1813–1825.
Puri, V., Konda, S., Ranjit, S., Aouadi, M., Chawla, A., Chouinard, M., et al. (2007). Fat-specific protein 27, a novel lipid droplet protein that enhances triglyceride storage. *Journal of Biological Chemistry, 282*, 34213–34218.
Ramanathan, N., Ahmed, M., Raffan, E., Stewart, C. L., O'Rahilly, S., Semple, R. K., et al. (2013). Identification and characterisation of a novel pathogenic mutation in the human lipodystrophy gene AGPAT2: C48R: A novel mutation in AGPAT2. *JIMD Reports, 9*, 73–80.
Razani, B., Combs, T. P., Wang, X. B., Frank, P. G., Park, D. S., Russell, R. G., et al. (2002). Caveolin-1-deficient mice are lean, resistant to diet-induced obesity, and show hypertriglyceridemia with adipocyte abnormalities. *Journal of Biological Chemistry, 277*, 8635–8647.
Rochford, J. J. (2010). Molecular mechanisms controlling human adipose tissue development: Insights from monogenic lipodystrophies. *Expert Reviews in Molecular Medicine, 12*, e24.
Rodriguez-Cuenca, S., Carobbio, S., Velagapudi, V. R., Barbarroja, N., Moreno-Navarrete, J. M., Tinahones, F. J., et al. (2012). Peroxisome proliferator-activated receptor gamma-dependent regulation of lipolytic nodes and metabolic flexibility. *Molecular and Cellular Biology, 32*, 1555–1565.
Rosen, E. D., & MacDougald, O. A. (2006). Adipocyte differentiation from the inside out. *Nature Reviews, 7*, 885–896.
Rosen, E. D., Sarraf, P., Troy, A. E., Bradwin, G., Moore, K., Milstone, D. S., et al. (1999). PPAR gamma is required for the differentiation of adipose tissue in vivo and in vitro. *Molecular Cell, 4*, 611–617.

Rosen, E. D., & Spiegelman, B. M. (2014). What we talk about when we talk about fat. *Cell, 156*, 20–44.
Rubio-Cabezas, O., Puri, V., Murano, I., Saudek, V., Semple, R. K., Dash, S., et al. (2009). Partial lipodystrophy and insulin resistant diabetes in a patient with a homozygous nonsense mutation in CIDEC. *EMBO Molecular Medicine, 1*, 280–287.
Safar Zadeh, E., Lungu, A. O., Cochran, E. K., Brown, R. J., Ghany, M. G., Heller, T., et al. (2013). The liver diseases of lipodystrophy: The long-term effect of leptin treatment. *Journal of Hepatology, 59*, 131–137.
Saha, P. K., Kojima, H., Martinez-Botas, J., Sunehag, A. L., & Chan, L. (2004). Metabolic adaptations in the absence of perilipin: Increased beta-oxidation and decreased hepatic glucose production associated with peripheral insulin resistance but normal glucose tolerance in perilipin-null mice. *Journal of Biological Chemistry, 279*, 35150–35158.
Savage, D. B., Tan, G. D., Acerini, C. L., Jebb, S. A., Agostini, M., Gurnell, M., et al. (2003). Human metabolic syndrome resulting from dominant-negative mutations in the nuclear receptor peroxisome proliferator-activated receptor-gamma. *Diabetes, 52*, 910–917.
Semple, R. K., Chatterjee, V. K., & O'Rahilly, S. (2006). PPAR gamma and human metabolic disease. *Journal of Clinical Investigation, 116*, 581–589.
Shackleton, S., Lloyd, D. J., Jackson, S. N., Evans, R., Niermeijer, M. F., Singh, B. M., et al. (2000). LMNA, encoding lamin A/C, is mutated in partial lipodystrophy. *Nature Genetics, 24*, 153–156.
Shastry, S., Delgado, M. R., Dirik, E., Turkmen, M., Agarwal, A. K., & Garg, A. (2010). Congenital generalized lipodystrophy, type 4 (CGL4) associated with myopathy due to novel PTRF mutations. *American Journal of Medical Genetics Part A, 152A*, 2245–2253.
Sim, M. F. M., Dennis, R. J., Aubry, E. M., Ramanathan, N., Sembongi, H., Saudek, V., et al. (2013). The human lipodystrophy protein seipin is an ER membrane adaptor for the adipogenic PA phosphatase lipin 1. *Molecular Metabolism, 2*, 38–46.
Sim, M. F. M., Talukder, M. M. U., Dennis, R. J., O'Rahilly, S., Edwardson, J. M., & Rochford, J. J. (2013). Analysis of naturally occurring mutations in the human lipodystrophy protein seipin reveals multiple potential pathogenic mechanisms. *Diabetologia, 56*(11), 2498–2506.
Skinner, J. R., Harris, L. A., Shew, T. M., Abumrad, N. A., & Wolins, N. E. (2013). Perilipin 1 moves between the fat droplet and the endoplasmic reticulum. *Adipocyte, 2*, 80–86.
Subauste, A. R., Das, A. K., Li, X., Elliott, B. G., Evans, C., El Azzouny, M., et al. (2012). Alterations in lipid signaling underlie lipodystrophy secondary to AGPAT2 mutations. *Diabetes, 61*, 2922–2931.
Sullivan, T., Escalante-Alcalde, D., Bhatt, H., Anver, M., Bhat, N., Nagashima, K., et al. (1999). Loss of A-type lamin expression compromises nuclear envelope integrity leading to muscular dystrophy. *Journal of Cell Biology, 147*, 913–920.
Sun, K., Kusminski, C. M., & Scherer, P. E. (2011). Adipose tissue remodeling and obesity. *Journal of Clinical Investigation, 121*, 2094–2101.
Szymanski, K. M., Binns, D., Bartz, R., Grishin, N. V., Li, W. P., Agarwal, A. K., et al. (2007). The lipodystrophy protein seipin is found at endoplasmic reticulum lipid droplet junctions and is important for droplet morphology. *Proceedings of the National Academy of Sciences of the United States of America, 104*, 20890–20895.
Tansey, J. T., Sztalryd, C., Gruia-Gray, J., Roush, D. L., Zee, J. V., Gavrilova, O., et al. (2001). Perilipin ablation results in a lean mouse with aberrant adipocyte lipolysis, enhanced leptin production, and resistance to diet-induced obesity. *Proceedings of the National Academy of Sciences of the United States of America, 98*, 6494–6499.
Tian, Y., Bi, J., Shui, G., Liu, Z., Xiang, Y., Liu, Y., et al. (2011). Tissue-autonomous function of *Drosophila* seipin in preventing ectopic lipid droplet formation. *PLoS Genetics, 7*, e1001364.

Toh, S. Y., Gong, J., Du, G., Li, J. Z., Yang, S., Ye, J., et al. (2008). Up-regulation of mitochondrial activity and acquirement of brown adipose tissue-like property in the white adipose tissue of fsp27 deficient mice. *PLoS One, 3*, e2890.
Tontonoz, P., & Spiegelman, B. M. (2008). Fat and beyond: The diverse biology of PPARgamma. *Annual Review of Biochemistry, 77*, 289–312.
Tsai, Y. S., Kim, H. J., Takahashi, N., Kim, H. S., Hagaman, J. R., Kim, J. K., et al. (2004). Hypertension and abnormal fat distribution but not insulin resistance in mice with P465L PPARgamma. *Journal of Clinical Investigation, 114*, 240–249.
Tsukahara, T., Tsukahara, R., Fujiwara, Y., Yue, J., Cheng, Y., Guo, H., et al. (2010). Phospholipase D2-dependent inhibition of the nuclear hormone receptor PPARgamma by cyclic phosphatidic acid. *Molecular Cell, 39*, 421–432.
Unger, R. H., Clark, G. O., Scherer, P. E., & Orci, L. (2010). Lipid homeostasis, lipotoxicity and the metabolic syndrome. *Biochimica et Biophysica Acta, 1801*, 209–214.
Unger, R. H., & Scherer, P. E. (2010). Gluttony, sloth and the metabolic syndrome: A roadmap to lipotoxicity. *Trends in Endocrinology and Metabolism, 21*(6), 345–352.
Van Maldergem, L., Magre, J., Khallouf, T. E., Gedde-Dahl, T., Jr., Delepine, M., Trygstad, O., et al. (2002). Genotype-phenotype relationships in Berardinelli-Seip congenital lipodystrophy. *Journal of Medical Genetics, 39*, 722–733.
Vantyghem, M. C., Balavoine, A. S., Douillard, C., Defrance, F., Dieudonne, L., Mouton, F., et al. (2012). How to diagnose a lipodystrophy syndrome. *Annales d'endocrinologie, 73*, 170–189.
Vatier, C., Bidault, G., Briand, N., Guenantin, A. C., Teyssieres, L., Lascols, O., et al. (2013). What the genetics of lipodystrophy can teach us about insulin resistance and diabetes. *Current Diabetes Reports, 13*, 757–767.
Victoria, B., Cabezas-Agricola, J. M., Gonzalez-Mendez, B., Lattanzi, G., Del Coco, R., Loidi, L., et al. (2010). Reduced adipogenic gene expression in fibroblasts from a patient with type 2 congenital generalized lipodystrophy. *Diabetic Medicine, 27*, 1178–1187.
Vigouroux, C., Caron-Debarle, M., Le Dour, C., Magre, J., & Capeau, J. (2011). Molecular mechanisms of human lipodystrophies: From adipocyte lipid droplet to oxidative stress and lipotoxicity. *International Journal of Biochemistry & Cell Biology, 43*(6), 862–876.
Virtue, S., & Vidal-Puig, A. (2010). Adipose tissue expandability, lipotoxicity and the metabolic syndrome—An allostatic perspective. *Biochimica et Biophysica Acta, 1801*, 338–349.
Vogel, P., Read, R., Hansen, G., Wingert, J., Dacosta, C. M., Buhring, L. M., et al. (2011). Pathology of congenital generalized lipodystrophy in Agpat2-/- mice. *Veterinary Pathology, 48*, 642–654.
Weedon, M. N., Ellard, S., Prindle, M. J., Caswell, R., Lango Allen, H., Oram, R., et al. (2013). An in-frame deletion at the polymerase active site of POLD1 causes a multisystem disorder with lipodystrophy. *Nature Genetics, 45*, 947–950.
Windpassinger, C., Auer-Grumbach, M., Irobi, J., Patel, H., Petek, E., Horl, G., et al. (2004). Heterozygous missense mutations in BSCL2 are associated with distal hereditary motor neuropathy and Silver syndrome. *Nature Genetics, 36*, 271–276.
Wojtanik, K. M., Edgemon, K., Viswanadha, S., Lindsey, B., Haluzik, M., Chen, W., et al. (2009). The role of LMNA in adipose: A novel mouse model of lipodystrophy based on the Dunnigan-type familial partial lipodystrophy mutation. *Journal of Lipid Research, 50*, 1068–1079.
Wolfrum, C., Shih, D. Q., Kuwajima, S., Norris, A. W., Kahn, C. R., & Stoffel, M. (2003). Role of Foxa-2 in adipocyte metabolism and differentiation. *Journal of Clinical Investigation, 112*, 345–356.
Wolinski, H., Kolb, D., Hermann, S., Koning, R. I., & Kohlwein, S. D. (2011). A role for seipin in lipid droplet dynamics and inheritance in yeast. *Journal of Cell Science, 124*, 3894–3904.

Worman, H. J., & Bonne, G. (2007). "Laminopathies": A wide spectrum of human diseases. *Experimental Cell Research*, *313*, 2121–2133.

Yamauchi, T., Kamon, J., Minokoshi, Y., Ito, Y., Waki, H., Uchida, S., et al. (2002). Adiponectin stimulates glucose utilization and fatty-acid oxidation by activating AMP-activated protein kinase. *Nature Medicine*, *8*, 1288–1295.

Yamauchi, T., Kamon, J., Waki, H., Terauchi, Y., Kubota, N., Hara, K., et al. (2001). The fat-derived hormone adiponectin reverses insulin resistance associated with both lipoatrophy and obesity. *Nature Medicine*, 7, 941–946.

Yang, J., Cron, P., Good, V. M., Thompson, V., Hemmings, B. A., & Barford, D. (2002). Crystal structure of an activated Akt/protein kinase B ternary complex with GSK3-peptide and AMP-PNP. *Nature Structural Biology*, *9*, 940–944.

Yang, W., Thein, S., Guo, X., Xu, F., Venkatesh, B., Sugii, S., et al. (2013). Seipin differentially regulates lipogenesis and adipogenesis through a conserved core sequence and an evolutionarily acquired C-terminus. *Biochemical Journal*, *452*(1), 37–44.

Yang, W., Thein, S., Wang, X., Bi, X., Ericksen, R. E., Xu, F., et al. (2014). BSCL2/seipin regulates adipogenesis through actin cytoskeleton remodelling. *Human Molecular Genetics*, *23*, 502–513.

Yeo, G. S., & Heisler, L. K. (2012). Unraveling the brain regulation of appetite: Lessons from genetics. *Nature Neuroscience*, *15*, 1343–1349.

Yun, S. J., Kim, E. K., Tucker, D. F., Kim, C. D., Birnbaum, M. J., & Bae, S. S. (2008). Isoform-specific regulation of adipocyte differentiation by Akt/protein kinase Balpha. *Biochemical and Biophysical Research Communications*, *371*, 138–143.

Zechner, R., Zimmermann, R., Eichmann, T. O., Kohlwein, S. D., Haemmerle, G., Lass, A., et al. (2012). Fat signals–lipases and lipolysis in lipid metabolism and signaling. *Cell Metabolism*, *15*, 279–291.

Zhou, Z., Yon Toh, S., Chen, Z., Guo, K., Ng, C. P., Ponniah, S., et al. (2003). Cidea-deficient mice have lean phenotype and are resistant to obesity. *Nature Genetics*, *35*, 49–56.

CHAPTER THREE

Mouse Models of Neurodevelopmental Disease of the Basal Ganglia and Associated Circuits

Samuel S. Pappas[*], **Daniel K. Leventhal**[*], **Roger L. Albin**[*,†,‡], **William T. Dauer**[*,§,1]

[*]Department of Neurology, University of Michigan, Ann Arbor, Michigan, USA
[†]Michigan Alzheimer Disease Center, University of Michigan, Ann Arbor, Michigan, USA
[‡]Geriatrics Research, Education, and Clinical Center, VAAAHS, Ann Arbor, Michigan, USA
[§]Department of Cell and Developmental Biology, University of Michigan, Ann Arbor, Michigan, USA
[1]Corresponding author: e-mail address: dauer@umich.edu

Contents

Current Topics in Developmental Biology, Volume 109
ISSN 0070-2153
http://dx.doi.org/10.1016/B978-0-12-397920-9.00001-9

Abstract

This chapter focuses on neurodevelopmental diseases that are tightly linked to abnormal function of the striatum and connected structures. We begin with an overview of three representative diseases in which striatal dysfunction plays a key role—Tourette syndrome and obsessive-compulsive disorder, Rett's syndrome, and primary dystonia. These diseases highlight distinct etiologies that disrupt striatal integrity and function during development, and showcase the varied clinical manifestations of striatal dysfunction. We then review striatal organization and function, including evidence for striatal roles in online motor control/action selection, reinforcement learning, habit formation, and action sequencing. A key barrier to progress has been the relative lack of animal models of these diseases, though recently there has been considerable progress. We review these efforts, including their relative merits providing insight into disease pathogenesis, disease symptomatology, and basal ganglia function.

1. INTRODUCTION

The basal ganglia (BG) are a group of subcortical brain structures conserved across vertebrate species (Grillner, Robertson, & Stephenson-Jones, 2013). These interconnected structures contribute to the control of movement performance and several learning-related motor functions, including habit, reinforcement, and motor sequence learning. The BG are able to accomplish these functions as participants in recurrent neural loops, through which information flows to the BG from distinct cortical areas, is outputted to thalamus, and subsequently relayed back to cortex (cortico-BG-thalamocortical loops). The striatum (composed of the caudate and putamen in humans) is the principal "input" nucleus of the BG, receiving massive excitatory projections from the cortex. Myriad neuronal connections within and between BG structures process this cortical input—perhaps performing a "gating" or "selection" function—and subsequently project this information to the thalamus, which then projects back to more restricted areas of the neocortex. Evidence for the involvement of the striatum and related BG structures in varied behavioral processes and developmental disorders is reviewed.

The BG are affected by a range of debilitating diseases that can manifest throughout an individuals lifespan. For example, Tourette's syndrome (TS) occurs primarily in school-aged children, while Parkinson disease (PD) primarily afflicts the aged. Treatment of these diseases has been challenging, in part because the function of these circuits remains poorly understood. Indeed, despite extensive research and clinical observation, a cohesive theory describing the functional roles of the striatum and connected BG circuits remains elusive.

Considerable focus has centered on diseases resulting from dysfunction or damage to the striatum. The diverse symptomatology of these diseases has strongly influenced hypotheses regarding striatal function. Broadly speaking, these hypotheses can be divided into "performance" and "learning" roles (Beeler et al., 2012; Leventhal et al., 2014). The short-duration response to levodopa is a striking example supporting a role for striatum in online motor performance. Evidence from diseases featuring pathological repetitive behaviors and thoughts such as TS and obsessive-compulsive disorder (OCD) has been critical to highlighting the importance of striatal function in learning and habit expression. A rapidly expanding literature supports a role for striatum in implicit learning processes, with implications for normal adaptive behavior, habit formation, and addiction.

This chapter focuses on neurodevelopmental diseases that are tightly linked to abnormal function of the striatum and connected structures. We begin with an overview of three representative diseases in which striatal dysfunction plays a key role—TS/OCD, Rett's syndrome (RTT), and primary dystonia. These diseases highlight distinct etiologies that disrupt striatal integrity and function during development, and showcase the varied clinical manifestations of striatal dysfunction. We then review striatal organization and function, including evidence for striatal roles in online motor control/action selection, reinforcement learning, habit formation, and action sequencing. A key barrier to progress has been the relative lack of animal models of these diseases, though there recently has been considerable progress. We review these efforts, including their relative merits providing insight into disease pathogenesis, disease symptomatology, and BG function.

2. DISEASES AND LINKS TO STRIATAL DYSFUNCTION

2.1. Tic disorders, Tourette syndrome, and obsessive-compulsive disorder

Although transient tic disorders, TS and OCD are distinguished clinically, the common comorbidity of tics and obsessive-compulsive behaviors in TS suggest that these disorders exist on a spectrum. In support of this view, imaging studies demonstrate similar abnormalities between these disorders, and implicate abnormal BG function as a pathological feature of many of their associated symptoms (Cavanna & Seri, 2013; Worbe et al., 2010). As such, we will consider the disorders individually and as a unit.

2.1.1 Tics and Tourette syndrome: Clinical features

Tics are sudden, repetitive, involuntary movements, or vocalizations that have no intended purpose and interfere with ongoing behavior (Bronfeld & Bar-Gad, 2013). They range from mild movements or utterances involving few muscle groups (e.g., eye blinking or grimacing) to more complex, coordinated sequential activation of multiple muscle groups (Cohen, Leckman, & Bloch, 2013). To distinguish tics from other types of brief, abnormal movements, several features are helpful. Tics are often associated with premonitory feelings, ranging from a psychic urge to physical sensations like itching or even pain (Cavanna & Rickards, 2013). Tics are suppressible, at least transiently. When suppressed, the pressure to perform tics can become overwhelming, so that they are almost always eventually expressed. Some have termed tics an "unvoluntary" movement disorder to emphasize the inevitability of tic expression despite the temporary ability to suppress them (Cohen et al., 2013), though "semivoluntary" may be more accurate. Finally, tics are stereotyped—that is, patients tend to express a limited range of tic-like movements at any given time.

Tic disorders are classified into transient versus chronic tic disorders. Transient tic disorders persist for less than 1 year and represent the mildest and most common form of tic disorder (Cohen et al., 2013). The repetitive behaviors in transient tic disorder may not be consistent and often change in severity over time. In contrast, chronic tic disorders involve a more stereotyped motor or vocal tic behavior that persists for more than 1 year. TS, the most severe form of tic disorder, is characterized by multiple motor and at least one vocal tic which begin before 18 years of age, increase in severity over time, and persist for at least 1 year (Kurlan, 2010). TS, first described by Gilles de la Tourette in 1885, affects approximately 1% of the population, but is three to four times more common in boys (Kurlan, 2010; Robertson, Eapen, & Cavanna, 2009). TS often begins between 3 and 9 years of age, with the worst symptoms during the early teens. Symptoms often improve in the late teens and into adulthood (Bloch et al., 2006). Caudate nucleus volumes in childhood predict later tic severity, consistent with both BG dysfunction in TS and abnormal trajectory of brain development (Bloch, Leckman, Zhu, & Peterson, 2005). Despite this typical pattern, tic disorders remain a lifelong disabling condition for many. Pedigrees and twin studies suggest that TS has a strong hereditary component, but specific genetic risk factors remain elusive (Deng, Gao, & Jankovic, 2012; Scharf et al., 2013).

2.1.2 Tics and Tourette syndrome: Evidence for striatal dysfunction

There is strong evidence that tic disorders have a basis in BG dysfunction. Secondary tics may arise from BG disease, including Huntington disease (Jankovic & Ashizawa, 1995), BG strokes (Kwak & Jankovic, 2002), and Wilson disease (Lorincz, 2012). The most effective treatment for tics is dopamine-blocking (or depleting) drugs, implicating nigrostriatal dopamine in their pathogenesis (Cavanna & Seri, 2013). Indeed, a syndrome of tardive tics after neuroleptic exposure has been reported (Bharucha & Sethi, 1995). Molecular imaging studies report abnormalities in the striatal dopaminergic system in patients with tics though results are not consistent (Albin et al., 2003, 2009; Denys et al., 2013; Singer et al., 2002; Wolf et al., 1996). Deep brain stimulation of thalamic nuclei tightly linked with the BG, or BG nuclei themselves, variably improves tics (Ackermans, Kuhn, Neuner, Temel, & Visser-Vandewalle, 2013; Priori et al., 2013; Vandewalle, van der Linden, Groenewegen, & Caemaert, 1999).

Consistent with these studies, several quantitative volumetric imaging studies have revealed differences in striatal size between patients with TS and controls (Felling & Singer, 2011; Hyde et al., 1995; Peterson et al., 2003; Roessner et al., 2011). Postmortem studies describe specific deficits in striatal and pallidal GABAergic and cholinergic neurons, but these studies are based on a small number of postmortem examinations (Kalanithi et al., 2005; Kataoka et al., 2010).

2.1.3 Obsessive-compulsive disorder: Clinical features

Up to 60% of patients with TS have comorbid OCD (Cavanna & Rickards, 2013). OCD is characterized by the presence of intrusive, unwanted thoughts (obsessions), and the performance of ritualized behaviors that are intended to neutralize the negative emotions resulting from obsessions (compulsions) (Sarvet, 2013). While these repeated behaviors do not provide pleasure or reward, they provide relief from obsessions. These distressing uncontrolled rituals vary in severity, leading to distress and diminished quality of life. OCD often begins during childhood, with a mean age at onset of ~20 years (Narayanaswamy et al., 2012), though a median onset age at onset of 10 has been reported (Sarvet, 2013). OCD is believed to affect 1–3% of the population, both in children and adults (Sarvet, 2013). Like TS, OCD has a strong genetic component, though specific genes have yet to be identified (Pauls, 2008).

2.1.4 Obsessive-compulsive disorder: Evidence for striatal dysfunction

As for tic disorders, dysfunction of corticostriatal circuits has been implicated in OCD. Structural abnormalities in frontal cortex and/or the BG have been observed in patients with secondary OCD (Berthier, Kulisevsky, Gironell, & Heras, 1996). Functional imaging has revealed enhanced connectivity between limbic cortical areas and the ventral striatum in patients with OCD compared to controls (Harrison et al., 2009), and decreased dopamine D2/3 receptor availability (Denys et al., 2013). DBS has also been attempted for OCD, targeting primarily ventral striatum or the internal capsule, with variable results (Blomstedt, Sjoberg, Hansson, Bodlund, & Hariz, 2013). DBS of the subthalamic nucleus for PD has been reported to both induce compulsive behaviors (Broen, Duits, Visser-Vandewalle, Temel, & Winogrodzka, 2011) and alleviate them (Fontaine et al., 2004). Finally, animals with OCD-like behavior (e.g., hyperdopaminergic mice, discussed in Section 4) have abnormalities of striatal function (Berridge, Aldridge, Houchard, & Zhuang, 2005; Welch et al., 2007).

While many similarities exist between tic disorders and OCD, including their frequent comorbidity, they exhibit distinct clinical pharmacology. Treatments for the two conditions differ. While dopamine-blocking (or depleting) agents are clearly effective for TS, they are at best adjunct therapy for OCD (Sarvet, 2013). Conversely, antidepressants, especially selective serotonin-reuptake inhibitors, are the pharmacologic treatment of choice for OCD (Sarvet, 2013), but are not effective for TS.

2.2. Rett syndrome

2.2.1 Rett syndrome—Clinical features

RTT is a childhood-onset neurodevelopmental disorder caused by mutations in the MECP2 gene encoding the X-linked methyl-CpG-binding protein 2 (MeCP2) protein (Amir et al., 1999). RTT is a leading cause of intellectual disability in girls (Jellinger, 2003). Subjects with RTT develop normally for 6–18 months, followed by loss of cognitive function and speech, regression of fine and gross motor skills, social withdrawal, and development of stereotypic hand movements (Hagberg, Hanefeld, Percy, & Skjeldal, 2002). Autism, ataxia, and seizures (Dolce, Ben-Zeev, Naidu, & Kossoff, 2013) are also characteristic following developmental regression, with Parkinsonism and dystonia occurring in older patients (FitzGerald, Jankovic, Glaze, Schultz, & Percy, 1990). While RTT occurs almost exclusively in females, in boys, MECP2 mutations can cause neonatal lethality (Villard et al., 2000) or, on rare occasion, an

RTT-like phenotype (Armstrong, Pineda, Aibar, Gean, & Monros, 2001; Dayer et al., 2007; Masuyama et al., 2005), likely due to somatic mosaicism (Armstrong et al., 2001).

2.2.2 Rett syndrome: Evidence for striatal dysfunction

On purely clinical grounds, the highly stereotyped hand movements that occur in RTT are reminiscent of tic/compulsive-like behavior, and similar to the stereotypies that result from striatal dysfunction in experimental animals (Berridge et al., 2005; Welch et al., 2007). The occurrence of dystonia and Parkinsonism in RTT also point to BG involvement (FitzGerald, Jankovic, Glaze, et al., 1990; FitzGerald, Jankovic, & Percy, 1990).

Beyond mere pattern recognition, however, many lines of evidence indicate that striatal dysfunction is a key feature in the pathogenesis of RTT. Imaging studies of RTT subjects demonstrate decreased cerebral volumes (Carter et al., 2008; Jellinger, 2003; Murakami, Courchesne, Haas, Press, & Yeung-Courchesne, 1992), with reductions in caudate nucleus, cortical, and midbrain volumes disproportionately reduced compared to other structures (Casanova et al., 1991; Dunn et al., 2002; Reiss et al., 1993; Subramaniam, Naidu, & Reiss, 1997). PET imaging studies demonstrate specific changes in BG transmitter function. Several lines of evidence demonstrate significant dopaminergic deficits in RTT, including reduced fluorodopa uptake (Dunn et al., 2002), decreased dopamine transporter (DAT) levels (Wenk, 1995; Wong et al., 1998), and lower density of D2 receptors (Harris et al., 1986) in caudate nucleus and putamen. Other studies demonstrate increased striatal D2 receptor binding, believed to reflect receptor upregulation in response to dopaminergic deficits (Chiron et al., 1993). These differences are likely due to the difference in the ages of the patients that were examined (Dunn & MacLeod, 2001), and demonstrate a possible age-dependent change with early D2 receptor increases followed by later reductions (Cordes et al., 1994). In contrast to specific DAT and D2 receptor changes, caudate nucleus D1 receptors and dopamine reuptake sites in cortical regions remain unchanged (Wenk, 1995). These results suggest a degree of specificity in affected dopaminergic systems. There is also evidence for striatal cholinergic dysfunction in RTT (Wenk & Mobley, 1996). For example, striatal vesicular acetylcholine transporter densities are reduced in women with RTT, and their clinical abilities are correlated with these levels (Brasic et al., 2012).

Postmortem studies of humans with RTT reveal abnormalities of the striatum and its connections. Consistent with functional imaging, choline acetyl-transferase (ChAT) activity is decreased specifically in the hippocampus, caudate nucleus, and thalamus (Wenk et al., 1993). The nigrostriatal dopaminergic system is affected, with decreased pigmentation of nigral neurons and decreased immunoreactivity for tyrosine hydroxylase (TH) (the rate-limiting enzyme in dopamine synthesis) (Jellinger, Armstrong, Zoghbi, & Percy, 1988; Kitt & Wilcox, 1995). Other catecholamines and their metabolites are similarly decreased in the substantia nigra of RTT subjects (Lekman et al., 1989). AMPA and NMDA receptor density are decreased and GABA receptor density is increased, but no changes are observed in mGluR receptor density in the BG (Blue, Naidu, & Johnston, 1999). While it is impossible to disentangle definitively changes that reflect the underlying disease process from compensatory or medication effects, these data support strongly a primary role BG dysfunction in the behavioral features of RTT.

2.3. Primary dystonia

2.3.1 Primary dystonia—Clinical features

Dystonia is a movement disorder characterized by abnormal, sustained, or intermittent, usually twisting postures maintained by agonist/antagonist co-contraction (Albanese et al., 2013). Dystonia is classified as "primary" if it occurs in isolation without neuropathological changes. In contrast, "secondary" dystonic movements occur consequent to CNS damage (e.g., from stroke, trauma, or neurodegeneration) and are typically accompanied by additional neurological signs and symptoms. Among the primary dystonias, there is a bimodal age at onset with peaks during childhood (school age to teenage years) and the fifth to sixth decade of life (Bressman et al., 2000). Age at symptom onset is strongly associated with etiology and pattern of involvement. Childhood-onset dystonia typically begins in an arm or leg, spreads to other body parts and generally reflects an inherited mutation. In contrast, adult-onset dystonia is almost always sporadic and remains confined to the cranio-cervical region or, less commonly, the upper extremities (e.g., writer's cramp) (Albanese et al., 2013). The pattern of involvement and time at symptom onset have been most extensively studied for DYT1 dystonia, the most common inherited form of primary dystonia. The DYT1 mutation (described in detail in Section 6) is only ~30% penetrant. Interestingly, manifesting subjects develop symptoms almost uniformly during the school-age years. Those that remain symptom free into their early twenties typically remain unaffected for life, strongly implying the presence of a "critical period" of vulnerability during CNS development as a key feature of disease pathogenesis.

2.3.2 Dystonia: Evidence for striatal dysfunction

A combination of clinical, neuroimaging, and basic scientific observations support a central role for striatal dysfunction in the pathophysiology of primary dystonia. Damage to the striatum and associated structures (e.g., pallidum, thalamus) are a common cause of secondary dystonia (Marsden, Obeso, Zarranz, & Lang, 1985; Mehanna & Jankovic, 2013). Metabolic diseases that disrupt striatal function (e.g., Wilson's disease, pantothenate kinase-associated neurodegeneration) also cause secondary dystonia (Gordon, 2002; Lorincz, 2012).

There are strong links between dopaminergic dysfunction and dystonia. Dystonia is commonly an early manifestation of PD, or a complication of its treatment with levodopa. Among drugs that induce dystonia, dopamine-receptor blockers are the most prominent. An interesting age-dependent feature of these medications is that acute dystonic reactions are much more common in young patients, while tardive syndromes tend to emerge in older patients (Dayalu & Chou, 2008). Furthermore, patients with dopa-responsive dystonia are, as the name implies, exquisitely responsive to levodopa. These patients also have an age-dependent phenotype, with young onset associated with dystonia and older onset associated with Parkinsonism (Trender-Gerhard et al., 2009). Together with the clinical features of DYT1 dystonia reviewed above, this relationship between age and symptom onset further strengthens the link between abnormal striatal development and dystonia. Commonly used treatments for dystonia include anticholinergic and GABAergic agents, which strongly influence striatal circuitry (though are admittedly not specific to striatal function). Pallidal or subthalamic DBS is highly effective for dystonia (Vidailhet, Jutras, Grabli, & Roze, 2013), further implicating these circuits in disease pathophysiology.

Behavioral functions typically attributed to striatal circuitry are disrupted in patients with dystonia, as well as carriers of dystonia mutations. A prominent theory of dystonia pathophysiology suggests that the action selection functions of the BG are deficient in dystonia, leading to the simultaneous activation of competing motor programs (Mink, 2003). A note of caution is in order, however, as the involvement of the BG in dystonia (and other movement disorders) is often cited as support for the action selection hypothesis. Motor sequence learning, which is more convincingly dependent on striatal function (see Section 3), is impaired in patients with DYT1 dystonia. On the other hand, similar impairments were found in nonmanifesting carriers of the DYT1 mutation, and impairments were not found in manifesting or nonmanifesting DYT6 mutation carriers

(Carbon et al., 2011). It is therefore not clear if impairments in motor sequencing are sensitive or specific for primary dystonia.

Imaging studies strongly implicate BG circuits in the pathophysiology of dystonia. Though generally considered a disorder of abnormal function of motor circuits (as opposed to structure), consistent changes in the volume of motor-related areas (motor cortex, BG, and cerebellum) have been found in both focal and generalized primary dystonias (Delmaire et al., 2007; Draganski, Thun-Hohenstein, Bogdahn, Winkler, & May, 2003; Egger et al., 2007; Garraux et al., 2004). Some of these changes are specific to patients manifesting dystonia, while others also occur in mutation carriers. Other subtle structural changes have been found using tractography techniques. Patients with cervical dystonia have abnormalities of BG-pontine tracts (Blood et al., 2012). Reduced cerebellothalamic connectivity has been found in both manifesting and nonmanifesting DYT1 mutation carriers, but reduced thalamocortical connectivity was present only in nonmanifesting carriers (Argyelan et al., 2009).

There are very few neuropathological studies in primary dystonia patients (Paudel, Hardy, Revesz, Holton, & Houlden, 2012), making it difficult to know what, if any, histopathological abnormalities are responsible for the neuroimaging findings. Most neuropathological studies have found no pathological changes (Furukawa, Hornykiewicz, Fahn, & Kish, 2000; Gibb, Lees, & Marsden, 1988; Walker, Brin, Sandu, Good, & Shashidharan, 2002; Zweig et al., 1988). In those that did find changes, it is not clear if the findings were incidental (Holton et al., 2008). For example, two patients with Meige syndrome (Kulisevsky, Marti, Ferrer, & Tolosa, 1988; Mark et al., 1994) had Lewy Body pathology in the brainstem. These patients were ages 69 and 72, however, ages at which Lewy Body pathology is relatively common in people without clinical evidence of Parkinsonism. Another study found tau and ubiquitin-positive inclusions in the brainstems of patients with DYT1 dystonia (McNaught et al., 2004). While one of these patients was 33, the other three were ages 78–83, again raising the possibility that these were incidental findings. Overall, neuropathological studies in primary dystonia patients are simply too limited to draw any conclusions.

Functional imaging of subjects with dystonia demonstrates clear abnormalities of BG function. BG metabolic activity is altered in dystonia patients, with basal ganglia fMRI signals remaining elevated after finger tapping compared to controls (Blood et al., 2004). Some of the most consistent findings are in functional imaging of the dopaminergic system, though the specific abnormalities vary with the type of dystonia. For example, striatal

D2-like receptor binding is decreased in patients with focal dystonia (Perlmutter et al., 1997) or nonmanifesting DYT1 mutations (Asanuma et al., 2005), but increased in patients with dopa-responsive dystonia (Rinne et al., 2004). A recent study suggests that at least some of these changes may be in the density of striatal D3 receptors (a member of the D2 receptor family) specifically (Karimi et al., 2011). D1 receptor density, at least in primary focal dystonia, however, seems to be unchanged from controls (Karimi et al., 2013).

3. STRIATAL ORGANIZATION AND FUNCTION

3.1. Summary of basic features of striatal organization

The striatum (composed of the caudate nucleus and putamen) is a major input nucleus for the BG, a group of subcortical gray matter structures that form recurrent loops with the thalamus and cortex (Fig. 3.1). It receives glutamatergic excitatory input from the whole neocortical mantle (Mathai & Smith, 2011), related structures such as the hippocampal formation and the amygdala (McGeorge & Faull, 1989), as well as thalamic intralaminar (centromedian and parafascicular nuclei—CM/Pf) and BG-recipient relay nuclei (Sadikot & Rymar, 2009). Some GABAergic striatal inputs come from within the BG, including from the globus pallidus, pars externa (GPe; Mallet et al., 2012). As detailed below, many of these afferents project preferentially to specific subpopulations of striatal neurons. The striatum sends a massive GABAergic efferent projection to GPe and the BG output nuclei (substantia nigra, pars reticulata (SNr) and globus pallidus, pars interna (GPi)).

The organization of corticostriate projections is complex. Lesion studies and, more recently, functional imaging (Choi, Yeo, & Buckner, 2012; Kemp & Powell, 1970) demonstrate a topographic organization, suggesting that different striatal regions are functionally specialized depending on their cortical inputs. For example, dorsolateral striatum in rodents (primarily dorsolateral putamen in primates) receives afferents from primary sensorimotor cortex. Similarly, dorsomedial striatum in rodents (primarily caudate nucleus in primates) receives afferents from cortical association areas, and ventral striatum (nucleus accumbens) receives limbic afferents (Parent & Hazrati, 1995). This functional specialization is at least partially maintained throughout the BG and its thalamic projections (Alexander, DeLong, & Strick, 1986). Cortico-BG-thalamocortical loops tend to project back to their

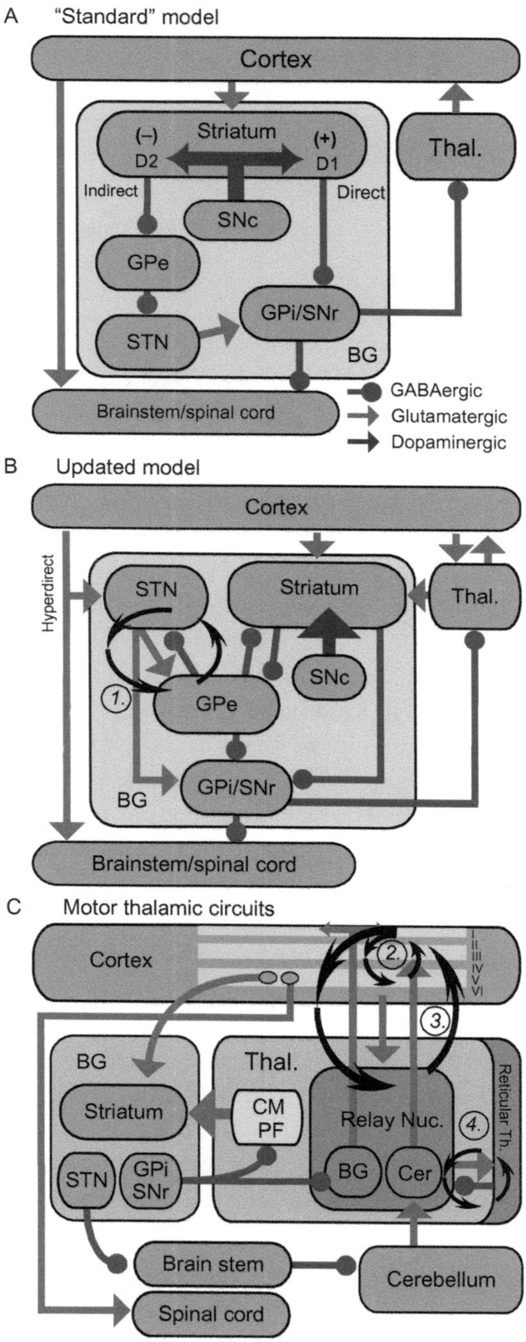

A "Standard" model
Cortex
Striatum
(–)
D2
(+)
D1
Indirect
Direct
SNc
GPe
STN
GPi/SNr
BG
Thal.
Brainstem/spinal cord
GABAergic
Glutamatergic
Dopaminergic
B Updated model
Cortex
Hyperdirect
STN
Striatum
Thal.
SNc
1.
GPe
GPi/SNr
BG
Brainstem/spinal cord
C Motor thalamic circuits
Cortex
I
II
III
IV
V
VI
2.
3.
BG
Thal.
Reticular Th.
Striatum
CM
PF
Relay Nuc.
4.
STN
GPi
SNr
BG
Cer
Brain stem
Cerebellum
Spinal cord

cortical site of origin, but also project to at least one other cortical area (Joel & Weiner, 1994). Thus, motor, associative, and limbic loops are not fully segregated. Sensitive tract-tracing studies also indicate that cortical regions project to the striatum in overlapping parasagittally arranged zones of considerable rostro-caudal extent, suggesting some interdigitation of corticostriate projections (Selemon & Goldman-Rakic, 1985). Other work suggests some convergence of corticostriate projections from cortical regions that are functionally linked by corticocortical connections (Yeterian & Van Hoesen, 1978).

The striatum is composed primarily of GABAergic medium spiny neurons (MSNs) and several interneuron types (Table 3.1). These cells exist in two interdigitated compartments known as "patch" (or striosomes) and "matrix" (Crittenden & Graybiel, 2011). MSNs represent 90–95% of neurons depending on the species, with lower percentages in primates (Graveland & DiFiglia, 1985; Tepper, Tecuapetla, Koos, & Ibanez-Sandoval, 2010). MSNs are characterized by the presence of multiple dendritic spines (small mushroom-shaped protuberances) that are the site of most excitatory synapses. They fire sparsely, requiring coordinated excitatory synaptic input to initiate spiking (Kreitzer, 2009). MSNs are divided into two groups of roughly equal proportions based on projection patterns (Gerfen & Surmeier, 2011). The monosynaptic connection from a subset of MSNs (dMSNs) to GPi/SNr is termed the "direct" pathway, but also sends collaterals to GPe. "Indirect" pathway MSNs (iMSNs), on the other hand, synapse only in GPe. dMSNs express D1 dopamine receptors with relatively low affinity for dopamine, while iMSNs express higher affinity D2 dopamine receptors (Gerfen & Surmeier, 2011). Thus, phasic increases or decreases in striatal dopamine concentration likely have distinct effects on dMSNs and iMSNs.

Figure 3.1 Cortical-BG-thalamic models. (A) The "standard" model. Green arrows, excitatory (glutamatergic) projections; Red circles, inhibitory (GABAergic) projections; Blue arrows, dopaminergic projections. (+) and (−) indicate excitatory and inhibitory effects of dopamine on MSNs, respectively. Abbreviations are defined in the text. (B) Updated model. Not all connections described in the text are illustrated for clarity. (C) Model emphasizing thalamic circuits. Roman numerals indicate cortical layers. Ovals in layer V indicate distinct populations of IT and PT neurons. Note the distinct projection patterns of the cerebellar and BG-recipient thalamus. In the thalamus, "BG" and "Cer" indicate the BG and cerebellar-recipient regions, respectively. Circular arrows in (B) and (C) indicate candidate oscillation generators. *Figure and caption reprinted with modification from Ellens and Leventhal (2013), copyright 2013, with permission from IOS Press.* (See the color plate.)

Table 3.1 Striatal cell types

Cell type	Transmitter	Nomenclature	Afferent connections	Efferent connections	Specific markers	Major electrophysiological properties
Projection neurons	GABA	MSN—Direct pathway	Cortex, thalamus, interneurons, GPe	GPi, SNr, collaterals to GPe	DARPP-32, D1R	Fire sparsely. Require coordinated excitatory synaptic input to initiate spike
		MSN—Indirect pathway	Cortex, thalamus, interneurons, GPe	GPe	DARPP-32, D2R, A2A receptors	Fire sparsely. Require coordinated excitatory synaptic input to initiate spike
Interneurons	GABA	FSI	Cortex, CM/Pf thalamus, GPe, interneurons	Proximal MSN dendrites/soma	Parvalbumin	Strong monosynaptic inhibition of MSN *in vitro*
		PLTS		MSN distal dendrite	NPY, SST, nNOS	Small IPSP on postsynaptic neuron
		CR+			CR	
		TH+			TH	GABA-dependent IPSP on postsynaptic neuron
	Cholinergic	TAN	CM/pf thalamus, cortex, interneurons	MSN, interneurons, volume transmission	ChAT, VAChT, TrkA	Spontaneously fire at low frequencies (<10 Hz). Pause in response to salient behavioral events

The remaining neurons in the striatum are aspiny interneurons, including medium-sized GABAergic and large cholinergic neurons (Kreitzer, 2009). GABAergic interneurons can be further classified into at least four subtypes based on distinctive electrophysiological properties and expression of different neurochemical markers (Tepper, 2010; Tepper et al., 2010). Fast-spiking interneurons (FSIs) express the calcium-binding protein parvalbumin and represent approximately 1% of the neurons in the striatum (Luk & Sadikot, 2001) in a lateral to medial gradient (higher density laterally). This suggests a role in sensorimotor functions. These cells receive afferents from cortex (Ramanathan, Hanley, Deniau, & Bolam, 2002), the thalamic CM nucleus (Sidibe & Smith, 1999), other interneurons (Chang & Kita, 1992), and GPe (Berke, 2011; Bevan, Booth, Eaton, & Bolam, 1998; Kreitzer, 2009). One FSI may innervate hundreds of MSNs (Koos & Tepper, 1999). Single FSI discharges are capable of inhibiting action potential generation in MSNs (or at least delaying their onset) due to potent, proximal synapses (Bennett & Bolam, 1994; Kubota & Kawaguchi, 2000). Despite *in vitro* evidence of strong monosynaptic FSI–MSN inhibition, however, *in vivo* studies have not consistently found a sharp depression in MSN firing immediately after FSI spikes (Adler, Katabi, Finkes, Prut, & Bergman, 2013; Gage, Stoetzner, Wiltschko, & Berke, 2010; Lansink, Goltstein, Lankelma, & Pennartz, 2010). FSIs are linked with each other via dendritic gap junctions (Fukuda, 2009; Kita, Kosaka, & Heizmann, 1990), creating a syncytium hypothesized to synchronize FSI firing on a very fine timescale. More recent studies suggest that this is true only when FSIs receive coincident cortical input (Hjorth, Blackwell, & Kotaleski, 2009), and only in specific frequency ranges (Russo, Nieus, Maggi, & Taverna, 2013). *In vivo* experiments seem to bear this out, with coordinated FSI activity only during specific, temporally precise behavioral events (Adler et al., 2013; Berke, 2008; Gage et al., 2010). Nonetheless, the small increase in FSI coordination mediated via gap junctions may be critical for coordinating striatal cell assemblies and synchronizing corticostriatal oscillations.

The other GABAergic interneurons are less well-characterized, especially calretinin-positive interneurons, which are quite sparse in rodents (Gittis, Nelson, Thwin, Palop, & Kreitzer, 2010; Tepper, 2010). Another class of interneurons expresses neuropeptide Y, somatostatin, and nitric oxide synthase (hereafter referred to as NPY neurons). These cells synapse sparsely on more distal dendrites of MSNs, and generate small IPSPs. They are characterized physiologically by persistent low-threshold spikes (Kawaguchi, Wilson, Augood, & Emson, 1995; Tepper, 2010), leading to their other descriptor—PLTS neurons. Whether their effects are mediated primarily via GABAergic transmission or release of neuromodulatory peptides remains unknown (Tepper, 2010). Finally, there is a class of TH (the rate-limiting enzyme in dopamine and noradrenaline synthesis) positive interneurons that evoke GABA-dependent IPSPs in postsynaptic cells. The functional significance of the non-FSI GABAergic interneurons remains unclear, as they have been difficult to identify in awake, behaving subjects.

Cholinergic interneurons probably correspond to tonically active neurons (TANs) recorded *in vivo*, which fire spontaneously at low frequencies (<10 Hz) (Schulz & Reynolds, 2013). They receive excitatory inputs predominantly from CM/Pf (Ding, Guzman, Peterson, Goldberg, & Surmeier, 2010; Matsumoto, Minamimoto, Graybiel, & Kimura, 2001), minor excitatory input from the cortex (Thomas, Smith, Levey, & Hersch, 2000), and inhibitory input from GABAergic interneurons and possibly MSNs (Goldberg & Reynolds, 2011; Sullivan, Chen, & Morikawa, 2008). Striatal cholinergic signaling may be primarily via "volume neurotransmission" since there are relatively few typical cholinergic synapses within the striatum (Pisani, Bernardi, Ding, & Surmeier, 2007). Striatal cholinergic neurons directly influence GABAergic interneuron and MSN activity, and interact with dopamine to regulate synaptic plasticity (Gerfen & Surmeier, 2011). Their axons arborize densely and diffusely in the striatal matrix, with sparse crossings into the patch compartment (Graybiel, Baughman, & Eckenstein, 1986; Kawaguchi, 1992). Unlike most other neurons, TAN firing is correlated in awake, behaving animals, and may transmit coordinated signals to large striatal regions. An example is the "TAN pause" (Aosaki et al., 1994; Morris, Arkadir, Nevet, Vaadia, & Bergman, 2004), a transient, coordinated decrease in TAN firing in response to salient behavioral events dependent on afferents from the CM/Pf nucleus (Ding et al., 2010; Matsumoto et al., 2001). This pause is generally preceded and immediately followed by an increase in TAN firing, in a burst-pause-burst pattern. GABAergic interneurons are activated by nicotinic stimulation, and there is reason to believe that NPY neurons, in particular, may mediate the TAN pause by a feedback mechanism (Sullivan et al., 2008). Striatal output is transiently suppressed by the coordinated activity of TANs *in vitro* (English et al., 2012), and the TAN pause may provide a temporal window for corticostriatal plasticity in coordination with dopamine transients (Cragg, 2006). Interestingly, the TAN pause is lost after striatal dopamine depletion (Schulz & Reynolds, 2013).

Midbrain dopaminergic projections to striatum are especially strong, and likely play a major role in the pathophysiology of multiple disorders including Parkinsonism, tardive drug reactions, tic disorders, dystonia, and schizophrenia. The substantia nigra, pars compacta (SNc) innervates dorsal (motor and associative) regions of striatum, while the ventral tegmental area (VTA) innervates the ventral nucleus accumbens. These two cell populations have distinct physiologic features, but even this simple dichotomy may not capture their full diversity (Morikawa & Paladini, 2011). Indeed, there is a lateral (SNc) to medial (VTA) gradient of physiologic properties and anatomic connectivity. More medial cells project to ventral striatum, which project back to more lateral midbrain dopaminergic neurons, which project to more dorsal striatal regions. In this way, an "ascending spiral" of midbrain-striatal circuits is created, providing a mechanism by which distinct BG circuits can interact with each other (Belin & Everitt, 2008; Haber, Fudge, & McFarland, 2000). Midbrain dopaminergic projections feature a large number of terminals diverging from a relatively small number of neurons; each dopaminergic nigral neuron innervates many striatal neurons. The relative lack of specificity of nigrostriatal projections suggests that this pathway performs a diffuse "biasing" function for striatal neurons, though nigral dopamine neuron activity may not be as uniform as previously assumed (Matsumoto & Hikosaka, 2009). Dopaminergic terminals tend to synapse on the necks of MSN spines, with cortico- and thalamostriatal terminals on spine heads. These "triadic" synapse complexes allow tight regulation of the contributions of specific cortical and thalamic afferents to MSN firing.

Dopamine modulates striatal output on multiple timescales via pre- and postsynaptic mechanisms. Presynaptic D2 receptors on multiple cell types regulate dopamine, acetylcholine, glutamate, and GABA release (Gerfen & Surmeier, 2011). Postsynaptic D1 receptor stimulation acutely enhances MSN excitability, while D2 receptor activation has opposite effects (Gerfen & Surmeier, 2011). The differential expression of D1 and D2 receptors by MSNs (D1-dMSN; D2-iMSN) results in contrasting effects of acute changes in dopamine levels on the direct and indirect BG pathways. Striatal interneurons also express multiple dopamine-receptor subtypes with complex responses to dopaminergic stimulation. In addition to direct effects on striatal neuronal activity, dopamine influences synaptic plasticity at cortico- and thalamostriatal synapses of dMSNs and iMSNs in distinct ways (Gerfen & Surmeier, 2011; Shen, Flajolet, Greengard, & Surmeier, 2008).

According to the prevailing "rate" model of BG physiology (Albin, Young, & Penney, 1989), striatal dopamine increases dMSN firing, decreases iMSN firing, and, ultimately, suppresses BG output to release thalamocortical circuits from tonic inhibition. In Parkinsonism, dMSN activity decreases and iMSN activity increases, resulting in excessive BG inhibition of thalamocortical activity. Strong experimental support for this model comes from selective activation of dMSNs or iMSNs using optogenetic techniques, which dramatically alters locomotor activity and BG output in a manner consistent with the rate model (Freeze, Kravitz, Hammack, Berke, & Kreitzer, 2013; Kravitz et al., 2010). Though the rate model successfully predicts many clinical and physiological phenomena, it is also clearly incomplete. For one thing, there are multiple types of hyperkinetic movement disorders (e.g., dystonia, chorea), yet the rate model generically predicts increased movement from excessive direct, or insufficient indirect, pathway activity. More telling, the rate model predicts that lesions of BG output should provoke abnormal movements, but such lesions are used to treat chorea and dystonia. Finally, direct recordings from BG-thalamocortical circuits in PD and other movement disorders show only partial agreement with rate model predictions, instead emphasizing correlations between neuronal activity patterns and behavioral output (Cleary et al., 2013; Ellens & Leventhal, 2013; Rubin, McIntyre, Turner, & Wichmann, 2012; Tang et al., 2009).

3.2. Overview of striatal function

3.2.1 Online motor control/action selection

Based largely on observations of patients with PD and Huntington disease, it has been long assumed (at least among neurologists) that the primary function of the striatum is to regulate moment-to-moment motor function (Albin et al., 1989; DeLong, 1990). The "short-duration response" of PD patients to levodopa, in which motor function improves on approximately the same timescale that levodopa enters the brain (Chan, Nutt, & Holford, 2004), provides strong support for this hypothesis. There is also experimental evidence that the striatum regulates online motor performance: motor output is acutely altered by intrastriatal drug infusions (Gittis et al., 2011; Leventhal et al., 2014; Worbe et al., 2009), electrical stimulation (Watanabe & Munoz, 2010), or optogenetic stimulation (Kravitz et al., 2010).

It has been suggested that the BG, in general, and striatum specifically, participate in action selection and initiation (Mink, 2003; Ratcliff & Frank, 2012; Redgrave, Vautrelle, & Reynolds, 2011). In part, this hypothesis grows out of the observation that decorticate rats can properly sequence stereotyped movement patterns and associate specific cues with preferred behaviors (Berridge & Whishaw, 1992; Whishaw & Kolb, 1984). It is often stated that the BG, which are highly conserved from invertebrates through primates, provide blanket suppression of brainstem motor programs (Grillner et al., 2013) given the high tonic firing of BG output neurons. Though details of these models vary, the basic idea is that the striatum uses convergent cortical and thalamic input regarding the current state of the animal to determine the next optimal motor program, which is released for execution by focused inhibition of BG output via the direct pathway. Meanwhile, the indirect pathway suppresses all alternative options so that the motor system is accessed by only one "action" at a time (Cui et al., 2013). Several computational models of the striatum (Humphries, Wood, & Gurney, 2009) and full BG circuits (Lo & Wang, 2006; Ratcliff & Frank, 2012) demonstrate that striatal circuitry could implement action selection. These models make implicit assumptions that neuronal assemblies turned off or on by model circuits represent individual "action plans." Further, they assume that the primary function of GABAergic BG output is to suppress neuronal activity in target structures. Recent studies, however, suggest that BG output plays a subtler role in manipulating the fine timing of spikes generated by target neurons (Goldberg, Farries, & Fee, 2013; Goldberg & Fee, 2012; Rubin et al., 2012).

Physiologic studies of the BG are broadly consistent with the action selection hypothesis. For example, many BG output (GPi/SNr) neurons abruptly decrease their firing rates just prior to movement initiation (Gulley, Kuwajima, Mayhill, & Rebec, 1999; Schmidt, Leventhal, Mallet, Chen, & Berke, 2013; but see Mink & Thach, 1991a). Indeed, firing rate changes in this subset of neurons strongly predicts movement onset (Freeze et al., 2013). Many BG output neurons increase their firing rates at movement onset, however, and both dMSNs and iMSNs fire at movement onset (Cui et al., 2013). These observations have been interpreted as suppression of competing motor programs by the indirect pathway.

While the studies described above are consistent with the concept that action selection is a primary role for the striatum, they are not proof. During tasks in which decisions are indicated with orienting movements to the left or right (i.e., saccades in primates or nose pokes in rodents), intrastriatal manipulations strongly bias decisions. For example, caudate microstimulation suppresses contralateral saccades (Watanabe & Munoz, 2010), and inactivation of dorsolateral striatum biases rats to move toward the inactivated side (Leventhal et al., 2014). Further, unilateral blockade of dopamine receptors or dopamine depletion biases rats toward responses ipsilateral to the lesion (Brown & Robbins, 1989; Carli, Evenden, & Robbins, 1985; Leventhal et al., 2014). In some sense, this biasing is similar to the rapid rotation observed in hemidopamine lesioned rodents exposed to amphetamine or apomorphine (Ungerstedt & Arbuthnott, 1970). Thus, there is strong evidence that the striatum participates online in the decision to orient to the left or right. Since the striatum is a bilateral structure, however, decisions regarding head/neck movements cannot be determined by the output of one striatum alone. Further computations must take place in downstream structures, perhaps by comparing BG output from both hemispheres. The superior colliculus, which receives bilateral afferents from SNr, likely plays a key role in decisions regarding eye/head/neck movements (Holmes et al., 2012; Lo & Wang, 2006). Regardless of the underlying neural mechanisms, it is clear that for orienting type movements, the striatum participates in action selection processes.

A natural question is whether this function is generalizable to limb movements. Since striatal "decisions" must be conveyed to the rest of the brain via BG output nuclei, lesion/inactivation studies of sensorimotor regions of GPi are informative. Disruption of BG output slows limb movements and causes hypometria (target undershoot), but with relatively preserved reaction times (Horak & Anderson, 1984; Inase, Buford, & Anderson, 1996; Limousin et al., 1999; Mink & Thach, 1991b; Wenger, Musch, & Mink, 1999). Critically, previously acquired motor sequences are completed in the correct order (Desmurget & Turner, 2010; Turner & Desmurget, 2010), but patients with therapeutic pallidotomies have difficulty acquiring new sequences (Brown et al., 2003; Obeso et al., 2009). These results argue against a primary role for the striatum in online decision-making processes, at least for some types of movements.

3.2.2 Reinforcement learning

An impressive convergence of theoretical and experimental results accumulating in the past couple of decades suggest that nigrostriatal dopamine carries signals critical to reward-based striatal learning processes (for a concise review, see Glimcher, 2011). The basic concept is that nigrostriatal dopaminergic neurotransmission mediates reward prediction errors (RPEs) for reinforcement learning. In these models, an error signal indicates the difference between the expected and actual value of rewards. This error signal is used to refine predictions regarding the expected reward given a specific action performed in a specific context. In seminal experiments performed by Schultz and colleagues, it became apparent that phasic nigrostriatal dopaminergic neuron signaling exhibits the properties of an RPE signal (Schultz, Apicella, & Ljungberg, 1993). Phasic activity of these neurons is characterized by high firing rates with resulting marked elevation of striatal dopamine release. Schultz's group studied SNc dopaminergic neuron spiking rates as monkeys learned associations between visual cues and rewards. These cells have a baseline tonic firing rate of about 5–10 Hz that is modulated by behavioral events. In untrained animals, unpredictable reward presentation was followed rapidly by a phasic increase in nigral activity. As animals learned that specific cues predict rewards, the burst of SN activity associated with the primary reward subsided. Instead, the phasic increase in dopamine neuron firing became associated with the (unpredictably timed) reward-predicting cue. Migration of the positive RPE signal from previously unexpected reward to the predicting cue is an explicit prediction of computational temporal difference models of reinforcement learning, which were formulated years prior to Schultz's experiments.

Just as an unexpected reward results in a burst of nigral activity, omission of an expected reward results in transient depression nigral activity (Schultz, Dayan, & Montague, 1997), compatible with a negative RPE signal. More recent experiments investigating the fine timing of phasic dopamine release in rodent striatum using fast-scan cyclic voltammetry have confirmed rapid changes in striatal dopamine concentration that correlate well with the phasic changes in dopamine neuron firing rates observed by Schultz and others (Brown, McCutcheon, Cone, Ragozzino, & Roitman, 2011; Flagel et al., 2011; Kiyatkin & Gratton, 1994). It should be noted, however, that cholinergic activity intrinsic to the striatum also modulates dopamine release independently of ascending activity in midbrain dopaminergic neurons (Threlfell et al., 2012). The details of intrinsic striatal microcircuitry notwithstanding, selective activation of VTA dopaminergic neurons is still sufficient to induce conditioned place preference (Tsai et al., 2009). Activating dMSNs or iMSNs independently of dopamine signaling is sufficient to reinforce or diminish, respectively, instrumental responding (Kravitz, Tye, & Kreitzer, 2012), further supporting a role for the striatum in reinforcement learning.

Though many dopamine neuron responses are consistent with an RPE signal, this is not uniformly true (Matsumoto & Hikosaka, 2009; Schultz, 2013). Many putative midbrain dopamine neurons fire in response to aversive as well-rewarding events. Others respond to novel stimuli, even if not previously associated with rewarding outcomes (Schultz, 2013). This may in part be attributable to contamination of putative dopaminergic neuronal populations with recordings from GABAergic interneurons (Cohen, Haesler, Vong, Lowell, & Uchida, 2012). These experiments suggest, however, that midbrain dopaminergic neurons do not code a pure RPE signal.

It should be stressed that while an associative learning role for striatal dopamine is the predominant view of nigrostriatal dopaminergic signaling, this is not the only cogent interpretation of these data. Berridge and Robinson (1998) put forward an alternative explanation in which nigrostriatal dopaminergic signaling provides what they term incentive salience (Berridge, 2007), essentially a mapping of the motivational significance of rewards and stimuli, as opposed to learning *per se*. This interpretation leads to many of the same predictions as interpreting dopaminergic signaling as an example of temporal difference learning, but may account for "dopamine-free" associative learning observed under special circumstances (Cannon & Palmiter, 2003; Flagel et al., 2011; Hnasko, Sotak, & Palmiter, 2005). Thus, it seems that midbrain dopaminergic signaling is sufficient, but not necessary, for certain types of implicit learning.

Specific hypotheses have been advanced regarding the cellular mechanisms mediating the effects of striatal dopamine (Robinson, Sotak, During, & Palmiter, 2006). The so-called "3-factor rule" postulates that cortico- and thalamostriatal synapses active when phasic changes in dopamine (increases or decreases) occur are selectively modified via long-term potentiation or depression. Transient changes in cholinergic activity (i.e., TAN pauses) also likely modulate dopamine-mediated synaptic plasticity, perhaps by providing a temporal window in which cortico- and thalamostriatal plasticity is most effective (Morris et al., 2004). The direction in which synaptic strength is modified depends on the precise timing of afferent input, as well as the dopamine-receptor profile of striatal neurons (Shen et al., 2008). D2 receptors have a higher affinity for dopamine than D1 receptors, raising the possibility that, at tonic dopamine levels, D2 receptors are largely occupied while D1 receptors are not. Phasic increases in dopamine would then leave the indirect pathway relatively unchanged, while modifying corticostriatal synapses in the direct pathway. Conversely, drops in dopamine concentration due to negative RPEs would leave already-vacant D1 receptors unoccupied, but D2 receptors would go from occupied to unoccupied. Thus, reinforcement learning from positive RPEs may take place primarily along the direct pathway, while independent learning from negative RPEs may take place along the indirect pathway. Support for this hypothesis comes from studies of implicit learning in patients with

PD. When off medications, patients tend to learn more from omitted rewards; on medications, they tend to learn more from rewarded events (Frank, Seeberger, & O'reilly, 2004). While an attractive hypothesis, independent manipulation of the direct and indirect pathways by phasic dopamine increases and decreases remains to be proven (Dreyer, Herrik, Berg, & Hounsgaard, 2010; Wall et al., 2011).

3.2.3 Habit formation

Habits are motor or cognitive behaviors that are performed automatically, without regard to the expected outcome of the behavior. From a practical standpoint, there are two primary assays used to identify actions as habitual in experimental animals (Yin & Knowlton, 2006). In the first, the anticipated value of a specific action is modified. A subject may be given free access to reward prior to a "probe session," devaluing successful task completion. For example, a rat trained to press a lever for a food reward may be sated prior to testing. If the lever-press behavior is "goal-directed," the sated rat will respond less often than a hungry rat. A rat that responds habitually will press the lever whether hungry or full. Alternatively, a previously desired outcome may be associated with an aversive stimulus. For example, pairing lithium chloride injections (which induce nausea) with chocolate milk renders the previously rewarding outcome undesirable (Smith & Graybiel, 2013a). If the subject avoids (or at least does not actively pursue) a devalued outcome, the behavior is interpreted as "goal-directed." If the subject continues to respond, the behavior is identified as a habit.

The second assay involves changing the "action–outcome" contingency of a task. In this case, the probability of reward given a response relative to the probability without that response is manipulated. This may be accomplished by increasing the probability of a rewarding event during free behavior, or decreasing the probability of reward after a response. Again, if subjects continue to respond to a cue in a manner that no longer increases the probability of reward ("stimulus–response" vs. "action–outcome" behavior), the behavior is considered habitual.

Initial stages of habit learning involve a goal-directed behavior, which becomes habitual following extended experience or training. The switch from goal-directed behavior to habit involves a transition in the neural activity of circuits controlling these behaviors (Smith & Graybiel, 2013b). Once a behavior (such as maze walking or lever pressing in rodents, or sequential eye movements in primates) becomes habitual, a "task-bracketing" activity pattern emerges such that striatal firing occurs prominently at the beginning and end of the task (Barnes, Kubota, Hu, Jin, & Graybiel, 2005; Jin, Tecuapetla, & Costa, 2014; Jin & Costa, 2010; Jog et al., 1999, #40056). This bracketing activity is thought to allow the striatum to "chunk" or crystallize the habitual action so it can be performed as a unit (Graybiel, 1998), allowing each sequence of actions within the habit to be marked as valuable and performed without extra oversight. Striatal task bracketing, however, is readily broken by "extinction" trials, when the behavior is extinguished in the absence of reward. Similar "task-bracketing" activity also evolves in infralimbic cortex, which is more tightly associated with the transition from goal-directed to habitual behavior (Smith & Graybiel, 2013a).

Experiments which lesion or inactivate specific brain regions demonstrate that both the dorsolateral striatum and prefrontal cortex are critical for habit formation. Lesions of dorsolateral striatum prior to training cause behavior that would otherwise have become habitual to remain goal-directed (Yin, Knowlton, & Balleine, 2004). Similar results were obtained by transiently inactivating dorsolateral striatum only during devaluation sessions (Yin, Knowlton, & Balleine, 2006). Conversely, lesions of dorsomedial striatum force animals from goal-directed to habitual behaviors (Yin, Ostlund, Knowlton, & Balleine, 2005). Similar dissociations have been found in cortex, where lesions or inactivation of infralimbic cortex promote goal-directedness (Coutureau & Killcross, 2003). The combination of task-bracketing activity and lesion/inactivation studies have led to a "dual operator" hypothesis of habit formation, in which infralimbic cortex and dorsolateral striatum cooperate to generate habitual behavior (Smith & Graybiel, 2013a). Interestingly, inactivation of infralimbic cortex after a new habit has replaced an old one caused rats to revert to the original habit, instead of abolishing habitual behavior altogether (Smith, Virkud, Deisseroth, & Graybiel, 2012).

In summary, dorsolateral striatum is critical to the formation and expression of habitual behavior, which may serve the adaptive purpose of freeing valuable cerebral resources. How the brain decides whether to operate in habitual or goal-directed mode remains an area of active research, but it is easy to imagine how dysfunction of such a habit system could result in tics or other compulsive behaviors.

3.2.4 Behavioral sequencing

Related to the concepts of action selection and habit formation is that of motor sequencing. Sequential movements of complex behavioral tasks have been studied in many species and have focused on tasks such as ordered button pressing (Kermadi & Joseph, 1995; Matsumoto, Hanakawa, Maki, Graybiel, & Kimura, 1999), joystick movements (Desmurget & Turner, 2010), and sequential eye movements in monkeys (Kermadi & Joseph, 1995; Lu, Matsuzawa, & Hikosaka, 2002). In rats, lever pressing (Yin, 2010) and grooming behavior (Berridge & Whishaw, 1992) have been commonly used. These studies, along with imaging studies in humans, have implicated the striatum in both the learning and execution of sequential movements. One must be careful, however, not to confuse a general deficit in motor function from a specific deficit in action sequencing.

One approach to assessing the performance of motor sequences independently of their acquisition is to test naturally occurring sequential behaviors, for example, rodent grooming. One grooming pattern, termed the "syntactic chain," has received particular attention due to its highly stereotyped organization (Berridge & Whishaw, 1992). The syntactic chain is made up of at least 25 individual actions that are performed in a stereotypic and predictable order, which can be broken into constituent parts. Rats with striatal, but not cortical or cerebellar lesions, consistently fail to complete syntactic chains, and do not recover this ability over time (Berridge & Whishaw, 1992). Similarly, inactivating caudate/putamen in monkeys caused incorrect button-presses when monkeys attempted to reproduce previously learned sequences (Miyachi, Hikosaka, Miyashita, Karadi, & Rand, 1997).

Dopamine also is implicated in the execution of sequential movements. Monkeys treated with MPTP move slower to complete previously learned motor sequences (Matsumoto et al., 1999), and mice lacking D1 receptors are impaired in completing grooming chains (Cromwell, Berridge, Drago, & Levine, 1998). Patients with PD are also impaired in executing movement sequences, though in this case it is virtually impossible to separate learning from performance effects (Benecke, Rothwell, Dick, Day, & Marsden, 1987). In contrast, hyperdopaminergic mice (a model of OCD reviewed in Section 4) perform more frequent, behaviorally rigid grooming chains (Berridge et al., 2005). Further, cocaine and amphetamine, which both increase catecholamines including dopamine, induce repetitive, stereotypic behaviors in rodents (Wolgin, 2012). This is likely via a D1 receptor-dependent process, as D1 agonists induce stereotypic movements, but D2 agonists do not (Berridge & Aldridge, 2000a, 2000b; Taylor, Rajbhandari, Berridge, & Aldridge, 2010).

The above evidence is suggestive of a role for the striatum, in particular, nigrostriatal dopamine, in the expression of stereotyped movement sequences. If so, the striatum must convey sequence-specific signals to the rest of the brain via BG output nodes (GPi and SNr), which has been investigated in several studies. Silencing GPi with muscimol slows the performance of learned sequences, and this finding is consistent across studies (Turner & Desmurget, 2010). However, this appears to be a property of movement in general after pallidal inactivation, not just sequenced movements (Inase et al., 1996; Wenger et al., 1999). In a key experiment, Desmurget and Turner compared the performance of well-learned and random movement sequences after pallidal inactivation. Both types of sequence were performed slowly, but the overlearned sequence was still performed faster compared to the random sequence. Thus, the monkeys were still able to anticipate the next movement, suggesting that the BG are not required for the execution of learned motor sequences (Desmurget & Turner, 2010).

We have thus far addressed the nuanced role of the striatum in executing motor sequences. Much less controversial is the idea that the striatum is critical for acquiring motor sequences. Some of this evidence comes from the study of songbirds, which learn a highly stereotyped song early in life that ultimately becomes crystallized and relatively immutable. Lesions to the songbird homolog of the BG markedly impair song learning, but have minimal effect on performance of crystallized songs (Andalman & Fee, 2009; Brainard & Doupe, 2013). Lesion studies in mammals reveal similar results. Lesions of dorsolateral striatum impair sequence learning in mice, while lesions of dorsomedial striatum do not (Yin, 2010). Further, functional imaging studies have found enhanced striatal activity as motor sequences are learned (Doyon, 2008; Grafton, Hazeltine, & Ivry, 1995; Lehericy et al., 2005; Rauch et al., 1997; Seidler et al., 2005). Nigrostriatal dopamine specifically seems to play a role in sequence acquisition, as both humans (Frank et al., 2004; Kwak, Bohnen, Muller, Dayalu, & Seidler, 2013) and nonhuman primates (Matsumoto et al., 1999) with striatal dopamine depletion are impaired in the acquisition of new motor sequences. Perhaps most convincing, disruptions of BG output impair sequence acquisition (Brown et al., 2003; Obeso et al., 2009), but, as reviewed above, do not eliminate knowledge of prelearned sequences.

In summary, the BG, in general, and nigrostriatal dopamine specifically, play a central role in the acquisition of movement sequences. While it is clear that these structures also influence the expression of learned sequences, the sequences themselves do not seem to be stored within BG circuits.

4. DISEASE MECHANISMS AND MOUSE MODELS

There are three broad categories of criteria used to assess the relevance of animal models to human disease (Table 3.2). Face validity refers to the similarity between disease features and those exhibited by the animal model. Construct validity refers to similarity in pathophysiological mechanisms, and predictive validity refers to similarity of treatment efficacy (Albelda & Joel, 2012). The study of a model with any one type of validity may lead to new insights into disease mechanism, though the more categories the model matches, the easier it is to extrapolate findings to humans. Early OCD models provide an example in which the study of a model with face validity ultimately led to previously unknown pathophysiological mechanisms (Campbell et al., 1999; Greer & Capecchi, 2002; Powell, Newman, Pendergast, & Lewis, 1999). In the following section, we describe current

Table 3.2 Validity

Disorder	**Validity**	
OCD-spectrum	Face	Ability to induce behaviors resembling compulsions or stereotypies. Repetitive, self-injurious grooming. Perseverative burying
	Construct	Involvement of orbitofrontal cortex and basal ganglia
	Predictive	Response to SSRIs. Response to high-frequency stimulation of subthalamic nucleus or ventromedial striatum
Rett syndrome	Face	Normal early development followed by developmental regression and motor abnormalities
	Construct	Neuropathological abnormalities: smaller, more densely packed neurons
	Predictive	Unknown
Dystonia	Face	Abnormal twisting movements, hyperactivity, action-induced abnormal movements
	Construct	Deficient cortical and basal ganglia inhibition, maladaptive plasticity, sensory abnormalities, targeting genes known to be involved in human disease
	Predictive	Response to trihexyphenidyl or DBS

efforts at modeling OCD-spectrum disorders, RTT, and primary dystonia, highlighting new findings as they relate to human disease.

5. OCD-SPECTRUM DISORDERS AND RODENT MODELS

As outlined in Section 2, dysfunction of a circuit including the orbitofrontal and anterior cingulate cortices, and ventral striatum are demonstrated repeatedly in functional imaging studies of OCD subjects (Graybiel & Rauch, 2000). Dysfunction of the orbitofrontal-BG network is considered the construct validity hallmark for OCD models. Other potential standards for construct validity include changes in neurochemical systems observed in imaging studies (e.g., dopaminergic, glutamatergic, and serotoninergic), though many of the imaging findings between different research groups are not consistent. Deep brain stimulation of the subthalamic nucleus or ventromedial striatum has been demonstrated effective in controlling OCD symptoms (Burdick, Goodman, & Foote, 2009; Greenberg et al., 2010, 2006; Lipsman, Neimat, & Lozano, 2007). Manipulation of these structures, for example, with optogenetic strategies, is therefore a strategy for demonstrating predictive validity, and may enable future investigations of the type of circuit modulation most effective in symptom control. Suppression of OCD-like behaviors by fluoxetine is the most commonly used pharmacological assay for predictive validity, and the specificity of this response is often addressed by showing that the norepinephrine transporter inhibitor desipramine is ineffective. Face validity is more difficult to assess. Behavioral assays include repetitive grooming or leaping, and perseverative object burying, which are not common human OCD behaviors. Measures of anxiety are useful ancillary methods, as anxiety is a common feature of OCD.

5.1. Spontaneous and circuit manipulation models

5.1.1 Optogenetic modeling of OCD

A study employing optogenetic methods provides direct evidence linking hyperactivity in the OCD circuit to persistent, repetitive behaviors (Ahmari et al., 2013). These investigators repeatedly activated ventromedial striatal terminals originating in orbitofrontal cortex and documented progressive increases in repetitive grooming. Acute stimulation of this circuit had no effect, nor did optogenetic stimulation of terminals originating from the motor or prelimbic cortices. The lack of effects of acute stimulation and persistence of repetitive grooming for up to 2 weeks following cessation of stimulation suggest activity in this circuit *per se* is not driving the OCD-like behavior. More durable alterations of synaptic strength or persistent alterations of function in a downstream structure may be involved. Regardless, this study provides strong causal evidence for hyperactivity of orbitofrontal striatal afferents in the generation of OCD-like behaviors.

5.1.2 Spontaneous stereotypy in deer mice

A large percentage of deer mice housed in laboratory colonies develop persistent stereotypies, including repetitive jumping, patterned running, and backward somersaulting. These repetitive behaviors occur spontaneously. Enriched housing conditions reduce the frequency of these behaviors (Powell et al., 1999). These deer mouse-specific stereotypies are distinct from analogous behaviors induced by dopamine agonists and are not exacerbated by treatment with these agents (Presti, Gibney, & Lewis, 2004; Presti, Powell, & Lewis, 2002). Several findings implicate abnormal corticostriatal activity in these behaviors. During repetitive rearing, striatal glutamate concentrations are elevated (Presti, Watson, Kennedy, Yang, & Lewis, 2004), striatal opioid concentrations are altered (Presti & Lewis, 2005), and intrastriatal administration of NMDA and D1 antagonists reduces these behaviors (Presti, Mikes, & Lewis, 2003). Metabolic imaging studies employing cytochrome oxidase histochemistry also implicate BG dysfunction in this model. Deer mice exhibiting high levels of stereotypies show decreased subthalamic nucleus cytochrome oxidase activity (Tanimura, King, Williams, & Lewis, 2011; Tanimura, Vaziri, & Lewis, 2010). Serotoninergic alterations are suggested by reduced density of striatal serotonin transporters (Wolmarans de, Brand, Stein, & Harvey, 2013), and selective serotonin-reuptake inhibitors (SSRIs) suppress these stereotypies (Korff, Stein, & Harvey, 2008). Deer mouse stereotypies exhibit significant face and predictive validity with suggestions of construct validity for OCD. The etiologic trigger causing these behaviors is unknown, however, limiting the utility of this model.

5.1.3 Sequential super-stereotypy in hyperdopaminergic mice

As described in Section 3, rodents exhibit a well-characterized stereotypic chain of grooming behaviors (Berridge, Fentress, & Parr, 1987), and investigators have dissected the neurobiological basis of these stereotypies by analyzing interventions that disrupt (Cromwell & Berridge, 1996; Cromwell et al., 1998) or enhance (Berridge & Aldridge, 2000a, 2000b; Deveney & Waddington, 1997) their expression. Grooming sequences exhibit an ordered chain of distinct movement phases (syntax) that is species specific. An important point is that such analyses not only assess the occurrence of grooming sequences but also quantify the rate of completion of the whole sequence after expression of the initial components. Enhanced sequence completion (super-stereotypy) is the putative homologue of OCD. Such studies strongly support a role for nigrostriatal neurotransmission in sequential behavior (Berridge et al., 2005; Cromwell et al., 1998). Pharmacologic studies indicated that super-stereotypy is driven by activation of D1 but not D2 receptors (Berridge & Aldridge, 2000a, 2000b). Berridge and colleagues (Berridge et al., 2005) examined "hyperdopaminergic" mice genetically modified to express ~10% of normal levels of the DAT, causing ~170% excess extracellular striatal dopamine concentrations (Zhuang et al., 2001) and altered corticostriatal neurotransmission (Wu, Cepeda, Zhuang, & Levine, 2007). These mice exhibit a clear super-stereotypy. This supports a role for nigrostriatal dysfunction in the pathogenesis of OCD-spectrum disorders and provides a potential mechanism underlying the development of overly rigid sequential behaviors. The results of this study are consistent with experiments that disrupt grooming behaviors (Cromwell et al., 1998) and those that induce stereotypies via pharmacological manipulation (Berridge & Aldridge, 2000a, 2000b; Deveney & Waddington, 1997). Considered together, these features mark DAT knockdown mice as a model with strong face validity for OCD-spectrum disorders.

5.1.4 "Genetic neurostimulatory" model

Functional imaging suggests that abnormal activity of corticostriate afferents is linked to OCD-spectrum behaviors. By selectively expressing a cholera toxin subunit enhancing G-protein signaling in cortical inputs to ventral striatum, Campbell and colleagues produced mice that exhibit a variety of repetitive behaviors, including repetitive leaping and nonaggressive biting of littermates. These mice expressed the "neurostimulatory" transgene selectively in layer 2 of piriform cortex, layers 2–3 of somatosensory cortex, and the intercalated nucleus of the amygdala. These cortical regions preferentially innervate ventral striatum and, in turn, the intercalated nucleus regulates outputs to the ventral striatum and prefrontal cortices (Campbell et al., 1999; McGrath, Campbell, & Burton, 1999). In this model, drugs that activate corticostriatal glutamate release exacerbate the OCD-like behaviors. This model demonstrates that chronic potentiation of cortical and limbic neurons, which likely increases excitatory inputs to the striatum, causes repetitive behavior (McGrath, Campbell, Parks, & Burton, 2000).

5.2. Gene-targeted models

5.2.1 Hoxb8 mutant mice

Mice null for the Hoxb8 homeobox transcription factor were the first genetic model of excessive grooming and self-injury (Greer & Capecchi, 2002). Subsequent reports of Hoxb8 inactivation confirmed this striking phenotype (Chen, Tvrdik, et al., 2010; Holstege et al., 2008). These mice spend at least twice as much time grooming, and initiate more syntactic grooming sequences under natural conditions, as well as following induction of grooming via water drops or misting.

Analysis of different lines of Hoxb8-null mice indicates varied mechanisms for the observed behaviors, highlighting the potential pitfalls of relying on face validity alone. The repetitive behaviors and hairless patches observed in an independent line of Hoxb8 mutants generated by Holstege and colleagues (Holstege et al., 2008) differ from those observed in the original line analyzed (Greer & Capecchi, 2002), as they appear to result primarily from a sensory abnormality (i.e., excessive itch). This important distinction derives from the fact that local lidocaine treatment alleviates excessive grooming in one line (Holstege et al., 2008) but has no effect on the original null mutant (Chen, Tvrdik, et al., 2010). The difference between these models is probably related to the different genetic strategies used for Hoxb8 inactivation. The original line (Chen, Tvrdik, et al., 2010; Greer & Capecchi, 2002) contains a nonsense codon in the first exon, while the later mutant was generated with a LacZ reporter knockin into the first exon (Holstege et al., 2008; van den Akker et al., 1999). The LacZ knockin allele may lead to a more complex phenotype by interfering with closely neighboring Hox genes (Chen, Tvrdik, et al., 2010). Both models demonstrate the previously unsuspected importance of Hoxb8 in the function of the adult nervous system. Indeed, the presence of hairless patches in the line that does not exhibit sensory abnormalities (Greer & Capecchi, 2002) suggests that Hoxb8 dysfunction may participate in the pathogenesis of trichotillomania, an OC-spectrum disorder characterized by compulsive pulling out of one's own hair (but does not feature sensory abnormalities such as excessive itch; Franklin, Zagrabbe, & Benavides, 2011).

The behavioral phenotype of Hoxb8-null mice arises from an unexpected source: CNS microglia (Chen, Tvrdik, et al., 2010). While Hoxb8 is expressed widely throughout the adult CNS, including OCD circuit regions, expression is limited to microglia. Definitive proof comes from experiments in which grooming behavior abnormalities are recapitulated by conditionally deleting Hoxb8 from the hematopoietic system. Rescue of behavioral abnormalities in constitutive knockout mice follows transplants of wild-type bone marrow. Conversely, some control mice receiving marrow transplants from Hoxb8 mutant mice develop hairless patches (Chen, Tvrdik, et al., 2010). Chen et al. suggest several possible mechanisms for microglial dysfunction disrupting OCD circuit function, including regulation of developmental neuronal death, cytokine-mediated stimulation or inhibition of neuronal activity, and the stabilization and maintenance of neural networks. Studies to date do not demonstrate directly any abnormalities of the putative frontostriatal OCD circuit. These findings fall within a larger literature linking immune dysfunction to neuropsychiatric syndromes (Hyman, 2010) and illustrate the unexpected and complex mechanisms that may occur in models identified by face validity.

5.2.2 SAPAP3 mutant mice

SAPAP3 (SAP90/PSD95-associated protein 3, or DLGAP3, or GKAP) is a member of the SAPAP family of postsynaptic scaffolding proteins interacting with PSD95 and Shank protein families (Kim et al., 1997; Takeuchi et al., 1997) at the postsynaptic excitatory synapse. These proteins form scaffolding complexes providing key regulation of postsynaptic excitatory signaling via trafficking and targeting of interacting proteins in the postsynaptic membrane. SAPAP proteins are widely expressed in the brain, and SAPAP3 is particularly enriched in the striatum (Welch, Wang, & Feng, 2004).

SAPAP3-null mice exhibit a striking phenotype. These animals are initially indistinguishable from littermate controls, but by 4–6 months develop skin lesions of the head, neck, and snout (Welch et al., 2007) (Fig. 3.2).

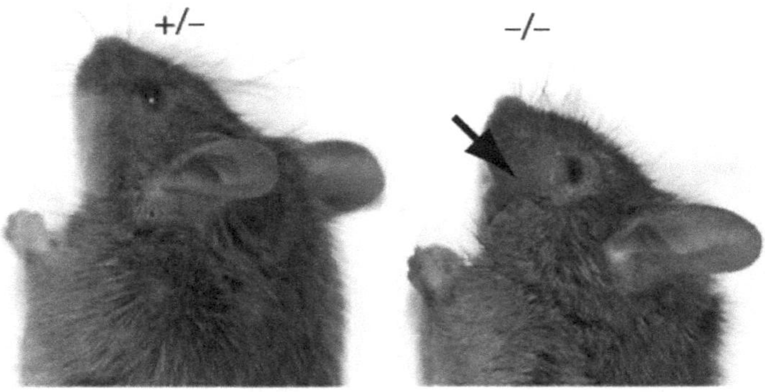

Figure 3.2 SAPAP3 –/– mice exhibit facial lesions due to self-injurious grooming. *Figure reprinted with permission from Macmillan Publishers Ltd.: [Nature] (Welch et al., 2007), copyright 2007.* (See the color plate.)

Increased number and duration of grooming bouts and increased anxiety characterize the 100% penetrant behavioral phenotype. All aspects of this OCD-like syndrome are suppressed by chronic treatment with fluoxetine. Further, pathological grooming was significantly reduced following striatal reexpression of SAPAP3 using lentivirus vectors. Consistent with a critical role for corticostriatal neurotransmission, SAPAP3-null mice exhibit specific reductions in excitatory postsynaptic field potentials and subtle defects in the structure of excitatory synapse postsynaptic complexes (Welch et al., 2007). Loss of SAPAP3 reduces AMPA receptor-mediated corticostriatal excitatory transmission in MSNs by enhancing receptor endocytosis through an mGluR5-dependent mechanism (Wan, Feng, & Calakos, 2011), suggesting that SAPAP3 normally functions to inhibit mGluR5-mediated endocytosis of AMPA receptors (Welch et al., 2007). Importantly, these effects appear relatively specific for the corticostriatal circuit, as thalamostriatal synaptic function remains intact in SAPAP3-null mice, possibly because of the expression of SAPAP4 in thalamostriatal synapses (Wan et al., 2013). These findings further support a role for dysfunctional excitatory corticostriatal neurotransmission in the pathogenesis of OCD.

Work from Xu and colleagues adds another piece to the SAPAP3 puzzle, demonstrating a functional interaction between SAPAP3 and the melanocortin 4 receptor (MC4R) (Xu et al., 2013). MC4R activation causes AMPAR endocytosis, and MC4R may also function in ventral striatal D1-bearing neurons to mediate behavioral responses to stress and procedural learning (Cui et al., 2012; Lim, Huang, Grueter, Rothwell, & Malenka, 2012). Loss of MC4R suppresses the compulsive behavior of SAPAP3 KO mice and, conversely, deletion of SAPAP3 reverses the metabolic abnormalities of MC4R-null mice. These effects were accompanied by normalization of striatal AMPAR-mediated excitatory field potentials. Whereas antagonism of MCR4 function did not affect grooming behavior in wild-type mice, chronic MC4R antagonism reduced grooming time and the number of grooming bouts in SAPAP3 KO mice, establishing a role for this receptor in the mature CNS. By injecting Cre-expressing viruses into SAPAP3-null mice carrying floxed MC4R alleles, Xu and colleagues confirmed that these effects resulted from MC4R function in the orbitofrontal cortical—ventral striatal circuit.

Additional work implicates defects in inhibitory neurotransmission as well, showing that SAPAP3-null mice have significantly fewer parvalbumin-positive FSIs, and lack the learning-related tuning of MSNs that requires FSI function. The loss of FSIs in this mouse model mimics the observed loss of FSI in TS (Kalanithi et al., 2005; Kataoka et al.,

2010). Optogenetic stimulation of orbitofrontal cortical fibers that terminate on FSIs rescues the electrophysiological and behavioral phenotypes, further supporting a role for these inhibitory interneurons (Burguiere, Monteiro, Feng, & Graybiel, 2013).

Considered together, these findings support the face, construct, and predictive validity of SAPA3-null mice as an OCD model. The multiple strengths of SAPAP3 as a model of OCD led human subject investigators to explore a role for SAPAP3 dysfunction as a cause of human OCD. Several preliminary studies have identified sequence variants in SAPAP3 in patients with OCD and related grooming disorders (Bienvenu et al., 2009; Boardman et al., 2011; Crane et al., 2011; Zuchner et al., 2009). Future studies in larger patient populations will be required to determine the contribution of SAPAP3 as a cause or risk factor for human OCD-spectrum disorders.

5.2.3 SLITRK-null mice

Two models of OCD are based on findings of an association of sequence variants in SLITRK1, a member of the SLITRK family of integral membrane proteins. The SLITRK family is so named because of similarities with the SLIT family (extracellular LRR repeats) and Trk tyrosine kinase receptors (intracellular tyrosine kinase domain). SLITRK1 sequence variants are associated with both TS and OCD (Abelson et al., 2005; Karagiannidis et al., 2012; Ozomaro et al., 2013), though some have questioned this association (e.g., Keen-Kim et al., 2006). Although the association of SLITRK variants with TS and OCD is highly controversial, these proteins are rational candidates for neurodevelopmental disease, as they are developmentally regulated, linked to neurite outgrowth (Aruga & Mikoshiba, 2003), and are widely expressed in CNS neurons (Beaubien & Cloutier, 2009; Stillman et al., 2009).

Slitrk1-null mice exhibit significantly increased "anxiety-like behaviors" in the elevated plus maze and open-field tasks, and increased freezing in a fear-conditioning task (Katayama et al., 2010). Norepinephrine levels in prefrontal cortex and norepinephrine metabolite levels in nucleus accumbens are modestly elevated, but the relationship of these findings to the behavioral phenotype is unclear. The administration of clonidine attenuates the mutant phenotype while not affecting the behavior of wild type mice. Clinical reports suggest that clonidine is successful in treating TS (Cohen, Detlor, Young, & Shaywitz, 1980; Cohen, Young, Nathanson, & Shaywitz, 1979; Robertson, 2000) and comorbid ADHD

(Rizzo, Gulisano, Cali, & Curatolo, 2013; Robertson, 2006), but no conclusive clinical trial demonstrates its effectiveness. The strength of this model remains uncertain due to its limited characterization and the absence of stronger evidence of construct and predictive validity.

While no associations with TS or OCD are published for SLITRK5, deletion of this gene in mice generates a model with face, construct, and predictive validity. Slitrk5-null mice appear normal and perform similarly to littermate controls on several motor tasks, but show increased measures of anxiety. At approximately 3 months of age, these mice exhibit pathologically increased grooming and resultant facial lesions reminiscent of those seen in SAPAP3 and Hoxb8-null mice. Heterozygous null mice also develop increased grooming and facial lesions, but approximately 4–6 months later than in the null animals. The repetitive and anxiety-related behavior is nearly extinguished following chronic treatment with fluoxetine. In null mice, FosB immunoreactivity is selectively decreased in the orbitofrontal cortex, indicating a relatively specific functional defect in the OCD circuit. Electrophysiological, morphological, and protein abundance studies of glutamatergic receptors revealed additional abnormalities in this circuit. Considered together, these studies provide strong support for the role of Slitrk5, and perhaps other family members, in the pathogenesis of OCD spectrum symptoms. The construct and predictive validity of this model suggest that it would be fruitful to search for Slitrk5 sequence alterations in patients with OCD and Tourettism.

5.2..4 Histidine decarboxylase-deficient mice

Genome-wide association studies have not identified risk factors that apply broadly to TS (Scharf et al., 2013). A small number of families exist that transmit TS with Mendelian frequency, however, allowing for an alternative strategy for the identification of potential pathways and disease mechanisms. A family carrying the nonsense mutation (W317X) in the HDC gene encoding histidine decarboxylase shows autosomal dominant transmission of TS (Ercan-Sencicek et al., 2010). Histidine decarboxylase is responsible for the synthesis of histamine from histidine (Schwartz, Arrang, Garbarg, Pollard, & Ruat, 1991). This mutation has not been identified outside of this index family, but additional studies identify HDC polymorphisms (Karagiannidis et al., 2013) and implicate histaminergic pathway abnormalities (Fernandez et al., 2012) in TS.

The W317X mutation truncates the HDC protein, abolishing its enzymatic activity (Ercan-Sencicek et al., 2010). While this suggests haploinsufficiency (Fleming & Wang, 2003), enzymatic assays indicate a dominant negative effect of the mutant protein (Ercan-Sencicek et al., 2010). In a recent study, Castellan Baldan et al. (2014) studied Hdc-null mice, which contained undetectable tissue histamine levels. These mice appear normal in open-field testing and do not exhibit motor stereotypies. Following a single dose of amphetamine, however, Hdc-null mice develop a significantly greater number of motor stereotypies, such as repetitive focused sniffing and orofacial movements. Hdc heterozygotes developed similarly abnormal stereotypies when administered a higher dose of amphetamine. Intracerebroventricular histamine pretreatment fully eliminated amphetamine-induced stereotypies in Hdc-null mice, confirming a direct role for histamine deficiency in the behavioral abnormalities. Similar to patients carrying the HDC W317X mutation, Hdc-null mice showed reduced auditory prepulse inhibition. No abnormalities of anxiety or excessive grooming were observed in these mice, though syntactic grooming was not examined.

Several experiments suggest that the behavioral effects of histamine deficiency act via altered dopamine function. Histamine infusion into Hdc-null mice significantly decreased striatal dopamine levels, consistent with a previously hypothesized negative regulation of dopamine by histamine (Schlicker, Malinowska, Kathmann, & Gothert, 1994). Fos staining of amphetamine-treated animals showed a pattern of abnormality consistent with enhanced dopamine signaling, and receptor-binding studies demonstrated dysregulation of D2 and D3 receptors. These findings are similar to those seen in PET studies of patients carrying the HDC W317X mutation, using a high-affinity D2 and D3 receptor agonist tracer (11C-PHNO). Amphetamine-induced stereotypies were attenuated by pretreatment with dopamine D2 antagonists. While this implies predictive validity for TS, D2 antagonists reduce total movement, which may confound these results.

Considering the recent associations of the histamine pathway with TS (Ercan-Sencicek et al., 2010; Fernandez et al., 2012; Karagiannidis et al., 2013), the Hdc mutant mouse is an attractive model of TS. While overt stereotypies are not observed, altered prepulse inhibition and psychostimulant-induced repetitive movements represent a moderate level of face validity (Castellan Baldan et al., 2014). Importantly, these mice contain a high degree of construct and potential predictive validity, supporting their relevance to TS. These findings warrant future studies of histaminergic alterations in a larger population of TS patients.

Despite the high heritability of TS and the large number of genetic studies of OC-spectrum disorders, there is limited success in reproducibly finding disease-related genes. This highlights the complex nature of OC-spectrum genetics. Future analysis of animal models with high face, construct, and predictive validity may shed light on important disease genes of interest. Like the present models employing optogenetic targeting, spontaneous behaviors, or gene-targeted approaches, keen -observation of repetitive behaviors and systems-level pathologies will be important. A potential translation from animal models to human disorders in the context of BG dysfunction may allow for further understanding of these complex disorders.

6. RETT SYNDROME AND MOUSE MODELS

RTT is a devastating neurobehavioral syndrome of early childhood. As described in Section 2, the illness affects girls essentially exclusively who, among other symptoms, develop characteristic stereotypic hand-wringing movements. The causative protein, MeCP2 (Amir et al., 1999; Van den Veyver & Zoghbi, 2000; Wan et al., 1999) was little studied but known to be a transcriptional repressor prior to its connection to RTT (Jones et al., 1998; Lewis et al., 1992; Nan et al., 1998). As reviewed below, our understanding of MeCP2 function has vastly increased since the initial reports linking it to RTT.

6.1. Loss of MeCP2 function in neurons recapitulates the Rett syndrome phenotype

A mouse genetic study prior to these discoveries (Tate, Skarnes, & Bird, 1996) reported that Mecp2-null chimeric mice exhibited developmental defects and embryonic lethality, concluding that MeCP2 was essential for embryonic development. Subsequent investigators succeeded in generating viable Mecp2-null mice, leading to the suggestion that the original strategy also disrupted genes surrounding Mecp2 (Guy, Hendrich, Holmes, Martin, & Bird, 2001). The initial two groups reporting viable MeCP2-null mice recovered them in Mendelian frequencies (Chen, Akbarian, Tudor, & Jaenisch, 2001; Guy et al., 2001). These mutants do not differ from their littermate controls at birth, but develop progressive neurological deterioration leading to death by 7–10 weeks of age. Behavioral abnormalities include stiff, uncoordinated gait, reduced spontaneous movement, and both hindlimb and forelimb clasping. No histopathological abnormalities were identified in the brain or other organs. Both groups also demonstrated that conditional deletion of MeCP2 from the CNS caused a phenotype indistinguishable from the germline mutants, establishing distinctive CNS requirements for MeCP2 function.

Both groups also observed that heterozygous null female mice—a genocopy of the human disease—display a milder behavioral phenotype and histopathological abnormalities. These mice appeared normal until 3–4 months of age, after which they became hypoactive and ataxic (Chen et al., 2001; Guy et al., 2001), developing respiratory abnormalities similar to those observed in the human disease (Guy et al., 2001). Chen and colleagues found the behavioral abnormalities were accompanied by subtle histochemical abnormalities, including smaller, more densely packed neurons in the cortex, cerebellum, and hippocampus. Female mice with heterozygous deletions confined to the CNS (using Nestin-Cre) displayed a similar phenotype, in agreement with the complete loss-of-function studies. Nestin-Cre causes deletion from neurons and glia, so Chen and colleagues further dissected the responsible cell type by using CamKII-Cre to delete MeCP2 selectively from postmitotic neurons. These mice developed behavioral and histopathological abnormalities similar to heterozygous null females. These findings established MeCP2 mutant mice as a robust model of RTT with construct and face validity, and identified neuronal cell autonomous dysfunction of MeCP2 as key to disease pathogenesis. The unique neuronal requirement for MeCP2 was surprising, considering its ubiquitous expression and previous work implicating its function as a general regulator of transcription.

A subsequent RTT model was developed by mimicking a human RTT mutation that results in a truncated MeCP2 protein (Shahbazian et al., 2002). These mice appear normal up until 6 weeks of age, when they develop progressive motor abnormalities similar to those described in previous studies (Chen et al., 2001; Guy et al., 2001). They also developed increased anxiety (McGill et al., 2006). In contrast to germline MeCP2-null mice, these animals typically live to at least 1 year of age, and the truncated protein localizes to heterochromatin domains normally, suggesting it retains some function (Shahbazian et al., 2002).

Mutations in MECP2 are linked to several other neurological diseases, including MECP2 duplication syndrome, severe neonatal encephalopathy, and PPM-X (X-linked syndrome of psychosis, pyramidal signs, and macroorchidism) syndrome (Gonzales & LaSalle, 2010; Villard, 2007). Of these, MECP2 duplication syndrome, characterized by seizures, intellectual disability, and developmental regression, has been modeled in mice (Collins et al., 2004). In this model, approximately twofold MECP2 overexpression initially improved performance in motor and contextual learning paradigms and enhanced synaptic plasticity. By 5–6 months of age, however, these animals became hypoactive and developed reduced seizure threshold and subtle histopathological abnormalities (Bodda et al., 2013; Jiang et al., 2013; Marshak, Meynard, De Vries, Kidane, & Cohen-Cory, 2012; McGill et al., 2006; Na et al., 2012). This work emphasizes that neurons are sensitive to even modest changes in MeCP2 function and is consistent with the distinct neurological syndromes linked to various MECP2 mutations.

The facts that RTT mice initially appear normal, and that an RTT syndrome can be modeled by MeCP2 deletion in postnatal neurons, argues against the notion that initial brain formation is abnormal in RTT. Rather, this work pointed to a role for MeCP2 in maturing neurons, and potentially in adult neurons—a possibility with important implications for potential treatment strategies. To explore this issue, several investigators examined the ability of MeCP2 reexpression at various ages to rescue RTT model phenotypes, and the effects of inducing MeCP2 loss-of-function in the mature CNS (Cheval et al., 2012; McGraw, Samaco, & Zoghbi, 2011; Nguyen et al., 2012). Reexpression of MeCP2 in early postnatal or adult RTT mice rescues behavioral (Giacometti, Luikenhuis, Beard, & Jaenisch, 2007; Guy, Gan, Selfridge, Cobb, & Bird, 2007) and electrophysiological (Guy et al., 2007) phenotypes of MeCP2-null mice, even when reexpression was induced following phenotype onset (Guy et al., 2007). This work strongly implied an ongoing role for MeCP2 in adult neurons. Several groups have confirmed the importance of MeCP2 in mature neuronal function, finding that reducing MeCP2 function in mature mice recapitulates the core RTT phenotypes seen following germline or early conditional deletion (Cheval et al., 2012; McGraw et al., 2011; Nguyen et al., 2012). Surprisingly, this requirement for MeCP2 was present even when deleted at 30–45 weeks of age, when a ~80% reduction was lethal (Cheval et al., 2012). Adult MeCP2 deletion causes brain shrinkage and increased neuronal density, characteristics of human RTT (Nguyen et al., 2012). Brain shrinkage is accompanied by dramatic morphological changes in the hippocampus, including severe retraction and reductions in dendritic spines of pyramidal neurons, and less complex astrocyte morphology. While mRNA levels for a number of inhibitory and excitatory synaptic proteins are normal in these mutants, the proteins themselves were significantly decreased. This led to the suggestion that MeCP2 loss-of-function disrupts posttranslational regulation of key synaptic components. A critical implication of these findings is that treatment of RTT will likely require lifetime therapy.

6.2. Gene Expression and electrophysiological abnormalities in RTT: Insights from mouse models

Considerable effort is focused on using the RTT models described above to identify key pathophysiological events of potential value for RTT therapy. Given the role of MeCP2 as a transcriptional repressor (Jones et al., 1998; Lewis et al., 1992; Nan et al., 1998), these studies have focused on identifying genes dysregulated by MeCP2 dysfunction, and the electrophysiological consequences of these changes.

Global transcriptional profiling of forebrains from asymptomatic, early symptomatic, or late symptomatic MeCP2-null mice does not show robust changes in gene expression, even in mice with overt phenotypes (Tudor, Akbarian, Chen, & Jaenisch, 2002). This lack of differential gene expression contrasts with postmortem studies of brains from RTT patients, which show striking changes in expression patterns, including increased levels of glial transcripts and decreased levels of presynaptic neuron-specific mRNAs (Colantuoni et al., 2001). These differences may reflect the fact that human postmortem studies of brains from RTT subject show gliosis and synapse loss, features not observed in mouse models. While it is possible that MeCP2-null mice have gene expression changes below the limits of detection, Tudor et al. suggest several possible biological explanations for their negative findings, including compensation by other methyl-binding proteins, transcriptional dysregulation in a small subset of cells, or the presence of only very subtle changes to expression levels in a highly vulnerable population of cells. These authors also raised the possibility that MeCP2 does not act solely as a transcriptional regulator, and that the observed changes in gene expression are secondary to some other function of MeCP2 (Tudor et al., 2002).

Other investigators identified specific transcriptional abnormalities in Mecp2 mutant mice. Both targeted and broad analyses have been performed. The gene encoding Uqcrc1, for example, a component of the mitochondrial respiratory chain, was over expressed in MeCP2-null mice during the period that abnormal behaviors developed (Kriaucionis et al., 2006). Because subtle region-specific changes may be masked in whole-tissue analysis, some groups have performed gene expression analysis in specific brain regions. Thousands of altered genes were identified in the hypothalamus or cerebellum of mice lacking or overexpressing MeCP2 (Ben-Shachar,

Chahrour, Thaller, Shaw, & Zoghbi, 2009; Chahrour et al., 2008). Approximately 85% of these genes were activated rather than repressed by MeCP2. This study highlights the heterogeneity of MeCP2 actions in different brain regions. In the hypothalamus and cerebellum, genes activated and repressed by MeCP2 fell into several functional categories, with enrichment for genes encoding neuropeptides and G protein-coupled receptors (Ben-Shachar et al., 2009; Chahrour et al., 2008).

Following the development of behavioral abnormalities, striata of MeCP2-null mice show reduced levels of genes encoding protein products involved in ion and chemical homeostasis, and increased levels of genes encoding proteins functioning in transcriptional regulation, development, and differentiation (Zhao, Goffin, Johnson, & Zhou, 2013). Bioinformatic analysis of differentially expressed genes in striatum, hippocampus, and cerebellum suggest a heterogeneous role for MeCP2 in these tissues (Zhao, Goffin, et al., 2013). This phenomenon is not specific to MeCP2 deletion during development, as alterations to gene expression also occur following deletion in adulthood (McGraw et al., 2011).

In different tissues, findings implicating alterations in both excitatory and inhibitory neurotransmission are described, but the specific transcriptional changes responsible for these effects are elusive. GABAergic neurotransmission, for example, was differentially affected in two different thalamic nuclei in MeCP2-null mice (Zhang, Zak, & Liu, 2010). MeCP2 appears to be enriched in cortical GABAergic compared to non-GABAergic neurons (Chao et al., 2010), and selective dysfunction of GABAergic neurotransmission is the most commonly reported electrophysiological abnormality in MeCP2-null mice. Decreases in GABA (but not glycine) release, reductions in the vesicular inhibitory transporter, and alterations in postsynaptic GABAergic receptors are found in the ventrolateral medulla of MeCP2-null mice (Medrihan et al., 2008). IPSP-based rhythmic activity is reduced in hippocampal slices from these animals (Zhang, He, Jugloff, & Eubanks, 2008). Conditional deletion of MeCP2 from all CNS GABAergic and glycine neurons (using a newly generated vesicular inhibitory amino acid transporter "Viaat"-Cre line) provided a particularly compelling link between disrupted inhibitory neurotransmission and phenotypes characteristic of germline MeCP2-null mice (Chao et al., 2010). Similar to the phenotype of germline MeCP2-null mice, deletion of MeCP2 with Viaat-Cre produced initially normal appearing mice that at ~5 weeks developed progressive motor dysfunction, repetitive stereotypies, and increased grooming and consequent skin lesions. These animals exhibited clear abnormalities in

GABAergic neurotransmission, including decreased levels of GAD1, GAD2, and GABA, accompanied by a decrease in GABA release and network hyperexcitability (Chao et al., 2010).

The repetitive stereotypies in "Viaat-Cre" mice are of particular interest, as compulsive-like repetitive behaviors and stereotypic hand wringing occur in RTT subjects. This aspect of the Viaat-cre/MeCP2 line indicated an overlap with OCD-spectrum mouse models and, along with motor dysfunction phenotypes, suggested that dysfunction of forebrain structures including ventral striatum could underlie the behaviors. Chao and colleagues tested this possibility by conditionally deleting MeCP2 with Dlx5/6-Cre mice that causes recombination in all forebrain GABAergic and cholinergic neurons. Consistent with a role in the OCD circuit, Dlx5/6 conditional MeCP2-null mice showed repetitive behaviors, impaired motor coordination, and increased prepulse inhibition.

While the effects of MeCP2 loss-of-function appear to predominantly affect GABAergic neurotransmission, disruption of glutamatergic signaling has also been observed. In MeCP2-null cortex, reductions of glutamatergic neurotransmission were identified, including reduced spontaneous firing of layer 5 somatosensory pyramidal neurons (Dani et al., 2005) and decreased numbers of hippocampal glutamatergic synapses (Chao, Zoghbi, & Rosenmund, 2007). The findings in cortex reflected a reduced excitatory and increased inhibitory total drive onto glutamatergic neurons, rather than changes in intrinsic excitability. The effects on glutamatergic synapse formation were bidirectional, with an increase in synapse number in MeCP2 overexpressing mice, consistent with the clinical consequences of MECP2 gene duplication.

7. PRIMARY DYSTONIA

Modeling primary dystonia is challenging, because manipulation of genes that cause the disease have not, in general, produced models with overt dystonic movements. Further complicating this situation, dystonia-like twisting may be caused by neural insults not thought relevant to causes of human dystonia. For example, abnormal twisting movements can be caused by disrupting the vestibular nuclei (Burke & Fahn, 1983), structures not believed to participate in the pathophysiology of human dystonia. While "behavioral" models of dystonia have face validity and may lead to novel insights into motor circuitry, particular care must be taken in interpreting the findings from models not based upon causes of human dystonia (i.e., models lacking construct validity).

Several models of secondary dystonia exhibit prolonged abnormal twisting movements consistent with human dystonias (Burke & Karanas, 1990). Thus far, however, none of the mouse models based on the manipulation of primary dystonia-causing genes exhibit dystonic-like movements. Instead, such models (most commonly Tor1a) have typically resulted in relatively subtle abnormalities on motor tasks (e.g., beam walking or rotarod) (Dang, Yokoi, Pence, & Li, 2006; Sciamanna, Hollis, et al., 2012; Yokoi, Dang, Li, Standaert, & Li, 2011; Yokoi, Dang, Mitsui, Li, & Li, 2008). In contrast, a recent rat TOR1A transgenic model is reported to exhibit hindlimb twisting during the tail-suspension test (Grundmann et al., 2012). Another feature suggested to establish face validity is the abnormal co-contraction of agonist and antagonist muscles that characterize dystonia. However, this measure is not specific for dystonic movements (Tanabe, Kim, Alagem, & Dauer, 2009).

Clinical electrophysiological studies of dystonic patients highlight several features potentially valuable for assessing the face and construct validity of dystonia models, including deficient intracortical inhibition, maladaptive cortical plasticity, and abnormal sensory function. Hallett presented an attractive model in which subcortical (e.g., BG) abnormalities result in sensorimotor cortex dysfunction that is the proximate cause of dystonic movements. Deficient inhibitory function is arguably the most consistently identified pathophysiological phenotype in dystonia (Hallett, 2011). This abnormality has been identified in genetic forms of childhood dystonia as well as idiopathic adult-onset disease. It is, however, not limited to brain regions subserving affected body parts, so appears to be a general phenomenon of the dystonic CNS that is perhaps necessary but not sufficient to produce dystonia (Berardelli, Rothwell, Day, & Marsden, 1985; Berardelli et al., 1998; Hallett, 2011; Nakashima et al., 1989; Panizza, Lelli, Nilsson, & Hallett, 1990; Tisch et al., 2006). Abnormal synaptic plasticity is another pathophysiological feature observed in a range of subjects with primary dystonia; this has been proposed as a key pathophysiological mechanism in dystonia (Edwards, Huang, Mir, Rothwell, & Bhatia, 2006; Peterson, Sejnowski, & Poizner, 2010; Quartarone, Classen, Morgante, Rosenkranz, & Hallett, 2009; Quartarone & Pisani, 2011; Quartarone, Siebner, & Rothwell, 2006). The occurrence of "task-specific" dystonias that arise in body parts of subjects used to perform highly trained stereotyped tasks, such as the fingers of musicians or telegraph operators, or the vocal cords of auctioneers, provides conceptual support for the notion that dystonia may represent maladaptive plasticity (Ferguson, 1971; Frucht, 2004; Scolding, Smith, Sturman, Brookes, & Lees, 1995). Indeed, impaired plasticity at striatal synapses has been a strong feature of several dystonia animal models (Bonsi et al., 2008; Martella et al., 2009; Peterson et al., 2010; Sciamanna et al., 2012). Dystonia patients show abnormalities of sensory processing and the organization of the somatosensory cortex is disrupted in the dystonic brain. (Bara-Jimenez, Catalan, Hallett, & Gerloff, 1998; Bara-Jimenez et al., 1998; Scontrini et al., 2009; Quartarone & Hallett, 2013). The demonstration of related sensory abnormalities in models of dystonia would therefore also serve as a measure of construct validity.

An emerging neuroimaging literature documenting relatively subtle structural abnormalities of the dystonic brain is challenging traditional views that primary dystonias lack structural pathologies and provides another opportunity to identify phenotypes shared by human subjects and mouse models (Ramdhani & Simonyan, 2013). This work includes voxel-based morphometry studies documenting differences in the size of various motor circuit components and diffusion tensor imaging pointing to microstructural abnormalities in key motor tracts. Neuroimaging has been used to provide support for the construct validity of a mouse model of DYT1 dystonia, as detailed below.

The most commonly used measure of predictive validity is the response of various phenotypes to the acute administration of the anticholinergic compound trihexyphenidyl. This is based on the fact that muscarinic anticholinergic medications are the most commonly used pharmacological treatment for primary dystonia (Burke & Fahn, 1985a, 1985b; Burke, Fahn, & Marsden, 1986; Fahn, 1983). Anticholinergic therapy is most practical and effective in younger patients, who can tolerate higher doses (Burke et al., 1986) and can be treated quickly after symptom onset (Greene, Shale, & Fahn, 1988). While undoubtedly effective in some patients, open trials and double-blind studies indicate that a substantial proportion of patients do not obtain long-term benefit, and dose-limiting side effects are common. Anticholinergic treatment was not beneficial in nearly 40% of subjects with childhood-onset dystonia, and only a minority of adult subjects experience long-term benefit from these medications (Burke & Fahn, 1985b; Fahn, 1983). Cognitive side effects such as memory loss and confusion are common in older adults (Burke et al., 1986; Taylor, Lang, Saint-Cyr, Riley, & Ranawaya, 1991). The use of anticholinergics as the "gold standard" of predictive validity for dystonia models may therefore be misleading. Other medications used to treat dystonia, including baclofen

($GABA_B$ receptor agonist) and benzodiazepines (enhance $GABA_A$-mediated neurotransmission), may also be valuable for predictive validity studies. However, the low efficacy of these compounds precludes their use as a sole measure of predictive validity.

Whatever pharmacological strategy is employed, it is important to recall that chronic administration of medication, and even deep brain stimulation, is typically required for symptomatic improvement of dystonia. For example, weeks are typically required to observe benefit following DBS (Krauss, 2002) and benefits may continue to accrue over months (Vercueil, Krack, & Pollak, 2002; Vidailhet et al., 2005). This pattern contrasts sharply to that seen in PD and essential tremor, which respond over seconds to days to deep brain stimulation (Wichmann & Delong, 2006). These observations underscore the importance of not relying on acute responses to treatment interventions when assessing the predictive validity of dystonia models.

7.1. Early "phenotypic" models

Prior to the discovery of mutations that cause primary dystonia in humans, investigators relied on mutants with abnormal twisting movements that arose spontaneously in rodent colonies. Perhaps, the earliest example is the dystonia musculorum mouse (Duchen et al., 1964). This model is of uncertain significance to primary dystonia, because it exhibits progressive peripheral axonal neuropathy including nerve fiber loss in sensory ganglia, and degenerative changes in the spinal cord and brainstem (Duchen et al., 1964). The most notable spontaneous models that, like primary dystonia, do not develop overt histopathology, are the dystonic (dt) rat (Lorden, McKeon, Baker, Cox, & Walkley, 1984), dystonic dt/sz hamster (Loscher et al., 1989), and tottering mice (Neychev, Fan, Mitev, Hess, & Jinnah, 2008). These models, recently reviewed elsewhere (Wilson & Hess, 2013), provide a wealth of information on motor circuit function, and, in the case of the dystonic rat and tottering mice, highlight the role of hindbrain structures in abnormal twisting movements. The mutation responsible for the dystonic hamster is unknown, while the dystonic rat is caused by loss-of-function of the atcay gene. In humans, mutations in this gene cause "Cayman" ataxia, a syndrome of apparently static cerebellar ataxia and psychomotor retardation that may include dystonia (Bomar et al., 2003; Xiao & Ledoux, 2005). The apparently complex phenotype of Cayman ataxia and the paucity of clinical description of the Cayman ataxia phenotype make it difficult to determine which features of the dystonic rat and dystonic mice are relevant to primary dystonia. This highlights the challenge of studying "phenotypic" models of dystonia identified from face validity alone. The remainder of this section will therefore focus on models based on genes that cause human primary dystonia.

7.2. DYT1 dystonia

DYT1 dystonia was the first form of primary dystonia for which a causative gene was identified. This disease is caused by a dominantly inherited mutation in the TOR1A gene encoding the protein torsinA (Ozelius et al., 1997). All patients have the same mutation: an in-frame 3 base-pair deletion that removes a single glutamic acid (hence, "ΔE"). TorsinA is a member of the AAA+ family of proteins (ATPases associated with a variety of cellular activities). These chaperone-like proteins typically function as oligomers and use ATP hydrolysis to disassemble protein complexes and alter protein conformations (Neuwald, Aravind, Spouge, & Koonin, 1999). TorsinA is localized to the endoplasmic reticular/nuclear membrane compartment (Kustedjo, Bracey, & Cravatt, 2000). While the function of torsinA in this compartment is not fully understood, several findings support its involvement in protein quality control. Biochemical studies show that the ΔE mutation impairs the ATPase activity of torsinA by inhibiting binding to proteins required for its enzymatic activity (LAP1 and LULL1) (Zhao, Brown, Chase, Eisele, & Schlieker, 2013). One consequence of inhibiting torsinA function is to enhance its concentration in the nuclear membrane where it interacts with several nuclear membrane-localized proteins (Goodchild & Dauer, 2004, 2005). There are numerous examples of cell biological effects observed when overexpressing wild-type torsinA that are reduced or abolished by the ΔE mutation (e.g., ;Caldwell et al., 2003 Chen, Burdette, et al., 2010; Hewett et al., 2007; Torres, Sweeney, Beaulieu, Shashidharan, & Caron, 2004). Among these is inhibited trafficking of polytopic membrane-bound proteins, including the DAT (Torres et al., 2004). Protein processing through the secretory pathway is defective in fibroblasts derived from subjects with DYT1 dystonia and torsinA-null mice (Hewett et al., 2007). TorsinA was reported to participate in the protein "quality control" functions of the ER (Nery et al., 2011) and to interact with synaptic vesicle membrane proteins (Granata, Watson, Collinson, Schiavo, & Warner, 2008; Kakazu, Koh, Ho, Gonzalez-Alegre, & Harata, 2012).

7.2.1 Transgenic models

Several transgenic rodent models of DYT1 dystonia were generated by broadly overexpressing DYT1 mutant human torsinA (Grundmann et al., 2012, 2007; Sharma et al., 2005; Shashidharan et al., 2005). Of these models, two are reported to display overt twisting behavior during tail suspension (Grundmann et al., 2012; Sharma et al., 2005), whereas others exhibit statistical differences in measures of beam walking, pawprint analysis, and motor learning on the rotarod (Grundmann et al., 2007; Sharma et al., 2005). Expressing DYT1 mutant human torsinA selectively in dopaminergic neurons recapitulates statistically significant abnormalities in beam walking seen in other models, but does not cause dystonic-like twisting movements (Page et al., 2010). The reasons for the discrepancies in the behavioral phenotypes of the broadly expressing models are not clear, but may result—at least in part—from differences in the pattern, expression levels, and insertion sites of the transgenes. For example, Sharma and colleagues did not generate independent lines of transgenic mice overexpressing levels of the DYT1 transgene similar to the hMT1 line (Sharma et al., 2005), as is typical in the generation of transgenic models. This raises the possibility that some features of this widely studied line (Balcioglu et al., 2007; Hewett, Johanson, Sharma, Standaert, & Balcioglu, 2010; Napolitano et al., 2010; Pisani et al., 2006; Sciamanna et al., 2009, 2011; Sharma et al., 2005; Zhao, DeCuypere, & LeDoux, 2008) may result from insertional effects of the transgene, rather than the actions of DYT1 mutant torsinA protein.

Most studies of these transgenic models focused on dopaminergic and cholinergic signaling, the disruption of which are implicated in other forms of dystonia (e.g., dopa-responsive dystonia, tardive dystonias, and dystonic dyskinesias in PD) and by the response of dystonic symptoms to anticholinergic medication. Studies assessing baseline levels and evoked release of dopamine report conflicting results. Some find abnormal basal levels of dopamine but no changes to transporter or receptor activity (Grundmann et al., 2007; Shashidharan et al., 2005), while others find normal levels of dopamine at baseline, but abnormalities following dopamine-modulating drugs or differences in metabolite turnover (Balcioglu et al., 2007; Hewett et al., 2010; Page et al., 2010; Zhao et al., 2008). The study by Page and colleagues is notable for utilizing fast-scan cyclic voltammetry, a method capable of distinguishing neurotransmitter release and subsequent uptake. These authors found deficient dopamine release, whereas DAT-mediated dopamine uptake was unaffected. Because this model overexpresses

DYT1 mutant torsinA selectively in dopaminergic neurons, it is unclear whether the ability to disrupt neurotransmitter release is specific to these cells, or whether torsinA can similarly affect other neuronal classes.

A series of studies by Pisani and colleagues, primarily utilizing the hMT1 line of DYT1 transgenic mice (Sharma et al., 2005) implicates abnormal function of striatal cholinergic interneurons in the pathogenesis of primary dystonia. These investigators demonstrate a paradoxical effect of D2 receptor stimulation on cholinergic interneuron firing, causing these cells to increase their firing rate (Pisani et al., 2006). In contrast, D2 receptor stimulation in control or transgenic mice overexpressing wild-type human torsinA does not cause detectable changes in cholinergic interneuron firing. Subsequent work on this model (Napolitano et al., 2010) pointed to postsynaptic D2 receptors as the site of this phenomenon. Conditionally deleting torsinA in cholinergic neurons recapitulated this paradoxical D2 effect, suggesting that torsinA dysfunction disrupts D2 receptor function on cholinergic interneurons in a cell autonomous fashion (Sciamanna, Tassone, et al., 2012). While the exact mechanism of D2 receptor dysregulation is unclear, abnormal interactions with N-type calcium channels and impaired coupling with RGS9 was suggested (Pisani et al., 2006). Additional work suggests that the D2 receptor dysfunction is responsible for the deficient "pause" response of cholinergic interneurons in hMT1 mice; this deficient pause is proposed to cause an imbalance in muscarinic M1 and M2 receptor stimulation that disrupts corticostriatal neurotransmission (Martella et al., 2009; Sciamanna, Tassone, et al., 2012). Similar abnormalities in corticostriatal neurotransmission are found in the transgenic rat model of DYT1 dystonia (Grundmann et al., 2012). Features of disrupted corticostriatal neurotransmission, including impaired long-term depression and synaptic depotentiation, are rescued by anticholinergic treatment. However, the relationship of these abnormalities to dystonic movements remains unclear, as they are present long before behavioral abnormalities appear (Sciamanna et al., 2011), and are present in rodents with and without abnormal twisting movements (Grundmann et al., 2012; Sciamanna, Tassone, et al., 2012; Sharma et al., 2005).

7.2.2 Gene-targeted *Tor1a* mutant mice

Two groups (Dang et al., 2006; Goodchild, Kim, & Dauer, 2005) used gene targeting to inactivate or "knockin" the DYT1 mutation into the Tor1a locus, and report similar effects. Both groups found that torsinA null and homozygous knockin mice are born in Mendelian ratios and appear morphologically normal, but die for unclear reasons during the first postnatal day. Goodchild and colleagues (Goodchild et al., 2005) identified morphological abnormalities in both torsinA null and homozygous knockin mice. They also observed decreased steady-state levels of mutant torsinA protein in the knockin mice, as well as skin fibroblasts derived from DYT1 subjects. Additional findings in torsinA-null mice indicate that neurons have a unique requirement for torsinA function. The nuclear membrane abnormalities observed in knockout and homozygous knockin lines occur exclusively in neurons, even though torsinA is expressed widely. In addition, these nuclear membrane abnormalities emerge in postmigratory maturing neurons, pointing to a specific role for torsinA in neuronal maturation. These findings are evocative of the neurodevelopmental signature of DYT1 dystonia. Considered together, these observations demonstrate that the DYT1 mutation impairs torsinA function, and highlight the importance of torsinA loss-of-function studies in advancing understanding the pathogenesis of DYT1 dystonia.

In contrast to the above findings, heterozygous mice that carry one DYT1 allele (a genocopy of human DYT1 subjects) are normally viable and do not exhibit overt motor abnormalities (Dang et al., 2005; Goodchild et al., 2005; Tanabe, Martin, & Dauer, 2012). These mice show DTI abnormalities similar to those seen in nonmanifesting DYT1 carriers, suggesting that they model the "nonmanifesting" carrier state (Ulug et al., 2011). Nonmanifesting DYT1 subjects exhibit metabolic and psychophysical abnormalities (Carbon et al., 2011, 2008; Carbon, Trost, Ghilardi, & Eidelberg, 2004; Ghilardi et al., 2003), which may in part be caused by the DTI observable microstructural and PET observable metabolic changes seen in humans and in these mice. Potential neuropathological and electrophysiological substrates of these findings are described in DYT1 heterozygous mice. These include subtle changes in cell size and synaptic structure in the striatum and cerebellum (Song et al., 2013), as well as abnormalities in synaptic vesicle recycling and calcium homeostasis (Iwabuchi, Kakazu, Koh, & Harata, 2013; Kakazu, Koh, Ho, et al., 2012; Kakazu, Koh, Iwabuchi, Gonzalez-Alegre, & Harata, 2012).

The early lethality of germline null mutants led investigators to conditionally delete torsinA in brain regions and cell types implicated in the disease. This work has focused on conditional deletion in cortex, striatum, and cerebellum (Sciamanna, Hollis, et al., 2012; Yokoi et al., 2011, 2008; Zhang et al., 2011). None of these models exhibit overt twisting movements. Nevertheless, conditional deletion of torsinA did alter motor function and produce cellular abnormalities. Deletion from cholinergic neurons (using ChAT-Cre) caused the D2-related electrophysiological abnormalities described above (Sciamanna, Hollis, et al., 2012). Conditional deletion from cortical glutamatergic neurons (using Emx1-Cre) caused hyperactivity and altered performance on beam walking, but did not disrupt dopamine metabolism or cortical barrel field patterning (Yokoi et al., 2008). Striatal specific torsinA-null mice (using Rgs9-Cre) also showed beam-walking deficits but were normal in open-field testing (Yokoi et al., 2011). These animals had reductions in striatal D2 receptors but did not exhibit abnormalities in dopamine metabolism. Subtle abnormalities of dendritic morphology were seen in cerebellar Purkinje cells following conditional deletion using Pcp2-Cre, but neither behavioral nor neurochemical studies were performed on these mutants (Zhang et al., 2011). The lack of overt phenotype in these conditional mutants is consistent with the possibility that DYT1 dystonia is a disorder requiring the simultaneous disruption of multiple motor circuit nodes for phenotypic expression. This notion has been emphasized previously (Tanabe et al., 2009) and is supported by the occurrence of dystonia following lesions to several different brain regions (Neychev et al., 2008; Prudente, Hess, & Jinnah, 2013). DYT1 dystonia is a neurodevelopmental disease (Tanabe et al., 2009), so torsinA deletion in these models may have occurred after the critical period for torsinA function has passed.

8. CONCLUDING REMARKS

Assisted by progress in identifying disease-related mutations and their phenotypic features, the expanding power of targeted mutations in rodents, and more sophisticated understanding of BG function, there has been remarkable progress in animal modeling of BG neurodevelopmental diseases in recent years. A better understanding of BG circuits and functions has allowed for the development of a large variety of mouse models, including spontaneously arising, transgenic, gene-targeted, and circuit manipulation models. Each of these approaches exhibits different levels of face, construct, and predictive validity, and provides unique benefits for understanding striatal-based disorders. Unfortunately, each model also exhibits specific drawbacks as they relate to human pathophysiological conditions. Considering each model in the context of the overall field will provide further insight into the function of these circuits in health and disease, and potential insight into novel therapeutic targets.

REFERENCES

Abelson, J. F., Kwan, K. Y., O'Roak, B. J., Baek, D. Y., Stillman, A. A., Morgan, T. M., et al. (2005). Sequence variants in SLITRK1 are associated with Tourette's syndrome. *Science*, *310*, 317–320.

Ackermans, L., Kuhn, J., Neuner, I., Temel, Y., & Visser-Vandewalle, V. (2013). Surgery for Tourette syndrome. *World Neurosurgery*, *80*, S29.e15–S29.e22.

Adler, A., Katabi, S., Finkes, I., Prut, Y., & Bergman, H. (2013). Different correlation patterns of cholinergic and GABAergic interneurons with striatal projection neurons. *Frontiers in Systems Neuroscience*, *7*, 47.

Ahmari, S. E., Spellman, T., Douglass, N. L., Kheirbek, M. A., Simpson, H. B., Deisseroth, K., et al. (2013). Repeated cortico-striatal stimulation generates persistent OCD-like behavior. *Science*, *340*, 1234–1239.

Albanese, A., Bhatia, K., Bressman, S. B., Delong, M. R., Fahn, S., Fung, V. S., et al. (2013). Phenomenology and classification of dystonia: A consensus update. *Movement Disorders*, *28*, 863–873.

Albelda, N., & Joel, D. (2012). Animal models of obsessive-compulsive disorder: Exploring pharmacology and neural substrates. *Neuroscience & Biobehavioral Reviews*, *36*, 47–63.

Albin, R. L., Koeppe, R. A., Bohnen, N. I., Nichols, T. E., Meyer, P., Wernette, K., et al. (2003). Increased ventral striatal monoaminergic innervation in Tourette syndrome. *Neurology*, *61*, 310–315.

Albin, R. L., Koeppe, R. A., Wernette, K., Zhuang, W., Nichols, T., Kilbourn, M. R., et al. (2009). Striatal [11C]dihydrotetrabenazine and [11C]methylphenidate binding in Tourette syndrome. *Neurology*, *72*, 1390–1396.

Albin, R. L., Young, A. B., & Penney, J. B. (1989). The functional anatomy of basal ganglia disorders. *Trends in Neurosciences*, *12*, 366–375.

Alexander, G. E., DeLong, M. R., & Strick, P. L. (1986). Parallel organization of functionally segregated circuits linking basal ganglia and cortex. *Annual Review of Neuroscience*, *9*, 357–381.

Amir, R. E., Van den Veyver, I. B., Wan, M., Tran, C. Q., Francke, U., & Zoghbi, H. Y. (1999). Rett syndrome is caused by mutations in X-linked MECP2, encoding methyl-CpG-binding protein 2. *Nature Genetics*, *23*, 185–188.

Andalman, A. S., & Fee, M. S. (2009). A basal ganglia-forebrain circuit in the songbird biases motor output to avoid vocal errors. *Proceedings of the National Academy of Sciences of the United States of America*, *106*, 12518–12523.

Aosaki, T., Tsubokawa, H., Ishida, A., Watanabe, K., Graybiel, A. M., & Kimura, M. (1994). Responses of tonically active neurons in the primate's striatum undergo systematic changes during behavioral sensorimotor conditioning. *The Journal of Neuroscience*, *14*, 3969–3984.

Argyelan, M., Carbon, M., Niethammer, M., Ulug, A. M., Voss, H. U., Bressman, S. B., et al. (2009). Cerebellothalamocortical connectivity regulates penetrance in dystonia. *The Journal of Neuroscience*, *29*, 9740–9747.

Armstrong, J., Pineda, M., Aibar, E., Gean, E., & Monros, E. (2001). Classic Rett syndrome in a boy as a result of somatic mosaicism for a MECP2 mutation. *Annals of Neurology*, *50*, 692.

Aruga, J., & Mikoshiba, K. (2003). Identification and characterization of Slitrk, a novel neuronal transmembrane protein family controlling neurite outgrowth. *Molecular and Cellular Neurosciences*, *24*, 117–129.

Asanuma, K., Ma, Y., Okulski, J., Dhawan, V., Chaly, T., Carbon, M., et al. (2005). Decreased striatal D2 receptor binding in non-manifesting carriers of the DYT1 dystonia mutation. *Neurology*, *64*, 347–349.

Balcioglu, A., Kim, M. O., Sharma, N., Cha, J. H., Breakefield, X. O., & Standaert, D. G. (2007). Dopamine release is impaired in a mouse model of DYT1 dystonia. *Journal of Neurochemistry*, *102*, 783–788.

Bara-Jimenez, W., Catalan, M. J., Hallett, M., & Gerloff, C. (1998). Abnormal somatosensory homunculus in dystonia of the hand. *Annals of Neurology*, *44*, 828–831.
Barnes, T. D., Kubota, Y., Hu, D., Jin, D. Z., & Graybiel, A. M. (2005). Activity of striatal neurons reflects dynamic encoding and recoding of procedural memories. *Nature*, *437*, 1158–1161.
Beaubien, F., & Cloutier, J. F. (2009). Differential expression of Slitrk family members in the mouse nervous system. *Developmental Dynamics*, *238*, 3285–3296.
Beeler, J. A., Frank, M. J., McDaid, J., Alexander, E., Turkson, S., Bernandez, M. S., et al. (2012). A role for dopamine-mediated learning in the pathophysiology and treatment of Parkinson's disease. *Cell Reports*, *2*, 1747–1761.
Belin, D., & Everitt, B. J. (2008). Cocaine seeking habits depend upon dopamine-dependent serial connectivity linking the ventral with the dorsal striatum. *Neuron*, *57*, 432–441.
Benecke, R., Rothwell, J. C., Dick, J. P., Day, B. L., & Marsden, C. D. (1987). Disturbance of sequential movements in patients with Parkinson's disease. *Brain*, *110*, 361–379.
Bennett, B. D., & Bolam, J. P. (1994). Synaptic input and output of parvalbumin-immunoreactive neurons in the neostriatum of the rat. *Neuroscience*, *62*, 707–719.
Ben-Shachar, S., Chahrour, M., Thaller, C., Shaw, C. A., & Zoghbi, H. Y. (2009). Mouse models of MeCP2 disorders share gene expression changes in the cerebellum and hypothalamus. *Human Molecular Genetics*, *18*, 2431–2442.
Berardelli, A., Rothwell, J. C., Day, B. L., & Marsden, C. D. (1985). Pathophysiology of blepharospasm and oromandibular dystonia. *Brain*, *108*, 593–608.
Berardelli, A., Rothwell, J. C., Hallett, M., Thompson, P. D., Manfredi, M., & Marsden, C. D. (1998). The pathophysiology of primary dystonia. *Brain*, *121*, 1195–1212.
Berke, J. D. (2008). Uncoordinated firing rate changes of striatal fast-spiking interneurons during behavioral task performance. *The Journal of Neuroscience*, *28*, 10075–10080.
Berke, J. D. (2011). Functional properties of striatal fast-spiking interneurons. *Frontiers in Systems Neuroscience*, *5*, 45.
Berridge, K. C. (2007). The debate over dopamine's role in reward: The case for incentive salience. *Psychopharmacology*, *191*, 391–431.
Berridge, K. C., & Aldridge, J. W. (2000a). Super-stereotypy I: Enhancement of a complex movement sequence by systemic dopamine D1 agonists. *Synapse*, *37*, 194–204.
Berridge, K. C., & Aldridge, J. W. (2000b). Super-stereotypy II: Enhancement of a complex movement sequence by intraventricular dopamine D1 agonists. *Synapse*, *37*, 205–215.
Berridge, K. C., Aldridge, J. W., Houchard, K. R., & Zhuang, X. (2005). Sequential super-stereotypy of an instinctive fixed action pattern in hyper-dopaminergic mutant mice: A model of obsessive compulsive disorder and Tourette's. *BMC Biology*, *3*, 4.
Berridge, K. C., Fentress, J. C., & Parr, H. (1987). Natural syntax rules control action sequence of rats. *Behavioural Brain Research*, *23*, 59–68.
Berridge, K. C., & Robinson, T. E. (1998). What is the role of dopamine in reward: Hedonic impact, reward learning, or incentive salience? *Brain Research Brain Research Reviews*, *28*, 309–369.
Berridge, K. C., & Whishaw, I. Q. (1992). Cortex, striatum and cerebellum: Control of serial order in a grooming sequence. *Experimental Brain Research*, *90*, 275–290.
Berthier, M. L., Kulisevsky, J., Gironell, A., & Heras, J. A. (1996). Obsessive-compulsive disorder associated with brain lesions: Clinical phenomenology, cognitive function, and anatomic correlates. *Neurology*, *47*, 353–361.
Bevan, M. D., Booth, P. A., Eaton, S. A., & Bolam, J. P. (1998). Selective innervation of neostriatal interneurons by a subclass of neuron in the globus pallidus of the rat. *The Journal of Neuroscience*, *18*, 9438–9452.
Bharucha, K. J., & Sethi, K. D. (1995). Tardive tourettism after exposure to neuroleptic therapy. *Movement Disorders*, *10*, 791–793.

Bienvenu, O. J., Wang, Y., Shugart, Y. Y., Welch, J. M., Grados, M. A., Fyer, A. J., et al. (2009). Sapap3 and pathological grooming in humans: Results from the OCD collaborative genetics study. *American Journal of Medical Genetics Part B: Neuropsychiatric Genetics, 150B*, 710–720.

Bloch, M. H., Leckman, J. F., Zhu, H., & Peterson, B. S. (2005). Caudate volumes in childhood predict symptom severity in adults with Tourette syndrome. *Neurology, 65*, 1253–1258.

Bloch, M. H., Peterson, B. S., Scahill, L., Otka, J., Katsovich, L., Zhang, H., et al. (2006). Adulthood outcome of tic and obsessive-compulsive symptom severity in children with Tourette syndrome. *Archives of Pediatrics & Adolescent Medicine, 160*, 65–69.

Blomstedt, P., Sjoberg, R. L., Hansson, M., Bodlund, O., & Hariz, M. I. (2013). Deep brain stimulation in the treatment of obsessive-compulsive disorder. *World Neurosurgery, 80*, e245–e253.

Blood, A. J., Flaherty, A. W., Choi, J. K., Hochberg, F. H., Greve, D. N., Bonmassar, G., et al. (2004). Basal ganglia activity remains elevated after movement in focal hand dystonia. *Annals of Neurology, 55*, 744–748.

Blood, A. J., Kuster, J. K., Woodman, S. C., Kirlic, N., Makhlouf, M. L., Multhaupt-Buell, T. J., et al. (2012). Evidence for altered basal ganglia-brainstem connections in cervical dystonia. *PLoS One, 7*, e31654.

Blue, M. E., Naidu, S., & Johnston, M. V. (1999). Altered development of glutamate and GABA receptors in the basal ganglia of girls with Rett syndrome. *Experimental Neurology, 156*, 345–352.

Boardman, L., van der Merwe, L., Lochner, C., Kinnear, C. J., Seedat, S., Stein, D. J., et al. (2011). Investigating SAPAP3 variants in the etiology of obsessive-compulsive disorder and trichotillomania in the South African white population. *Comprehensive Psychiatry, 52*, 181–187.

Bodda, C., Tantra, M., Mollajew, R., Arunachalam, J. P., Laccone, F. A., Can, K., et al. (2013). Mild overexpression of Mecp2 in mice causes a higher susceptibility toward seizures. *The American Journal of Pathology, 183*, 195–210.

Bomar, J. M., Benke, P. J., Slattery, E. L., Puttagunta, R., Taylor, L. P., Seong, E., et al. (2003). Mutations in a novel gene encoding a CRAL-TRIO domain cause human Cayman ataxia and ataxia/dystonia in the jittery mouse. *Nature Genetics, 35*, 264–269.

Bonsi, P., Martella, G., Cuomo, D., Platania, P., Sciamanna, G., Bernardi, G., et al. (2008). Loss of muscarinic autoreceptor function impairs long-term depression but not long-term potentiation in the striatum. *The Journal of Neuroscience, 28*, 6258–6263.

Brainard, M. S., & Doupe, A. J. (2013). Translating birdsong: Songbirds as a model for basic and applied medical research. *Annual Review of Neuroscience, 36*, 489–517.

Brasic, J. R., Bibat, G., Kumar, A., Zhou, Y., Hilton, J., Yablonski, M. E., et al. (2012). Correlation of the vesicular acetylcholine transporter densities in the striata to the clinical abilities of women with Rett syndrome. *Synapse, 66*, 471–482.

Bressman, S. B., Sabatti, C., Raymond, D., de Leon, D., Klein, C., Kramer, P. L., et al. (2000). The DYT1 phenotype and guidelines for diagnostic testing. *Neurology, 54*, 1746–1752.

Broen, M., Duits, A., Visser-Vandewalle, V., Temel, Y., & Winogrodzka, A. (2011). Impulse control and related disorders in Parkinson's disease patients treated with bilateral subthalamic nucleus stimulation: A review. *Parkinsonism & Related Disorders, 17*, 413–417.

Bronfeld, M., & Bar-Gad, I. (2013). Tic disorders: What happens in the basal ganglia? *The Neuroscientist, 19*, 101–108.

Brown, R. G., Jahanshahi, M., Limousin-Dowsey, P., Thomas, D., Quinn, N. P., & Rothwell, J. C. (2003). Pallidotomy and incidental sequence learning in Parkinson's disease. *Neuroreport, 14*, 21–24.

Brown, H. D., McCutcheon, J. E., Cone, J. J., Ragozzino, M. E., & Roitman, M. F. (2011). Primary food reward and reward-predictive stimuli evoke different patterns of phasic dopamine signaling throughout the striatum. *The European Journal of Neuroscience*, *34*, 1997–2006.

Brown, V. J., & Robbins, T. W. (1989). Elementary processes of response selection mediated by distinct regions of the striatum. *The Journal of Neuroscience*, *9*, 3760–3765.

Burdick, A., Goodman, W. K., & Foote, K. D. (2009). Deep brain stimulation for refractory obsessive-compulsive disorder. *Frontiers in Bioscience (Landmark Edition)*, *14*, 1880–1890.

Burguiere, E., Monteiro, P., Feng, G., & Graybiel, A. M. (2013). Optogenetic stimulation of lateral orbitofronto-striatal pathway suppresses compulsive behaviors. *Science*, *340*, 1243–1246.

Burke, R. E., & Fahn, S. (1983). Chlorpromazine methiodide acts at the vestibular nuclear complex to induce barrel rotation in the rat. *Brain Research*, *288*, 273–281.

Burke, R. E., & Fahn, S. (1985a). Pharmacokinetics of trihexyphenidyl after short-term and long-term administration to dystonic patients. *Annals of Neurology*, *18*, 35–40.

Burke, R. E., & Fahn, S. (1985b). Serum trihexyphenidyl levels in the treatment of torsion dystonia. *Neurology*, *35*, 1066–1069.

Burke, R. E., Fahn, S., & Marsden, C. D. (1986). Torsion dystonia: A double-blind, prospective trial of high-dosage trihexyphenidyl. *Neurology*, *36*, 160–164.

Burke, R. E., & Karanas, A. L. (1990). Quantitative morphological analysis of striatal cholinergic neurons in perinatal asphyxia. *Annals of Neurology*, *27*, 81–88.

Caldwell, G. A., Cao, S., Sexton, E. G., Gelwix, C. C., Bevel, J. P., & Caldwell, K. A. (2003). Suppression of polyglutamine-induced protein aggregation in Caenorhabditis elegans by torsin proteins. *Human Molecular Genetics*, *12*, 307–319.

Campbell, K. M., de Lecea, L., Severynse, D. M., Caron, M. G., McGrath, M. J., Sparber, S. B., et al. (1999). OCD-Like behaviors caused by a neuropotentiating transgene targeted to cortical and limbic D1+ neurons. *The Journal of Neuroscience*, *19*, 5044–5053.

Cannon, C. M., & Palmiter, R. D. (2003). Reward without dopamine. *The Journal of Neuroscience*, *23*, 10827–10831.

Carbon, M., Argyelan, M., Ghilardi, M. F., Mattis, P., Dhawan, V., Bressman, S., et al. (2011). Impaired sequence learning in dystonia mutation carriers: A genotypic effect. *Brain*, *134*, 1416–1427.

Carbon, M., Ghilardi, M. F., Argyelan, M., Dhawan, V., Bressman, S. B., & Eidelberg, D. (2008). Increased cerebellar activation during sequence learning in DYT1 carriers: An equiperformance study. *Brain*, *131*, 146–154.

Carbon, M., Trost, M., Ghilardi, M. F., & Eidelberg, D. (2004). Abnormal brain networks in primary torsion dystonia. *Advances in Neurology*, *94*, 155–161.

Carli, M., Evenden, J. L., & Robbins, T. W. (1985). Depletion of unilateral striatal dopamine impairs initiation of contralateral actions and not sensory attention. *Nature*, *313*, 679–682.

Carter, J. C., Lanham, D. C., Pham, D., Bibat, G., Naidu, S., & Kaufmann, W. E. (2008). Selective cerebral volume reduction in Rett syndrome: A multiple-approach MR imaging study. *AJNR. American Journal of Neuroradiology*, *29*, 436–441.

Casanova, M. F., Naidu, S., Goldberg, T. E., Moser, H. W., Khoromi, S., Kumar, A., et al. (1991). Quantitative magnetic resonance imaging in Rett syndrome. *The Journal of Neuropsychiatry and Clinical Neurosciences*, *3*, 66–72.

Castellan Baldan, L., Williams, K. A., Gallezot, J. D., Pogorelov, V., Rapanelli, M., Crowley, M., et al. (2014). Histidine decarboxylase deficiency causes Tourette syndrome: Parallel findings in humans and mice. *Neuron*, *81*, 77–90.

Cavanna, A. E., & Rickards, H. (2013). The psychopathological spectrum of Gilles de la Tourette syndrome. *Neuroscience & Biobehavioral Reviews*, *37*, 1008–1015.

Cavanna, A. E., & Seri, S. (2013). Tourette's syndrome. *BMJ*, *347*, f4964.
Chahrour, M., Jung, S. Y., Shaw, C., Zhou, X., Wong, S. T., Qin, J., et al. (2008). MeCP2, a key contributor to neurological disease, activates and represses transcription. *Science*, *320*, 1224–1229.
Chan, P. L., Nutt, J. G., & Holford, N. H. (2004). Modeling the short- and long-duration responses to exogenous levodopa and to endogenous levodopa production in Parkinson's disease. *Journal of Pharmacokinetics and Pharmacodynamics*, *31*, 243–268.
Chang, H. T., & Kita, H. (1992). Interneurons in the rat striatum: Relationships between parvalbumin neurons and cholinergic neurons. *Brain Research*, *574*, 307–311.
Chao, H. T., Chen, H., Samaco, R. C., Xue, M., Chahrour, M., Yoo, J., et al. (2010). Dysfunction in GABA signalling mediates autism-like stereotypies and Rett syndrome phenotypes. *Nature*, *468*, 263–269.
Chao, H. T., Zoghbi, H. Y., & Rosenmund, C. (2007). MeCP2 controls excitatory synaptic strength by regulating glutamatergic synapse number. *Neuron*, *56*, 58–65.
Chen, R. Z., Akbarian, S., Tudor, M., & Jaenisch, R. (2001). Deficiency of methyl-CpG binding protein-2 in CNS neurons results in a Rett-like phenotype in mice. *Nature Genetics*, *27*, 327–331.
Chen, P., Burdette, A. J., Porter, J. C., Ricketts, J. C., Fox, S. A., Nery, F. C., et al. (2010). The early-onset torsion dystonia-associated protein, torsinA, is a homeostatic regulator of endoplasmic reticulum stress response. *Human Molecular Genetics*, *19*, 3502–3515.
Chen, S. K., Tvrdik, P., Peden, E., Cho, S., Wu, S., Spangrude, G., et al. (2010). Hematopoietic origin of pathological grooming in Hoxb8 mutant mice. *Cell*, *141*, 775–785.
Cheval, H., Guy, J., Merusi, C., De Sousa, D., Selfridge, J., & Bird, A. (2012). Postnatal inactivation reveals enhanced requirement for MeCP2 at distinct age windows. *Human Molecular Genetics*, *21*, 3806–3814.
Chiron, C., Bulteau, C., Loc'h, C., Raynaud, C., Garreau, B., Syrota, A., et al. (1993). Dopaminergic D2 receptor SPECT imaging in Rett syndrome: Increase of specific binding in striatum. *Journal of Nuclear Medicine*, *34*, 1717–1721.
Choi, E. Y., Yeo, B. T., & Buckner, R. L. (2012). The organization of the human striatum estimated by intrinsic functional connectivity. *Journal of Neurophysiology*, *108*, 2242–2263.
Cleary, D. R., Raslan, A. M., Rubin, J. E., Bahgat, D., Viswanathan, A., Heinricher, M. M., et al. (2013). Deep brain stimulation entrains local neuronal firing in human globus pallidus internus. *Journal of Neurophysiology*, *109*, 978–987.
Cohen, D. J., Detlor, J., Young, J. G., & Shaywitz, B. A. (1980). Clonidine ameliorates Gilles de la Tourette syndrome. *Archives of General Psychiatry*, *37*, 1350–1357.
Cohen, J. Y., Haesler, S., Vong, L., Lowell, B. B., & Uchida, N. (2012). Neuron-type-specific signals for reward and punishment in the ventral tegmental area. *Nature*, *482*, 85–88.
Cohen, S. C., Leckman, J. F., & Bloch, M. H. (2013). Clinical assessment of Tourette syndrome and tic disorders. *Neuroscience & Biobehavioral Reviews*, *37*, 997–1007.
Cohen, D. J., Young, J. G., Nathanson, J. A., & Shaywitz, B. A. (1979). Clonidine in Tourette's syndrome. *Lancet*, *2*, 551–553.
Colantuoni, C., Jeon, O. H., Hyder, K., Chenchik, A., Khimani, A. H., Narayanan, V., et al. (2001). Gene expression profiling in postmortem Rett syndrome brain: Differential gene expression and patient classification. *Neurobiology of Disease*, *8*, 847–865.
Collins, A. L., Levenson, J. M., Vilaythong, A. P., Richman, R., Armstrong, D. L., Noebels, J. L., et al. (2004). Mild overexpression of MeCP2 causes a progressive neurological disorder in mice. *Human Molecular Genetics*, *13*, 2679–2689.
Cordes, M., Snow, B. J., Cooper, S., Schulzer, M., Pate, B. D., Ruth, T. J., et al. (1994). Age-dependent decline of nigrostriatal dopaminergic function: A positron emission

tomographic study of grandparents and their grandchildren. *Annals of Neurology, 36*, 667–670.

Coutureau, E., & Killcross, S. (2003). Inactivation of the infralimbic prefrontal cortex reinstates goal-directed responding in overtrained rats. *Behavioural Brain Research, 146*, 167–174.

Cragg, S. J. (2006). Meaningful silences: How dopamine listens to the ACh pause. *Trends in Neurosciences, 29*, 125–131.

Crane, J., Fagerness, J., Osiecki, L., Gunnell, B., Stewart, S. E., Pauls, D. L., et al. (2011). Family-based genetic association study of DLGAP3 in Tourette syndrome. *American Journal of Medical Genetics Part B: Neuropsychiatric Genetics, 156B*, 108–114.

Crittenden, J. R., & Graybiel, A. M. (2011). Basal ganglia disorders associated with imbalances in the striatal striosome and matrix compartments. *Frontiers in Neuroanatomy, 5*, 59.

Cromwell, H. C., & Berridge, K. C. (1996). Implementation of action sequences by a neostriatal site: A lesion mapping study of grooming syntax. *The Journal of Neuroscience, 16*, 3444–3458.

Cromwell, H. C., Berridge, K. C., Drago, J., & Levine, M. S. (1998). Action sequencing is impaired in D1A-deficient mutant mice. *The European Journal of Neuroscience, 10*, 2426–2432.

Cui, G., Jun, S. B., Jin, X., Pham, M. D., Vogel, S. S., Lovinger, D. M., et al. (2013). Concurrent activation of striatal direct and indirect pathways during action initiation. *Nature, 494*, 238–242.

Cui, H., Mason, B. L., Lee, C., Nishi, A., Elmquist, J. K., & Lutter, M. (2012). Melanocortin 4 receptor signaling in dopamine 1 receptor neurons is required for procedural memory learning. *Physiology & Behavior, 106*, 201–210.

Dang, M. T., Yokoi, F., McNaught, K. S., Jengelley, T. A., Jackson, T., Li, J., et al. (2005). Generation and characterization of Dyt1 DeltaGAG knock-in mouse as a model for early-onset dystonia. *Experimental Neurology, 196*, 452–463.

Dang, M. T., Yokoi, F., Pence, M. A., & Li, Y. (2006). Motor deficits and hyperactivity in Dyt1 knockdown mice. *Neuroscience Research, 56*, 470–474.

Dani, V. S., Chang, Q., Maffei, A., Turrigiano, G. G., Jaenisch, R., & Nelson, S. B. (2005). Reduced cortical activity due to a shift in the balance between excitation and inhibition in a mouse model of Rett syndrome. *Proceedings of the National Academy of Sciences of the United States of America, 102*, 12560–12565.

Dayalu, P., & Chou, K. L. (2008). Antipsychotic-induced extrapyramidal symptoms and their management. *Expert Opinion on Pharmacotherapy, 9*, 1451–1462.

Dayer, A. G., Bottani, A., Bouchardy, I., Fluss, J., Antonarakis, S. E., Haenggeli, C. A., et al. (2007). MECP2 mutant allele in a boy with Rett syndrome and his unaffected heterozygous mother. *Brain Dev, 29*, 47–50.

Delmaire, C., Vidailhet, M., Elbaz, A., Bourdain, F., Bleton, J. P., Sangla, S., et al. (2007). Structural abnormalities in the cerebellum and sensorimotor circuit in writer's cramp. *Neurology, 69*, 376–380.

DeLong, M. R. (1990). Primate models of movement disorders of basal ganglia origin. *Trends in Neurosciences, 13*, 281–285.

Deng, H., Gao, K., & Jankovic, J. (2012). The genetics of Tourette syndrome. *Nature Reviews Neurology, 8*, 203–213.

Denys, D., de Vries, F., Cath, D., Figee, M., Vulink, N., Veltman, D. J., et al. (2013). Dopaminergic activity in Tourette syndrome and obsessive-compulsive disorder. *European Neuropsychopharmacology, 23*(11), 1423–1431.

Desmurget, M., & Turner, R. S. (2010). Motor sequences and the basal ganglia: Kinematics, not habits. *The Journal of Neuroscience, 30*, 7685–7690.

Deveney, A. M., & Waddington, J. L. (1997). Psychopharmacological distinction between novel full-efficacy "D1-like" dopamine receptor agonists. *Pharmacology, Biochemistry and Behavior, 58*, 551–558.

Ding, J. B., Guzman, J. N., Peterson, J. D., Goldberg, J. A., & Surmeier, D. J. (2010). Thalamic gating of corticostriatal signaling by cholinergic interneurons. *Neuron, 67*, 294–307.
Dolce, A., Ben-Zeev, B., Naidu, S., & Kossoff, E. H. (2013). Rett syndrome and epilepsy: An update for child neurologists. *Pediatric Neurology, 48*, 337–345.
Doyon, J. (2008). Motor sequence learning and movement disorders. *Current Opinion in Neurology, 21*, 478–483.
Draganski, B., Thun-Hohenstein, C., Bogdahn, U., Winkler, J., & May, A. (2003). "Motor circuit" gray matter changes in idiopathic cervical dystonia. *Neurology, 61*, 1228–1231.
Dreyer, J. K., Herrik, K. F., Berg, R. W., & Hounsgaard, J. D. (2010). Influence of phasic and tonic dopamine release on receptor activation. *The Journal of Neuroscience, 30*, 14273–14283.
Duchen, L. W., Strich, S. J., & Falconer, D. S. (1964). Clinical and pathological studies of an hereditary neuropathy in mice (*dystonia musculorum*). *Brain, 87*, 367–378.
Dunn, H. G., & MacLeod, P. M. (2001). Rett syndrome: Review of biological abnormalities. *The Canadian Journal of Neurological Sciences, 28*, 16–29.
Dunn, H. G., Stoessl, A. J., Ho, H. H., MacLeod, P. M., Poskitt, K. J., Doudet, D. J., et al. (2002). Rett syndrome: Investigation of nine patients, including PET scan. *The Canadian Journal of Neurological Sciences, 29*, 345–357.
Edwards, M. J., Huang, Y. Z., Mir, P., Rothwell, J. C., & Bhatia, K. P. (2006). Abnormalities in motor cortical plasticity differentiate manifesting and nonmanifesting DYT1 carriers. *Movement Disorders, 21*, 2181–2186.
Egger, K., Mueller, J., Schocke, M., Brenneis, C., Rinnerthaler, M., Seppi, K., et al. (2007). Voxel based morphometry reveals specific gray matter changes in primary dystonia. *Movement Disorders, 22*, 1538–1542.
Ellens, D. J., & Leventhal, D. K. (2013). Review: Electrophysiology of basal ganglia and cortex in models of Parkinson disease. *Journal of Parkinson's Disease, 3*, 241–254.
English, D. F., Ibanez-Sandoval, O., Stark, E., Tecuapetla, F., Buzsaki, G., Deisseroth, K., et al. (2012). GABAergic circuits mediate the reinforcement-related signals of striatal cholinergic interneurons. *Nature Neuroscience, 15*, 123–130.
Ercan-Sencicek, A. G., Stillman, A. A., Ghosh, A. K., Bilguvar, K., O'Roak, B. J., Mason, C. E., et al. (2010). L-histidine decarboxylase and Tourette's syndrome. *The New England Journal of Medicine, 362*, 1901–1908.
Fahn, S. (1983). High dosage anticholinergic therapy in dystonia. *Neurology, 33*, 1255–1261.
Felling, R. J., & Singer, H. S. (2011). Neurobiology of tourette syndrome: Current status and need for further investigation. *The Journal of Neuroscience, 31*, 12387–12395.
Ferguson, D. (1971). An Australian study of telegraphists' cramp. *British Journal of Industrial Medicine, 28*, 280–285.
Fernandez, T. V., Sanders, S. J., Yurkiewicz, I. R., Ercan-Sencicek, A. G., Kim, Y. S., Fishman, D. O., et al. (2012). Rare copy number variants in Tourette syndrome disrupt genes in histaminergic pathways and overlap with autism. *Biological Psychiatry, 71*, 392–402.
FitzGerald, P. M., Jankovic, J., Glaze, D. G., Schultz, R., & Percy, A. K. (1990). Extrapyramidal involvement in Rett's syndrome. *Neurology, 40*, 293–295.
FitzGerald, P. M., Jankovic, J., & Percy, A. K. (1990). Rett syndrome and associated movement disorders. *Movement Disorders, 5*, 195–202.
Flagel, S. B., Clark, J. J., Robinson, T. E., Mayo, L., Czuj, A., Willuhn, I., et al. (2011). A selective role for dopamine in stimulus-reward learning. *Nature, 469*, 53–57.
Fleming, J. V., & Wang, T. C. (2003). The production of 53-55-kDa isoforms is not required for rat L-histidine decarboxylase activity. *The Journal of Biological Chemistry, 278*, 686–694.
Fontaine, D., Mattei, V., Borg, M., von Langsdorff, D., Magnie, M. N., Chanalet, S., et al. (2004). Effect of subthalamic nucleus stimulation on obsessive-compulsive disorder in a patient with Parkinson disease. Case report. *Journal of Neurosurgery, 100*, 1084–1086.

Frank, M. J., Seeberger, L. C., & O'reilly, R. C. (2004). By carrot or by stick: Cognitive reinforcement learning in parkinsonism. *Science, 306*, 1940–1943.
Franklin, M. E., Zagrabbe, K., & Benavides, K. L. (2011). Trichotillomania and its treatment: A review and recommendations. *Expert Review of Neurotherapeutics, 11*, 1165–1174.
Freeze, B. S., Kravitz, A. V., Hammack, N., Berke, J. D., & Kreitzer, A. C. (2013). Control of basal ganglia output by direct and indirect pathway projection neurons. *The Journal of Neuroscience, 33*, 18531–18539.
Frucht, S. J. (2004). Focal task-specific dystonia in musicians. *Advances in Neurology, 94*, 225–230.
Fukuda, T. (2009). Network architecture of gap junction-coupled neuronal linkage in the striatum. *The Journal of Neuroscience, 29*, 1235–1243.
Furukawa, Y., Hornykiewicz, O., Fahn, S., & Kish, S. J. (2000). Striatal dopamine in early-onset primary torsion dystonia with the DYT1 mutation. *Neurology, 54*, 1193–1195.
Gage, G. J., Stoetzner, C. R., Wiltschko, A. B., & Berke, J. D. (2010). Selective activation of striatal fast-spiking interneurons during choice execution. *Neuron, 67*, 466–479.
Garraux, G., Bauer, A., Hanakawa, T., Wu, T., Kansaku, K., & Hallett, M. (2004). Changes in brain anatomy in focal hand dystonia. *Annals of Neurology, 55*, 736–739.
Gerfen, C. R., & Surmeier, D. J. (2011). Modulation of striatal projection systems by dopamine. *Annual Review of Neuroscience, 34*, 441–466.
Ghilardi, M. F., Carbon, M., Silvestri, G., Dhawan, V., Tagliati, M., Bressman, S., et al. (2003). Impaired sequence learning in carriers of the DYT1 dystonia mutation. *Annals of Neurology, 54*, 102–109.
Giacometti, E., Luikenhuis, S., Beard, C., & Jaenisch, R. (2007). Partial rescue of MeCP2 deficiency by postnatal activation of MeCP2. *Proceedings of the National Academy of Sciences of the United States of America, 104*, 1931–1936.
Gibb, W. R., Lees, A. J., & Marsden, C. D. (1988). Pathological report of four patients presenting with cranial dystonias. *Movement Disorders, 3*, 211–221.
Gittis, A. H., Leventhal, D. K., Fensterheim, B. A., Pettibone, J. R., Berke, J. D., & Kreitzer, A. C. (2011). Selective inhibition of striatal fast-spiking interneurons causes dyskinesias. *The Journal of Neuroscience, 31*, 15727–15731.
Gittis, A. H., Nelson, A. B., Thwin, M. T., Palop, J. J., & Kreitzer, A. C. (2010). Distinct roles of GABAergic interneurons in the regulation of striatal output pathways. *The Journal of Neuroscience, 30*, 2223–2234.
Glimcher, P. W. (2011). Understanding dopamine and reinforcement learning: The dopamine reward prediction error hypothesis. *Proceedings of the National Academy of Sciences of the United States of America, 108*(Suppl. 3), 15647–15654.
Goldberg, J. H., Farries, M. A., & Fee, M. S. (2013). Basal ganglia output to the thalamus: Still a paradox. *Trends in Neurosciences, 36*, 695–705.
Goldberg, J. H., & Fee, M. S. (2012). A cortical motor nucleus drives the basal ganglia-recipient thalamus in singing birds. *Nature Neuroscience, 15*, 620–627.
Goldberg, J. A., & Reynolds, J. N. (2011). Spontaneous firing and evoked pauses in the tonically active cholinergic interneurons of the striatum. *Neuroscience, 198*, 27–43.
Gonzales, M. L., & LaSalle, J. M. (2010). The role of MeCP2 in brain development and neurodevelopmental disorders. *Current Psychiatry Reports, 12*, 127–134.
Goodchild, R. E., & Dauer, W. T. (2004). Mislocalization to the nuclear envelope: An effect of the dystonia-causing torsinA mutation. *Proceedings of the National Academy of Sciences of the United States of America, 101*, 847–852.
Goodchild, R. E., & Dauer, W. T. (2005). The AAA+ protein torsinA interacts with a conserved domain present in LAP1 and a novel ER protein. *The Journal of Cell Biology, 168*, 855–862.

Goodchild, R. E., Kim, C. E., & Dauer, W. T. (2005). Loss of the dystonia-associated protein torsinA selectively disrupts the neuronal nuclear envelope. *Neuron, 48*, 923–932.

Gordon, N. (2002). Pantothenate kinase-associated neurodegeneration (Hallervorden-Spatz syndrome). *European Journal of Paediatric Neurology, 6*, 243–247.

Grafton, S. T., Hazeltine, E., & Ivry, R. (1995). Functional mapping of sequence learning in normal humans. *Journal of Cognitive Neuroscience, 7*, 497–510.

Granata, A., Watson, R., Collinson, L. M., Schiavo, G., & Warner, T. T. (2008). The dystonia-associated protein torsinA modulates synaptic vesicle recycling. *The Journal of Biological Chemistry, 283*, 7568–7579.

Graveland, G. A., & DiFiglia, M. (1985). The frequency and distribution of medium-sized neurons with indented nuclei in the primate and rodent neostriatum. *Brain Research, 327*, 307–311.

Graybiel, A. M. (1998). The basal ganglia and chunking of action repertoires. *Neurobiology of Learning and Memory, 70*, 119–136.

Graybiel, A. M., Baughman, R. W., & Eckenstein, F. (1986). Cholinergic neuropil of the striatum observes striosomal boundaries. *Nature, 323*, 625–627.

Graybiel, A. M., & Rauch, S. L. (2000). Toward a neurobiology of obsessive-compulsive disorder. *Neuron, 28*, 343–347.

Greenberg, B. D., Gabriels, L. A., Malone, D. A. J., Rezai, A. R., Friehs, G. M., Okun, M. S., et al. (2010). Deep brain stimulation of the ventral internal capsule/ventral striatum for obsessive-compulsive disorder: Worldwide experience. *Molecular Psychiatry, 15*, 64–79.

Greenberg, B. D., Malone, D. A., Friehs, G. M., Rezai, A. R., Kubu, C. S., Malloy, P. F., et al. (2006). Three-year outcomes in deep brain stimulation for highly resistant obsessive-compulsive disorder. *Neuropsychopharmacology, 31*, 2384–2393.

Greene, P., Shale, H., & Fahn, S. (1988). Analysis of open-label trials in torsion dystonia using high dosages of anticholinergics and other drugs. *Movement Disorders, 3*, 46–60.

Greer, J. M., & Capecchi, M. R. (2002). Hoxb8 is required for normal grooming behavior in mice. *Neuron, 33*, 23–34.

Grillner, S., Robertson, B., & Stephenson-Jones, M. (2013). The evolutionary origin of the vertebrate basal ganglia and its role in action selection. *The Journal of Physiology, 591*, 5425–5431.

Grundmann, K., Glockle, N., Martella, G., Sciamanna, G., Hauser, T. K., Yu, L., et al. (2012). Generation of a novel rodent model for DYT1 dystonia. *Neurobiology of Disease, 47*, 61–74.

Grundmann, K., Reischmann, B., Vanhoutte, G., Hubener, J., Teismann, P., Hauser, T. K., et al. (2007). Overexpression of human wildtype torsinA and human DeltaGAG torsinA in a transgenic mouse model causes phenotypic abnormalities. *Neurobiology of Disease, 27*, 190–206.

Gulley, J. M., Kuwajima, M., Mayhill, E., & Rebec, G. V. (1999). Behavior-related changes in the activity of substantia nigra pars reticulata neurons in freely moving rats. *Brain Research, 845*, 68–76.

Guy, J., Gan, J., Selfridge, J., Cobb, S., & Bird, A. (2007). Reversal of neurological defects in a mouse model of Rett syndrome. *Science, 315*, 1143–1147.

Guy, J., Hendrich, B., Holmes, M., Martin, J. E., & Bird, A. (2001). A mouse Mecp2-null mutation causes neurological symptoms that mimic Rett syndrome. *Nature Genetics, 27*, 322–326.

Haber, S. N., Fudge, J. L., & McFarland, N. R. (2000). Striatonigrostriatal pathways in primates form an ascending spiral from the shell to the dorsolateral striatum. *The Journal of Neuroscience, 20*, 2369–2382.

Hagberg, B., Hanefeld, F., Percy, A., & Skjeldal, O. (2002). An update on clinically applicable diagnostic criteria in Rett syndrome. Comments to Rett Syndrome Clinical Criteria Consensus Panel Satellite to European Paediatric Neurology Society Meeting, Baden Baden, Germany, 11 September 2001. *European Journal of Paediatric Neurology*, *6*, 293–297.

Hallett, M. (2011). Neurophysiology of dystonia: The role of inhibition. *Neurobiology of Disease*, *42*, 177–184.

Harris, J. C., Wong, D. F., Wagner, H. N. J., Rett, A., Naidu, S., Dannals, R. F., et al. (1986). Positron emission tomographic study of D2 dopamine receptor binding and CSF biogenic amine metabolites in Rett syndrome. *American Journal of Medical Genetics Supplement*, *1*, 201–210.

Harrison, B. J., Soriano-Mas, C., Pujol, J., Ortiz, H., Lopez-Sola, M., Hernandez-Ribas, R., et al. (2009). Altered corticostriatal functional connectivity in obsessive-compulsive disorder. *Archives of General Psychiatry*, *66*, 1189–1200.

Hewett, J., Johanson, P., Sharma, N., Standaert, D., & Balcioglu, A. (2010). Function of dopamine transporter is compromised in DYT1 transgenic animal model in vivo. *Journal of Neurochemistry*, *113*, 228–235.

Hewett, J. W., Tannous, B., Niland, B. P., Nery, F. C., Zeng, J., Li, Y., et al. (2007). Mutant torsinA interferes with protein processing through the secretory pathway in DYT1 dystonia cells. *Proceedings of the National Academy of Sciences of the United States of America*, *104*, 7271–7276.

Hjorth, J., Blackwell, K. T., & Kotaleski, J. H. (2009). Gap junctions between striatal fast-spiking interneurons regulate spiking activity and synchronization as a function of cortical activity. *The Journal of Neuroscience*, *29*, 5276–5286.

Hnasko, T. S., Sotak, B. N., & Palmiter, R. D. (2005). Morphine reward in dopamine-deficient mice. *Nature*, *438*, 854–857.

Holmes, A. L., Forcelli, P. A., DesJardin, J. T., Decker, A. L., Teferra, M., West, E. A., et al. (2012). Superior colliculus mediates cervical dystonia evoked by inhibition of the substantia nigra pars reticulata. *The Journal of Neuroscience*, *32*, 13326–13332.

Holstege, J. C., de Graaff, W., Hossaini, M., Cardona Cano, S., Jaarsma, D., van den Akker, E., et al. (2008). Loss of Hoxb8 alters spinal dorsal laminae and sensory responses in mice. *Proceedings of the National Academy of Sciences of the United States of America*, *105*, 6338–6343.

Holton, J. L., Schneider, S. A., Ganesharajah, T., Gandhi, S., Strand, C., Shashidharan, P., et al. (2008). Neuropathology of primary adult-onset dystonia. *Neurology*, *70*, 695–699.

Horak, F. B., & Anderson, M. E. (1984). Influence of globus pallidus on arm movements in monkeys. I. Effects of kainic acid-induced lesions. *Journal of Neurophysiology*, *52*, 290–304.

Humphries, M. D., Wood, R., & Gurney, K. (2009). Dopamine-modulated dynamic cell assemblies generated by the GABAergic striatal microcircuit. *Neural Networks*, *22*, 1174–1188.

Hyde, T. M., Stacey, M. E., Coppola, R., Handel, S. F., Rickler, K. C., & Weinberger, D. R. (1995). Cerebral morphometric abnormalities in Tourette's syndrome: A quantitative MRI study of monozygotic twins. *Neurology*, *45*, 1176–1182.

Hyman, S. E. (2010). A bone to pick with compulsive behavior. *Cell*, *141*, 752–754.

Inase, M., Buford, J. A., & Anderson, M. E. (1996). Changes in the control of arm position, movement, and thalamic discharge during local inactivation in the globus pallidus of the monkey. *Journal of Neurophysiology*, *75*, 1087–1104.

Iwabuchi, S., Kakazu, Y., Koh, J. Y., & Harata, N. C. (2013). Abnormal cytoplasmic calcium dynamics in central neurons of a dystonia mouse model. *Neuroscience Letters*, *548*, 61–66.

Jankovic, J., & Ashizawa, T. (1995). Tourettism associated with Huntington's disease. *Movement Disorders*, *10*, 103–105.

Jellinger, K. A. (2003). Rett syndrome—An update. *Journal of Neural Transmission*, *110*, 681–701.

Jellinger, K., Armstrong, D., Zoghbi, H. Y., & Percy, A. K. (1988). Neuropathology of Rett syndrome. *Acta Neuropathologica*, *76*, 142–158.

Jiang, M., Ash, R. T., Baker, S. A., Suter, B., Ferguson, A., Park, J., et al. (2013). Dendritic arborization and spine dynamics are abnormal in the mouse model of MECP2 duplication syndrome. *The Journal of Neuroscience*, *33*, 19518–19533.

Jin, X., & Costa, R. M. (2010). Start/stop signals emerge in nigrostriatal circuits during sequence learning. *Nature*, *466*, 457–462.

Jin, X., Tecuapetla, F., & Costa, R. M. (2014). Basal ganglia subcircuits distinctively encode the parsing and concatenation of action sequences. *Nature Neuroscience*, *17*(3), 423–430.

Joel, D., & Weiner, I. (1994). The organization of the basal ganglia-thalamocortical circuits: Open interconnected rather than closed segregated. *Neuroscience*, *63*, 363–379.

Jog, M. S., Kubota, Y., Connolly, C. I., Hillegaart, V., & Graybiel, A. M. (1999). Building neural representations of habits. *Science*, *286*, 1745–1749.

Jones, P. L., Veenstra, G. J., Wade, P. A., Vermaak, D., Kass, S. U., Landsberger, N., et al. (1998). Methylated DNA and MeCP2 recruit histone deacetylase to repress transcription. *Nature Genetics*, *19*, 187–191.

Kakazu, Y., Koh, J. Y., Ho, K. W., Gonzalez-Alegre, P., & Harata, N. C. (2012). Synaptic vesicle recycling is enhanced by torsinA that harbors the DYT1 dystonia mutation. *Synapse*, *66*, 453–464.

Kakazu, Y., Koh, J. Y., Iwabuchi, S., Gonzalez-Alegre, P., & Harata, N. C. (2012). Miniature release events of glutamate from hippocampal neurons are influenced by the dystonia-associated protein torsinA. *Synapse*, *66*, 807–822.

Kalanithi, P. S., Zheng, W., Kataoka, Y., DiFiglia, M., Grantz, H., Saper, C. B., et al. (2005). Altered parvalbumin-positive neuron distribution in basal ganglia of individuals with Tourette syndrome. *Proceedings of the National Academy of Sciences of the United States of America*, *102*, 13307–13312.

Karagiannidis, I., Dehning, S., Sandor, P., Tarnok, Z., Rizzo, R., Wolanczyk, T., et al. (2013). Support of the histaminergic hypothesis in Tourette syndrome: Association of the histamine decarboxylase gene in a large sample of families. *Journal of Medical Genetics*, *50*, 760–764.

Karagiannidis, I., Rizzo, R., Tarnok, Z., Wolanczyk, T., Hebebrand, J., Nothen, M. M., et al. (2012). Replication of association between a SLITRK1 haplotype and Tourette syndrome in a large sample of families. *Molecular Psychiatry*, *17*, 665–668.

Karimi, M., Moerlein, S. M., Videen, T. O., Luedtke, R. R., Taylor, M., Mach, R. H., et al. (2011). Decreased striatal dopamine receptor binding in primary focal dystonia: A D2 or D3 defect? *Movement Disorders*, *26*, 100–106.

Karimi, M., Moerlein, S. M., Videen, T. O., Su, Y., Flores, H. P., & Perlmutter, J. S. (2013). Striatal dopamine D1-like receptor binding is unchanged in primary focal dystonia. *Movement Disorders*, *28*(14), 2002–2006.

Kataoka, Y., Kalanithi, P. S., Grantz, H., Schwartz, M. L., Saper, C., Leckman, J. F., et al. (2010). Decreased number of parvalbumin and cholinergic interneurons in the striatum of individuals with Tourette syndrome. *The Journal of Comparative Neurology*, *518*, 277–291.

Katayama, K., Yamada, K., Ornthanalai, V. G., Inoue, T., Ota, M., Murphy, N. P., et al. (2010). Slitrk1-deficient mice display elevated anxiety-like behavior and noradrenergic abnormalities. *Molecular Psychiatry*, *15*, 177–184.

Kawaguchi, Y. (1992). Large aspiny cells in the matrix of the rat neostriatum in vitro: Physiological identification, relation to the compartments and excitatory postsynaptic currents. *Journal of Neurophysiology*, *67*, 1669–1682.

Kawaguchi, Y., Wilson, C. J., Augood, S. J., & Emson, P. C. (1995). Striatal interneurones: Chemical, physiological and morphological characterization. *Trends in Neurosciences*, *18*, 527–535.

Keen-Kim, D., Mathews, C. A., Reus, V. I., Lowe, T. L., Herrera, L. D., Budman, C. L., et al. (2006). Overrepresentation of rare variants in a specific ethnic group may confuse interpretation of association analyses. *Human Molecular Genetics, 15*, 3324–3328.

Kemp, J. M., & Powell, T. P. (1970). The cortico-striate projection in the monkey. *Brain, 93*, 525–546.

Kermadi, I., & Joseph, J. P. (1995). Activity in the caudate nucleus of monkey during spatial sequencing. *Journal of Neurophysiology, 74*, 911–933.

Kim, E., Naisbitt, S., Hsueh, Y. P., Rao, A., Rothschild, A., Craig, A. M., et al. (1997). GKAP, a novel synaptic protein that interacts with the guanylate kinase-like domain of the PSD-95/SAP90 family of channel clustering molecules. *The Journal of Cell Biology, 136*, 669–678.

Kita, H., Kosaka, T., & Heizmann, C. W. (1990). Parvalbumin-immunoreactive neurons in the rat neostriatum: A light and electron microscopic study. *Brain Research, 536*, 1–15.

Kitt, C. A., & Wilcox, B. J. (1995). Preliminary evidence for neurodegenerative changes in the substantia nigra of Rett syndrome. *Neuropediatrics, 26*, 114–118.

Kiyatkin, E. A., & Gratton, A. (1994). Electrochemical monitoring of extracellular dopamine in nucleus accumbens of rats lever-pressing for food. *Brain Research, 652*, 225–234.

Koos, T., & Tepper, J. M. (1999). Inhibitory control of neostriatal projection neurons by GABAergic interneurons. *Nature Neuroscience, 2*, 467–472.

Korff, S., Stein, D. J., & Harvey, B. H. (2008). Stereotypic behaviour in the deer mouse: Pharmacological validation and relevance for obsessive compulsive disorder. *Progress in Neuro-Psychopharmacology & Biological Psychiatry, 32*, 348–355.

Krauss, J. K. (2002). Deep brain stimulation for dystonia in adults. Overview and developments. *Stereotactic and Functional Neurosurgery, 78*, 168–182.

Kravitz, A. V., Freeze, B. S., Parker, P. R., Kay, K., Thwin, M. T., Deisseroth, K., et al. (2010). Regulation of parkinsonian motor behaviours by optogenetic control of basal ganglia circuitry. *Nature, 466*, 622–626.

Kravitz, A. V., Tye, L. D., & Kreitzer, A. C. (2012). Distinct roles for direct and indirect pathway striatal neurons in reinforcement. *Nature Neuroscience, 15*, 816–818.

Kreitzer, A. C. (2009). Physiology and pharmacology of striatal neurons. *Annual Review of Neuroscience, 32*, 127–147.

Kriaucionis, S., Paterson, A., Curtis, J., Guy, J., Macleod, N., & Bird, A. (2006). Gene expression analysis exposes mitochondrial abnormalities in a mouse model of Rett syndrome. *Molecular and Cellular Biology, 26*, 5033–5042.

Kubota, Y., & Kawaguchi, Y. (2000). Dependence of GABAergic synaptic areas on the interneuron type and target size. *The Journal of Neuroscience, 20*, 375–386.

Kulisevsky, J., Marti, M. J., Ferrer, I., & Tolosa, E. (1988). Meige syndrome: Neuropathology of a case. *Movement Disorders, 3*, 170–175.

Kurlan, R. (2010). Clinical practice. Tourette's syndrome. *The New England Journal of Medicine, 363*, 2332–2338.

Kustedjo, K., Bracey, M. H., & Cravatt, B. F. (2000). Torsin A and its torsion dystonia-associated mutant forms are lumenal glycoproteins that exhibit distinct subcellular localizations. *The Journal of Biological Chemistry, 275*, 27933–27939.

Kwak, Y., Bohnen, N. I., Muller, M. L., Dayalu, P., & Seidler, R. D. (2013). Striatal denervation pattern predicts levodopa effects on sequence learning in Parkinson's disease. *Journal of Motor Behavior, 45*, 423–429.

Kwak, C. H., & Jankovic, J. (2002). Tourettism and dystonia after subcortical stroke. *Movement Disorders, 17*, 821–825.

Lansink, C. S., Goltstein, P. M., Lankelma, J. V., & Pennartz, C. M. (2010). Fast-spiking interneurons of the rat ventral striatum: Temporal coordination of activity with principal cells and responsiveness to reward. *The European Journal of Neuroscience, 32*, 494–508.

Lehericy, S., Benali, H., Van de Moortele, P. F., Pelegrini-Issac, M., Waechter, T., Ugurbil, K., et al. (2005). Distinct basal ganglia territories are engaged in early and advanced motor sequence learning. *Proceedings of the National Academy of Sciences of the United States of America, 102*, 12566–12571.

Lekman, A., Witt-Engerstrom, I., Gottfries, J., Hagberg, B. A., Percy, A. K., & Svennerholm, L. (1989). Rett syndrome: Biogenic amines and metabolites in postmortem brain. *Pediatric Neurology, 5*, 357–362.

Leventhal, D. K., Stoetzner, C. R., Abraham, R., Pettibone, J., DeMarco, K., & Berke, J. D. (2014). Dissociable effects of dopamine on learning and performance within sensorimotor striatum. *Basal Ganglia*. in press. http://dx.doi.org/10.1016/j.baga.2013.11.001.

Lewis, J. D., Meehan, R. R., Henzel, W. J., Maurer-Fogy, I., Jeppesen, P., Klein, F., et al. (1992). Purification, sequence, and cellular localization of a novel chromosomal protein that binds to methylated DNA. *Cell, 69*, 905–914.

Lim, B. K., Huang, K. W., Grueter, B. A., Rothwell, P. E., & Malenka, R. C. (2012). Anhedonia requires MC4R-mediated synaptic adaptations in nucleus accumbens. *Nature, 487*, 183–189.

Limousin, P., Brown, R. G., Jahanshahi, M., Asselman, P., Quinn, N. P., Thomas, D., et al. (1999). The effects of posteroventral pallidotomy on the preparation and execution of voluntary hand and arm movements in Parkinson's disease. *Brain, 122*, 315–327.

Lipsman, N., Neimat, J. S., & Lozano, A. M. (2007). Deep brain stimulation for treatment-refractory obsessive-compulsive disorder: The search for a valid target. *Neurosurgery, 61*, 1–11, discussion 11.

Lo, C. C., & Wang, X. J. (2006). Cortico-basal ganglia circuit mechanism for a decision threshold in reaction time tasks. *Nature Neuroscience, 9*, 956–963.

Lorden, J. F., McKeon, T. W., Baker, H. J., Cox, N., & Walkley, S. U. (1984). Characterization of the rat mutant dystonic (dt): A new animal model of dystonia musculorum deformans. *The Journal of Neuroscience, 4*, 1925–1932.

Lorincz, M. T. (2012). Recognition and treatment of neurologic Wilson's disease. *Seminars in Neurology, 32*, 538–543.

Loscher, W., Fisher, J. E. J., Schmidt, D., Fredow, G., Honack, D., & Iturrian, W. B. (1989). The sz mutant hamster: A genetic model of epilepsy or of paroxysmal dystonia? *Movement Disorders, 4*, 219–232.

Lu, X., Matsuzawa, M., & Hikosaka, O. (2002). A neural correlate of oculomotor sequences in supplementary eye field. *Neuron, 34*, 317–325.

Luk, K. C., & Sadikot, A. F. (2001). GABA promotes survival but not proliferation of parvalbumin-immunoreactive interneurons in rodent neostriatum: An in vivo study with stereology. *Neuroscience, 104*, 93–103.

Mallet, N., Micklem, B. R., Henny, P., Brown, M. T., Williams, C., Bolam, J. P., et al. (2012). Dichotomous organization of the external globus pallidus. *Neuron, 74*, 1075–1086.

Mark, M. H., Sage, J. I., Dickson, D. W., Heikkila, R. E., Manzino, L., Schwarz, K. O., et al. (1994). Meige syndrome in the spectrum of Lewy body disease. *Neurology, 44*, 1432–1436.

Marsden, C. D., Obeso, J. A., Zarranz, J. J., & Lang, A. E. (1985). The anatomical basis of symptomatic hemidystonia. *Brain, 108*, 463–483.

Marshak, S., Meynard, M. M., De Vries, Y. A., Kidane, A. H., & Cohen-Cory, S. (2012). Cell-autonomous alterations in dendritic arbor morphology and connectivity induced by overexpression of MeCP2 in Xenopus central neurons in vivo. *PLoS One, 7*, e33153.

Martella, G., Tassone, A., Sciamanna, G., Platania, P., Cuomo, D., Viscomi, M. T., et al. (2009). Impairment of bidirectional synaptic plasticity in the striatum of a mouse model of DYT1 dystonia: Role of endogenous acetylcholine. *Brain, 132*, 2336–2349.

Masuyama, T., Matsuo, M., Jing, J. J., Tabara, Y., Kitsuki, K., Yamagata, H., et al. (2005). Classic Rett syndrome in a boy with R133C mutation of MECP2. *Brain Dev*, *27*, 439–442.
Mathai, A., & Smith, Y. (2011). The corticostriatal and corticosubthalamic pathways: Two entries, one target. So what? *Frontiers in Systems Neuroscience*, *5*, 64.
Matsumoto, N., Hanakawa, T., Maki, S., Graybiel, A. M., & Kimura, M. (1999). Role of [corrected] nigrostriatal dopamine system in learning to perform sequential motor tasks in a predictive manner. *Journal of Neurophysiology*, *82*, 978–998.
Matsumoto, M., & Hikosaka, O. (2009). Two types of dopamine neuron distinctly convey positive and negative motivational signals. *Nature*, *459*, 837–841.
Matsumoto, N., Minamimoto, T., Graybiel, A. M., & Kimura, M. (2001). Neurons in the thalamic CM-Pf complex supply striatal neurons with information about behaviorally significant sensory events. *Journal of Neurophysiology*, *85*, 960–976.
McGeorge, A. J., & Faull, R. L. (1989). The organization of the projection from the cerebral cortex to the striatum in the rat. *Neuroscience*, *29*, 503–537.
McGill, B. E., Bundle, S. F., Yaylaoglu, M. B., Carson, J. P., Thaller, C., & Zoghbi, H. Y. (2006). Enhanced anxiety and stress-induced corticosterone release are associated with increased Crh expression in a mouse model of Rett syndrome. *Proceedings of the National Academy of Sciences of the United States of America*, *103*, 18267–18272.
McGrath, M. J., Campbell, K. M., & Burton, F. H. (1999). The role of cognitive and affective processing in a transgenic mouse model of cortical-limbic neuropotentiated compulsive behavior. *Behavioral Neuroscience*, *113*, 1249–1256.
McGrath, M. J., Campbell, K. M., Parks, C. R., & Burton, F. H. (2000). Glutamatergic drugs exacerbate symptomatic behavior in a transgenic model of comorbid Tourette's syndrome and obsessive-compulsive disorder. *Brain Research*, *877*, 23–30.
McGraw, C. M., Samaco, R. C., & Zoghbi, H. Y. (2011). Adult neural function requires MeCP2. *Science*, *333*, 186.
McNaught, K. S., Kapustin, A., Jackson, T., Jengelley, T. A., Jnobaptiste, R., Shashidharan, P., et al. (2004). Brainstem pathology in DYT1 primary torsion dystonia. *Annals of Neurology*, *56*, 540–547.
Medrihan, L., Tantalaki, E., Aramuni, G., Sargsyan, V., Dudanova, I., Missler, M., et al. (2008). Early defects of GABAergic synapses in the brain stem of a MeCP2 mouse model of Rett syndrome. *Journal of Neurophysiology*, *99*, 112–121.
Mehanna, R., & Jankovic, J. (2013). Movement disorders in cerebrovascular disease. *Lancet Neurology*, *12*, 597–608.
Mink, J. W. (2003). The Basal Ganglia and involuntary movements: Impaired inhibition of competing motor patterns. *Archives of Neurology*, *60*, 1365–1368.
Mink, J. W., & Thach, W. T. (1991a). Basal ganglia motor control. II. Late pallidal timing relative to movement onset and inconsistent pallidal coding of movement parameters. *Journal of Neurophysiology*, *65*, 301–329.
Mink, J. W., & Thach, W. T. (1991b). Basal ganglia motor control. III. Pallidal ablation: Normal reaction time, muscle cocontraction, and slow movement. *Journal of Neurophysiology*, *65*, 330–351.
Miyachi, S., Hikosaka, O., Miyashita, K., Karadi, Z., & Rand, M. K. (1997). Differential roles of monkey striatum in learning of sequential hand movement. *Experimental Brain Research*, *115*, 1–5.
Morikawa, H., & Paladini, C. A. (2011). Dynamic regulation of midbrain dopamine neuron activity: Intrinsic, synaptic, and plasticity mechanisms. *Neuroscience*, *198*, 95–111.
Morris, G., Arkadir, D., Nevet, A., Vaadia, E., & Bergman, H. (2004). Coincident but distinct messages of midbrain dopamine and striatal tonically active neurons. *Neuron*, *43*, 133–143.

Murakami, J. W., Courchesne, E., Haas, R. H., Press, G. A., & Yeung-Courchesne, R. (1992). Cerebellar and cerebral abnormalities in Rett syndrome: A quantitative MR analysis. *AJR. American Journal of Roentgenology*, *159*, 177–183.

Na, E. S., Nelson, E. D., Adachi, M., Autry, A. E., Mahgoub, M. A., Kavalali, E. T., et al. (2012). A mouse model for MeCP2 duplication syndrome: MeCP2 overexpression impairs learning and memory and synaptic transmission. *The Journal of Neuroscience*, *32*, 3109–3117.

Nakashima, K., Rothwell, J. C., Day, B. L., Thompson, P. D., Shannon, K., & Marsden, C. D. (1989). Reciprocal inhibition between forearm muscles in patients with writer's cramp and other occupational cramps, symptomatic hemidystonia and hemiparesis due to stroke. *Brain*, *112*, 681–697.

Nan, X., Ng, H. H., Johnson, C. A., Laherty, C. D., Turner, B. M., Eisenman, R. N., et al. (1998). Transcriptional repression by the methyl-CpG-binding protein MeCP2 involves a histone deacetylase complex. *Nature*, *393*, 386–389.

Napolitano, F., Pasqualetti, M., Usiello, A., Santini, E., Pacini, G., Sciamanna, G., et al. (2010). Dopamine D2 receptor dysfunction is rescued by adenosine A2A receptor antagonism in a model of DYT1 dystonia. *Neurobiology of Disease*, *38*, 434–445.

Narayanaswamy, J. C., Viswanath, B., Veshnal Cherian, A., Bada Math, S., Kandavel, T., & Janardhan Reddy, Y. C. (2012). Impact of age of onset of illness on clinical phenotype in OCD. *Psychiatry Research*, *200*, 554–559.

Nery, F. C., Armata, I. A., Farley, J. E., Cho, J. A., Yaqub, U., Chen, P., et al. (2011). TorsinA participates in endoplasmic reticulum-associated degradation. *Nature Communications*, *2*, 393.

Neuwald, A. F., Aravind, L., Spouge, J. L., & Koonin, E. V. (1999). AAA+: A class of chaperone-like ATPases associated with the assembly, operation, and disassembly of protein complexes. *Genome Research*, *9*, 27–43.

Neychev, V. K., Fan, X., Mitev, V. I., Hess, E. J., & Jinnah, H. A. (2008). The basal ganglia and cerebellum interact in the expression of dystonic movement. *Brain*, *131*, 2499–2509.

Nguyen, M. V., Du, F., Felice, C. A., Shan, X., Nigam, A., Mandel, G., et al. (2012). MeCP2 is critical for maintaining mature neuronal networks and global brain anatomy during late stages of postnatal brain development and in the mature adult brain. *The Journal of Neuroscience*, *32*, 10021–10034.

Obeso, J. A., Jahanshahi, M., Alvarez, L., Macias, R., Pedroso, I., Wilkinson, L., et al. (2009). What can man do without basal ganglia motor output? The effect of combined unilateral subthalamotomy and pallidotomy in a patient with Parkinson's disease. *Experimental Neurology*, *220*, 283–292.

Ozelius, L. J., Hewett, J. W., Page, C. E., Bressman, S. B., Kramer, P. L., Shalish, C., et al. (1997). The early-onset torsion dystonia gene (DYT1) encodes an ATP-binding protein. *Nature Genetics*, *17*, 40–48.

Ozomaro, U., Cai, G., Kajiwara, Y., Yoon, S., Makarov, V., Delorme, R., et al. (2013). Characterization of SLITRK1 variation in obsessive-compulsive disorder. *PLoS One*, *8*, e70376.

Page, M. E., Bao, L., Andre, P., Pelta-Heller, J., Sluzas, E., Gonzalez-Alegre, P., et al. (2010). Cell-autonomous alteration of dopaminergic transmission by wild type and mutant (DeltaE) TorsinA in transgenic mice. *Neurobiology of Disease*, *39*, 318–326.

Panizza, M., Lelli, S., Nilsson, J., & Hallett, M. (1990). H-reflex recovery curve and reciprocal inhibition of H-reflex in different kinds of dystonia. *Neurology*, *40*, 824–828.

Parent, A., & Hazrati, L. N. (1995). Functional anatomy of the basal ganglia. I. The cortico-basal ganglia-thalamo-cortical loop. *Brain Research Brain Research Reviews*, *20*, 91–127.

Paudel, R., Hardy, J., Revesz, T., Holton, J. L., & Houlden, H. (2012). Review: Genetics and neuropathology of primary pure dystonia. *Neuropathology and Applied Neurobiology, 38*, 520–534.

Pauls, D. L. (2008). The genetics of obsessive compulsive disorder: A review of the evidence. *American Journal of Medical Genetics Part C: Seminars in Medical Genetics, 148C*, 133–139.

Perlmutter, J. S., Stambuk, M. K., Markham, J., Black, K. J., McGee-Minnich, L., Jankovic, J., et al. (1997). Decreased [18F]spiperone binding in putamen in idiopathic focal dystonia. *The Journal of Neuroscience, 17*, 843–850.

Peterson, D. A., Sejnowski, T. J., & Poizner, H. (2010). Convergent evidence for abnormal striatal synaptic plasticity in dystonia. *Neurobiology of Disease, 37*, 558–573.

Peterson, B. S., Thomas, P., Kane, M. J., Scahill, L., Zhang, H., Bronen, R., et al. (2003). Basal Ganglia volumes in patients with Gilles de la Tourette syndrome. *Archives of General Psychiatry, 60*, 415–424.

Pisani, A., Bernardi, G., Ding, J., & Surmeier, D. J. (2007). Re-emergence of striatal cholinergic interneurons in movement disorders. *Trends in Neurosciences, 30*, 545–553.

Pisani, A., Martella, G., Tscherter, A., Bonsi, P., Sharma, N., Bernardi, G., et al. (2006). Altered responses to dopaminergic D2 receptor activation and N-type calcium currents in striatal cholinergic interneurons in a mouse model of DYT1 dystonia. *Neurobiology of Disease, 24*, 318–325.

Powell, S. B., Newman, H. A., Pendergast, J. F., & Lewis, M. H. (1999). A rodent model of spontaneous stereotypy: Initial characterization of developmental, environmental, and neurobiological factors. *Physiology & Behavior, 66*, 355–363.

Presti, M. F., Gibney, B. C., & Lewis, M. H. (2004). Effects of intrastriatal administration of selective dopaminergic ligands on spontaneous stereotypy in mice. *Physiology & Behavior, 80*, 433–439.

Presti, M. F., & Lewis, M. H. (2005). Striatal opioid peptide content in an animal model of spontaneous stereotypic behavior. *Behavioural Brain Research, 157*, 363–368.

Presti, M. F., Mikes, H. M., & Lewis, M. H. (2003). Selective blockade of spontaneous motor stereotypy via intrastriatal pharmacological manipulation. *Pharmacology, Biochemistry and Behavior, 74*, 833–839.

Presti, M. F., Powell, S. B., & Lewis, M. H. (2002). Dissociation between spontaneously emitted and apomorphine-induced stereotypy in Peromyscus maniculatus bairdii. *Physiology & Behavior, 75*, 347–353.

Presti, M. F., Watson, C. J., Kennedy, R. T., Yang, M., & Lewis, M. H. (2004). Behavior-related alterations of striatal neurochemistry in a mouse model of stereotyped movement disorder. *Pharmacology, Biochemistry and Behavior, 77*, 501–507.

Priori, A., Giannicola, G., Rosa, M., Marceglia, S., Servello, D., Sassi, M., et al. (2013). Deep brain electrophysiological recordings provide clues to the pathophysiology of Tourette syndrome. *Neuroscience & Biobehavioral Reviews, 37*, 1063–1068.

Prudente, C. N., Hess, E. J., & Jinnah, H. A. (2013). Dystonia as a network disorder: What is the role of the cerebellum? *Neuroscience, 260C*, 23–35.

Quartarone, A., Classen, J., Morgante, F., Rosenkranz, K., & Hallett, M. (2009). Consensus paper: Use of transcranial magnetic stimulation to probe motor cortex plasticity in dystonia and levodopa-induced dyskinesia. *Brain Stimulation, 2*, 108–117.

Quartarone, A., & Hallett, M. (2013). Emerging concepts in the physiological basis of dystonia. *Movement Disorders, 28*, 958–967.

Quartarone, A., & Pisani, A. (2011). Abnormal plasticity in dystonia: Disruption of synaptic homeostasis. *Neurobiology of Disease, 42*, 162–170.

Quartarone, A., Siebner, H. R., & Rothwell, J. C. (2006). Task-specific hand dystonia: Can too much plasticity be bad for you? *Trends in Neurosciences, 29*, 192–199.

Ramanathan, S., Hanley, J. J., Deniau, J. M., & Bolam, J. P. (2002). Synaptic convergence of motor and somatosensory cortical afferents onto GABAergic interneurons in the rat striatum. *The Journal of Neuroscience, 22*, 8158–8169.

Ramdhani, R. A., & Simonyan, K. (2013). Primary dystonia: Conceptualizing the disorder through a structural brain imaging lens. *Tremor and Other Hyperkinetic Movements (New York), 3*, 1–11.

Ratcliff, R., & Frank, M. J. (2012). Reinforcement-based decision making in corticostriatal circuits: Mutual constraints by neurocomputational and diffusion models. *Neural Computation, 24*, 1186–1229.

Rauch, S. L., Whalen, P. J., Savage, C. R., Curran, T., Kendrick, A., Brown, H. D., et al. (1997). Striatal recruitment during an implicit sequence learning task as measured by functional magnetic resonance imaging. *Human Brain Mapping, 5*, 124–132.

Redgrave, P., Vautrelle, N., & Reynolds, J. N. (2011). Functional properties of the basal ganglia's re-entrant loop architecture: Selection and reinforcement. *Neuroscience, 198*, 138–151.

Reiss, A. L., Faruque, F., Naidu, S., Abrams, M., Beaty, T., Bryan, R. N., et al. (1993). Neuroanatomy of Rett syndrome: A volumetric imaging study. *Annals of Neurology, 34*, 227–234.

Rinne, J. O., Iivanainen, M., Metsahonkala, L., Vainionpaa, L., Paakkonen, L., Nagren, K., et al. (2004). Striatal dopaminergic system in dopa-responsive dystonia: A multi-tracer PET study shows increased D2 receptors. *Journal of Neural Transmission, 111*, 59–67.

Rizzo, R., Gulisano, M., Cali, P. V., & Curatolo, P. (2013). Tourette syndrome and comorbid ADHD: Current pharmacological treatment options. *European Journal of Paediatric Neurology, 17*, 421–428.

Robertson, M. M. (2000). Tourette syndrome, associated conditions and the complexities of treatment. *Brain, 123*(Pt. 3), 425–462.

Robertson, M. M. (2006). Attention deficit hyperactivity disorder, tics and Tourette's syndrome: The relationship and treatment implications. A commentary. *European Child & Adolescent Psychiatry, 15*, 1–11.

Robertson, M. M., Eapen, V., & Cavanna, A. E. (2009). The international prevalence, epidemiology, and clinical phenomenology of Tourette syndrome: A cross-cultural perspective. *Journal of Psychosomatic Research, 67*, 475–483.

Robinson, S., Sotak, B. N., During, M. J., & Palmiter, R. D. (2006). Local dopamine production in the dorsal striatum restores goal-directed behavior in dopamine-deficient mice. *Behavioral Neuroscience, 120*, 196–200.

Roessner, V., Overlack, S., Schmidt-Samoa, C., Baudewig, J., Dechent, P., Rothenberger, A., et al. (2011). Increased putamen and callosal motor subregion in treatment-naive boys with Tourette syndrome indicates changes in the bihemispheric motor network. *The Journal of Child Psychology and Psychiatry, 52*, 306–314.

Rubin, J. E., McIntyre, C. C., Turner, R. S., & Wichmann, T. (2012). Basal ganglia activity patterns in parkinsonism and computational modeling of their downstream effects. *The European Journal of Neuroscience, 36*, 2213–2228.

Russo, G., Nieus, T. R., Maggi, S., & Taverna, S. (2013). Dynamics of action potential firing in electrically connected striatal fast-spiking interneurons. *Frontiers in Cellular Neuroscience, 7*, 209.

Sadikot, A. F., & Rymar, V. V. (2009). The primate centromedian-parafascicular complex: Anatomical organization with a note on neuromodulation. *Brain Research Bulletin, 78*, 122–130.

Sarvet, B. (2013). Childhood obsessive-compulsive disorder. *Pediatrics in Review, 34*, 19–27, quiz 28.

Scharf, J. M., Yu, D., Mathews, C. A., Neale, B. M., Stewart, S. E., Fagerness, J. A., et al. (2013). Genome-wide association study of Tourette's syndrome. *Molecular Psychiatry, 18*, 721–728.
Schlicker, E., Malinowska, B., Kathmann, M., & Gothert, M. (1994). Modulation of neurotransmitter release via histamine H3 heteroreceptors. *Fundamental & Clinical Pharmacology, 8*, 128–137.
Schmidt, R., Leventhal, D. K., Mallet, N., Chen, F., & Berke, J. D. (2013). Canceling actions involves a race between basal ganglia pathways. *Nature Neuroscience, 16*, 1118–1124.
Schultz, W. (2013). Updating dopamine reward signals. *Current Opinion in Neurobiology, 23*, 229–238.
Schultz, W., Apicella, P., & Ljungberg, T. (1993). Responses of monkey dopamine neurons to reward and conditioned stimuli during successive steps of learning a delayed response task. *The Journal of Neuroscience, 13*, 900–913.
Schultz, W., Dayan, P., & Montague, P. R. (1997). A neural substrate of prediction and reward. *Science, 275*, 1593–1599.
Schulz, J. M., & Reynolds, J. N. (2013). Pause and rebound: Sensory control of cholinergic signaling in the striatum. *Trends in Neurosciences, 36*, 41–50.
Schwartz, J. C., Arrang, J. M., Garbarg, M., Pollard, H., & Ruat, M. (1991). Histaminergic transmission in the mammalian brain. *Physiological Reviews, 71*, 1–51.
Sciamanna, G., Bonsi, P., Tassone, A., Cuomo, D., Tscherter, A., Viscomi, M. T., et al. (2009). Impaired striatal D2 receptor function leads to enhanced GABA transmission in a mouse model of DYT1 dystonia. *Neurobiology of Disease, 34*, 133–145.
Sciamanna, G., Hollis, R., Ball, C., Martella, G., Tassone, A., Marshall, A., et al. (2012). Cholinergic dysregulation produced by selective inactivation of the dystonia-associated protein torsinA. *Neurobiology of Disease, 47*, 416–427.
Sciamanna, G., Tassone, A., Mandolesi, G., Puglisi, F., Ponterio, G., Martella, G., et al. (2012). Cholinergic dysfunction alters synaptic integration between thalamostriatal and corticostriatal inputs in DYT1 dystonia. *The Journal of Neuroscience, 32*, 11991–12004.
Sciamanna, G., Tassone, A., Martella, G., Mandolesi, G., Puglisi, F., Cuomo, D., et al. (2011). Developmental profile of the aberrant dopamine D2 receptor response in striatal cholinergic interneurons in DYT1 dystonia. *PLoS One, 6*, e24261.
Scolding, N. J., Smith, S. M., Sturman, S., Brookes, G. B., & Lees, A. J. (1995). Auctioneer's jaw: A case of occupational oromandibular hemidystonia. *Movement Disorders, 10*, 508–509.
Scontrini, A., Conte, A., Defazio, G., Fiorio, M., Fabbrini, G., Suppa, A., et al. (2009). Somatosensory temporal discrimination in patients with primary focal dystonia. *Journal of Neurology, Neurosurgery and Psychiatry, 80*, 1315–1319.
Seidler, R. D., Purushotham, A., Kim, S. G., Ugurbil, K., Willingham, D., & Ashe, J. (2005). Neural correlates of encoding and expression in implicit sequence learning. *Experimental Brain Research, 165*, 114–124.
Selemon, L. D., & Goldman-Rakic, P. S. (1985). Longitudinal topography and interdigitation of corticostriatal projections in the rhesus monkey. *The Journal of Neuroscience, 5*, 776–794.
Shahbazian, M., Young, J., Yuva-Paylor, L., Spencer, C., Antalffy, B., Noebels, J., et al. (2002). Mice with truncated MeCP2 recapitulate many Rett syndrome features and display hyperacetylation of histone H3. *Neuron, 35*, 243–254.
Sharma, N., Baxter, M. G., Petravicz, J., Bragg, D. C., Schienda, A., Standaert, D. G., et al. (2005). Impaired motor learning in mice expressing torsinA with the DYT1 dystonia mutation. *The Journal of Neuroscience, 25*, 5351–5355.
Shashidharan, P., Sandu, D., Potla, U., Armata, I. A., Walker, R. H., McNaught, K. S., et al. (2005). Transgenic mouse model of early-onset DYT1 dystonia. *Human Molecular Genetics, 14*, 125–133.

Shen, W., Flajolet, M., Greengard, P., & Surmeier, D. J. (2008). Dichotomous dopaminergic control of striatal synaptic plasticity. *Science, 321*, 848–851.
Sidibe, M., & Smith, Y. (1999). Thalamic inputs to striatal interneurons in monkeys: Synaptic organization and co-localization of calcium binding proteins. *Neuroscience, 89*, 1189–1208.
Singer, H. S., Szymanski, S., Giuliano, J., Yokoi, F., Dogan, A. S., Brasic, J. R., et al. (2002). Elevated intrasynaptic dopamine release in Tourette's syndrome measured by PET. *The American Journal of Psychiatry, 159*, 1329–1336.
Smith, K. S., & Graybiel, A. M. (2013a). A dual operator view of habitual behavior reflecting cortical and striatal dynamics. *Neuron, 79*, 361–374.
Smith, K. S., & Graybiel, A. M. (2013b). Using optogenetics to study habits. *Brain Research, 1511*, 102–114.
Smith, K. S., Virkud, A., Deisseroth, K., & Graybiel, A. M. (2012). Reversible online control of habitual behavior by optogenetic perturbation of medial prefrontal cortex. *Proceedings of the National Academy of Sciences of the United States of America, 109*, 18932–18937.
Song, C. H., Bernhard, D., Bolarinwa, C., Hess, E. J., Smith, Y., & Jinnah, H. A. (2013). Subtle microstructural changes of the striatum in a DYT1 knock-in mouse model of dystonia. *Neurobiology of Disease, 54*, 362–371.
Stillman, A. A., Krsnik, Z., Sun, J., Rasin, M. R., State, M. W., Sestan, N., et al. (2009). Developmentally regulated and evolutionarily conserved expression of SLITRK1 in brain circuits implicated in Tourette syndrome. *The Journal of Comparative Neurology, 513*, 21–37.
Subramaniam, B., Naidu, S., & Reiss, A. L. (1997). Neuroanatomy in Rett syndrome: Cerebral cortex and posterior fossa. *Neurology, 48*, 399–407.
Sullivan, M. A., Chen, H., & Morikawa, H. (2008). Recurrent inhibitory network among striatal cholinergic interneurons. *The Journal of Neuroscience, 28*, 8682–8690.
Takeuchi, M., Hata, Y., Hirao, K., Toyoda, A., Irie, M., & Takai, Y. (1997). SAPAPs. A family of PSD-95/SAP90-associated proteins localized at postsynaptic density. *The Journal of Biological Chemistry, 272*, 11943–11951.
Tanabe, L. M., Kim, C. E., Alagem, N., & Dauer, W. T. (2009). Primary dystonia: Molecules and mechanisms. *Nature Reviews Neurology, 5*, 598–609.
Tanabe, L. M., Martin, C., & Dauer, W. T. (2012). Genetic background modulates the phenotype of a mouse model of DYT1 dystonia. *PLoS One, 7*, e32245.
Tang, C. C., Root, D. H., Duke, D. C., Zhu, Y., Teixeria, K., Ma, S., et al. (2009). Decreased firing of striatal neurons related to licking during acquisition and overtraining of a licking task. *The Journal of Neuroscience, 29*, 13952–13961.
Tanimura, Y., King, M. A., Williams, D. K., & Lewis, M. H. (2011). Development of repetitive behavior in a mouse model: Roles of indirect and striosomal basal ganglia pathways. *International Journal of Developmental Neuroscience, 29*, 461–467.
Tanimura, Y., Vaziri, S., & Lewis, M. H. (2010). Indirect basal ganglia pathway mediation of repetitive behavior: Attenuation by adenosine receptor agonists. *Behavioural Brain Research, 210*, 116–122.
Tate, P., Skarnes, W., & Bird, A. (1996). The methyl-CpG binding protein MeCP2 is essential for embryonic development in the mouse. *Nature Genetics, 12*, 205–208.
Taylor, A. E., Lang, A. E., Saint-Cyr, J. A., Riley, D. E., & Ranawaya, R. (1991). Cognitive processes in idiopathic dystonia treated with high-dose anticholinergic therapy: Implications for treatment strategies. *Clinical Neuropharmacology, 14*, 62–77.
Taylor, J. L., Rajbhandari, A. K., Berridge, K. C., & Aldridge, J. W. (2010). Dopamine receptor modulation of repetitive grooming actions in the rat: Potential relevance for Tourette syndrome. *Brain Research, 1322*, 92–101.
Tepper, J. M. (2010). GABAergic interneurons of the striatum. In H. Steiner & K. Tseng (Eds.), *Handbook of basal ganglia structure and function* (pp. 151–166). Oxford: Elsevier B.V.

Tepper, J. M., Tecuapetla, F., Koos, T., & Ibanez-Sandoval, O. (2010). Heterogeneity and diversity of striatal GABAergic interneurons. *Frontiers in Neuroanatomy*, *4*, 150.

Thomas, T. M., Smith, Y., Levey, A. I., & Hersch, S. M. (2000). Cortical inputs to m2-immunoreactive striatal interneurons in rat and monkey. *Synapse*, *37*, 252–261.

Threlfell, S., Lalic, T., Platt, N. J., Jennings, K. A., Deisseroth, K., & Cragg, S. J. (2012). Striatal dopamine release is triggered by synchronized activity in cholinergic interneurons. *Neuron*, *75*, 58–64.

Tisch, S., Limousin, P., Rothwell, J. C., Asselman, P., Quinn, N., Jahanshahi, M., et al. (2006). Changes in blink reflex excitability after globus pallidus internus stimulation for dystonia. *Movement Disorders*, *21*, 1650–1655.

Torres, G. E., Sweeney, A. L., Beaulieu, J. M., Shashidharan, P., & Caron, M. G. (2004). Effect of torsinA on membrane proteins reveals a loss of function and a dominant-negative phenotype of the dystonia-associated DeltaE-torsinA mutant. *Proceedings of the National Academy of Sciences of the United States of America*, *101*, 15650–15655.

Trender-Gerhard, I., Sweeney, M. G., Schwingenschuh, P., Mir, P., Edwards, M. J., Gerhard, A., et al. (2009). Autosomal-dominant GTPCH1-deficient DRD: Clinical characteristics and long-term outcome of 34 patients. *Journal of Neurology, Neurosurgery and Psychiatry*, *80*, 839–845.

Tsai, H. C., Zhang, F., Adamantidis, A., Stuber, G. D., Bonci, A., de Lecea, L., et al. (2009). Phasic firing in dopaminergic neurons is sufficient for behavioral conditioning. *Science*, *324*, 1080–1084.

Tudor, M., Akbarian, S., Chen, R. Z., & Jaenisch, R. (2002). Transcriptional profiling of a mouse model for Rett syndrome reveals subtle transcriptional changes in the brain. *Proceedings of the National Academy of Sciences of the United States of America*, *99*, 15536–15541.

Turner, R. S., & Desmurget, M. (2010). Basal ganglia contributions to motor control: A vigorous tutor. *Current Opinion in Neurobiology*, *20*, 704–716.

Ulug, A. M., Vo, A., Argyelan, M., Tanabe, L., Schiffer, W. K., Dewey, S., et al. (2011). Cerebellothalamocortical pathway abnormalities in torsinA DYT1 knock-in mice. *Proceedings of the National Academy of Sciences of the United States of America*, *108*, 6638–6643.

Ungerstedt, U., & Arbuthnott, G. W. (1970). Quantitative recording of rotational behavior in rats after 6-hydroxy-dopamine lesions of the nigrostriatal dopamine system. *Brain Research*, *24*, 485–493.

van den Akker, E., Reijnen, M., Korving, J., Brouwer, A., Meijlink, F., & Deschamps, J. (1999). Targeted inactivation of Hoxb8 affects survival of a spinal ganglion and causes aberrant limb reflexes. *Mechanisms of Development*, *89*, 103–114.

Van den Veyver, I. B., & Zoghbi, H. Y. (2000). Methyl-CpG-binding protein 2 mutations in Rett syndrome. *Current Opinion in Genetics & Development*, *10*, 275–279.

Vandewalle, V., van der Linden, C., Groenewegen, H. J., & Caemaert, J. (1999). Stereotactic treatment of Gilles de la Tourette syndrome by high frequency stimulation of thalamus. *Lancet*, *353*, 724.

Vercueil, L., Krack, P., & Pollak, P. (2002). Results of deep brain stimulation for dystonia: A critical reappraisal. *Movement Disorders*, *17*(Suppl. 3), S89–S93.

Vidailhet, M., Jutras, M. F., Grabli, D., & Roze, E. (2013). Deep brain stimulation for dystonia. *Journal of Neurology, Neurosurgery and Psychiatry*, *84*, 1029–1042.

Vidailhet, M., Vercueil, L., Houeto, J. L., Krystkowiak, P., Benabid, A. L., Cornu, P., et al. (2005). Bilateral deep-brain stimulation of the globus pallidus in primary generalized dystonia. *The New England Journal of Medicine*, *352*, 459–467.

Villard, L. (2007). MECP2 mutations in males. *Journal of Medical Genetics*, *44*, 417–423.

Villard, L., Kpebe, A., Cardoso, C., Chelly, P. J., Tardieu, P. M., & Fontes, M. (2000). Two affected boys in a Rett syndrome family: Clinical and molecular findings. *Neurology*, *55*, 1188–1193.

Walker, R. H., Brin, M. F., Sandu, D., Good, P. F., & Shashidharan, P. (2002). TorsinA immunoreactivity in brains of patients with DYT1 and non-DYT1 dystonia. *Neurology, 58*, 120–124.

Wall, V. Z., Parker, J. G., Fadok, J. P., Darvas, M., Zweifel, L., & Palmiter, R. D. (2011). A behavioral genetics approach to understanding D1 receptor involvement in phasic dopamine signaling. *Molecular and Cellular Neurosciences, 46*, 21–31.

Wan, Y., Ade, K. K., Caffall, Z., Ilcim Ozlu, M., Eroglu, C., Feng, G., et al. (2013). Circuit-selective striatal synaptic dysfunction in the Sapap3 knockout mouse model of obsessive-compulsive disorder. *Biological Psychiatry, 75*(8), 623–630.

Wan, Y., Feng, G., & Calakos, N. (2011). Sapap3 deletion causes mGluR5-dependent silencing of AMPAR synapses. *The Journal of Neuroscience, 31*, 16685–16691.

Wan, M., Lee, S. S., Zhang, X., Houwink-Manville, I., Song, H. R., Amir, R. E., et al. (1999). Rett syndrome and beyond: Recurrent spontaneous and familial MECP2 mutations at CpG hotspots. *The American Journal of Human Genetics, 65*, 1520–1529.

Watanabe, M., & Munoz, D. P. (2010). Saccade suppression by electrical microstimulation in monkey caudate nucleus. *The Journal of Neuroscience, 30*, 2700–2709.

Welch, J. M., Lu, J., Rodriguiz, R. M., Trotta, N. C., Peca, J., Ding, J. D., et al. (2007). Cortico-striatal synaptic defects and OCD-like behaviours in Sapap3-mutant mice. *Nature, 448*, 894–900.

Welch, J. M., Wang, D., & Feng, G. (2004). Differential mRNA expression and protein localization of the SAP90/PSD-95-associated proteins (SAPAPs) in the nervous system of the mouse. *The Journal of Comparative Neurology, 472*, 24–39.

Wenger, K. K., Musch, K. L., & Mink, J. W. (1999). Impaired reaching and grasping after focal inactivation of globus pallidus pars interna in the monkey. *Journal of Neurophysiology, 82*, 2049–2060.

Wenk, G. L. (1995). Alterations in dopaminergic function in Rett syndrome. *Neuropediatrics, 26*, 123–125.

Wenk, G. L., & Mobley, S. L. (1996). Choline acetyltransferase activity and vesamicol binding in Rett syndrome and in rats with nucleus basalis lesions. *Neuroscience, 73*, 79–84.

Wenk, G. L., O'Leary, M., Nemeroff, C. B., Bissette, G., Moser, H., & Naidu, S. (1993). Neurochemical alterations in Rett syndrome. *Brain Research. Developmental Brain Research, 74*, 67–72.

Whishaw, I. Q., & Kolb, B. (1984). Decortication abolishes place but not cue learning in rats. *Behavioural Brain Research, 11*, 123–134.

Wichmann, T., & Delong, M. R. (2006). Deep brain stimulation for neurologic and neuropsychiatric disorders. *Neuron, 52*, 197–204.

Wilson, B. K., & Hess, E. J. (2013). Animal models for dystonia. *Movement Disorders, 28*, 982–989.

Wolf, S. S., Jones, D. W., Knable, M. B., Gorey, J. G., Lee, K. S., Hyde, T. M., et al. (1996). Tourette syndrome: Prediction of phenotypic variation in monozygotic twins by caudate nucleus D2 receptor binding. *Science, 273*, 1225–1227.

Wolgin, D. L. (2012). Amphetamine stereotypy, the basal ganglia, and the "selection problem". *Behavioural Brain Research, 231*, 297–308.

Wolmarans de, W., Brand, L., Stein, D. J., & Harvey, B. H. (2013). Reappraisal of spontaneous stereotypy in the deer mouse as an animal model of obsessive-compulsive disorder (OCD): Response to escitalopram treatment and basal serotonin transporter (SERT) density. *Behavioural Brain Research, 256*, 545–553.

Wong, D. F., Ricaurte, G., Grunder, G., Rothman, R., Naidu, S., Singer, H., et al. (1998). Dopamine transporter changes in neuropsychiatric disorders. *Advances in Pharmacology, 42*, 219–223.

Worbe, Y., Baup, N., Grabli, D., Chaigneau, M., Mounayar, S., McCairn, K., et al. (2009). Behavioral and movement disorders induced by local inhibitory dysfunction in primate striatum. *Cerebral Cortex*, *19*, 1844–1856.

Worbe, Y., Mallet, L., Golmard, J. L., Behar, C., Durif, F., Jalenques, I., et al. (2010). Repetitive behaviours in patients with Gilles de la Tourette syndrome: Tics, compulsions, or both? *PLoS One*, *5*, e12959.

Wu, N., Cepeda, C., Zhuang, X., & Levine, M. S. (2007). Altered corticostriatal neurotransmission and modulation in dopamine transporter knock-down mice. *Journal of Neurophysiology*, *98*, 423–432.

Xiao, J., & Ledoux, M. S. (2005). Caytaxin deficiency causes generalized dystonia in rats. *Brain Research Molecular Brain Research*, *141*, 181–192.

Xu, P., Grueter, B. A., Britt, J. K., McDaniel, L., Huntington, P. J., Hodge, R., et al. (2013). Double deletion of melanocortin 4 receptors and SAPAP3 corrects compulsive behavior and obesity in mice. *Proceedings of the National Academy of Sciences of the United States of America*, *110*, 10759–10764.

Yeterian, E. H., & Van Hoesen, G. W. (1978). Cortico-striate projections in the rhesus monkey: The organization of certain cortico-caudate connections. *Brain Research*, *139*, 43–63.

Yin, H. H. (2010). The sensorimotor striatum is necessary for serial order learning. *The Journal of Neuroscience*, *30*, 14719–14723.

Yin, H. H., & Knowlton, B. J. (2006). The role of the basal ganglia in habit formation. *Nature Reviews Neuroscience*, 7, 464–476.

Yin, H. H., Knowlton, B. J., & Balleine, B. W. (2004). Lesions of dorsolateral striatum preserve outcome expectancy but disrupt habit formation in instrumental learning. *The European Journal of Neuroscience*, *19*, 181–189.

Yin, H. H., Knowlton, B. J., & Balleine, B. W. (2006). Inactivation of dorsolateral striatum enhances sensitivity to changes in the action-outcome contingency in instrumental conditioning. *Behavioural Brain Research*, *166*, 189–196.

Yin, H. H., Ostlund, S. B., Knowlton, B. J., & Balleine, B. W. (2005). The role of the dorsomedial striatum in instrumental conditioning. *The European Journal of Neuroscience*, *22*, 513–523.

Yokoi, F., Dang, M. T., Li, J., Standaert, D. G., & Li, Y. (2011). Motor deficits and decreased striatal dopamine receptor 2 binding activity in the striatum-specific Dyt1 conditional knockout mice. *PLoS One*, *6*, e24539.

Yokoi, F., Dang, M. T., Mitsui, S., Li, J., & Li, Y. (2008). Motor deficits and hyperactivity in cerebral cortex-specific Dyt1 conditional knockout mice. *Journal of Biochemistry*, *143*, 39–47.

Zhang, L., He, J., Jugloff, D. G., & Eubanks, J. H. (2008). The MeCP2-null mouse hippocampus displays altered basal inhibitory rhythms and is prone to hyperexcitability. *Hippocampus*, *18*, 294–309.

Zhang, L., Yokoi, F., Jin, Y. H., DeAndrade, M. P., Hashimoto, K., Standaert, D. G., et al. (2011). Altered dendritic morphology of Purkinje cells in Dyt1 DeltaGAG knock-in and purkinje cell-specific Dyt1 conditional knockout mice. *PLoS One*, *6*, e18357.

Zhang, Z. W., Zak, J. D., & Liu, H. (2010). MeCP2 is required for normal development of GABAergic circuits in the thalamus. *Journal of Neurophysiology*, *103*, 2470–2481.

Zhao, C., Brown, R. S., Chase, A. R., Eisele, M. R., & Schlieker, C. (2013). Regulation of Torsin ATPases by LAP1 and LULL1. *Proceedings of the National Academy of Sciences of the United States of America*, *110*, E1545–E1554.

Zhao, Y., DeCuypere, M., & LeDoux, M. S. (2008). Abnormal motor function and dopamine neurotransmission in DYT1 DeltaGAG transgenic mice. *Experimental Neurology*, *210*, 719–730.

Zhao, Y. T., Goffin, D., Johnson, B. S., & Zhou, Z. (2013). Loss of MeCP2 function is associated with distinct gene expression changes in the striatum. *Neurobiology of Disease, 59*, 257–266.

Zhuang, X., Oosting, R. S., Jones, S. R., Gainetdinov, R. R., Miller, G. W., Caron, M. G., et al. (2001). Hyperactivity and impaired response habituation in hyperdopaminergic mice. *Proceedings of the National Academy of Sciences of the United States of America, 98*, 1982–1987.

Zuchner, S., Wendland, J. R., Ashley-Koch, A. E., Collins, A. L., Tran-Viet, K. N., Quinn, K., et al. (2009). Multiple rare SAPAP3 missense variants in trichotillomania and OCD. *Molecular Psychiatry, 14*, 6–9.

Zweig, R. M., Hedreen, J. C., Jankel, W. R., Casanova, M. F., Whitehouse, P. J., & Price, D. L. (1988). Pathology in brainstem regions of individuals with primary dystonia. *Neurology, 38*, 702–706.

CHAPTER FOUR

Mouse Models of Heart Failure: Cell Signaling and Cell Survival

Lorna R. Fiedler[1], Evie Maifoshie, Michael D. Schneider[1]

British Heart Foundation Centre of Research Excellence, National Heart and Lung Institute, Imperial College London, London, UK

[1]Corresponding authors: e-mail address: l.fiedler@imperial.ac.uk; m.d.schneider@imperial.ac.uk

Contents

Abstract

Heart failure is one of the paramount global causes of morbidity and mortality. Despite this pandemic need, the available clinical counter-measures have not altered

ISSN 0070-2153
http://dx.doi.org/10.1016/B978-0-12-397920-9.00002-0

substantially in recent decades, most notably in the context of pharmacological interventions. Cell death plays a causal role in heart failure, and its inhibition poses a promising approach that has not been thoroughly explored. In previous approaches to target discovery, clinical failures have reflected a deficiency in mechanistic understanding, and in some instances, failure to systematically translate laboratory findings toward the clinic. Here, we review diverse mouse models of heart failure, with an emphasis on those that identify potential targets for pharmacological inhibition of cell death, and on how their translation into effective therapies might be improved in the future.

1. HEART FAILURE: THE UNMET NEED

The advent of technologies for gene manipulation has revolutionized mechanistic studies of disease systems and driven the discovery of new therapeutic strategies. Heart disease, a leading cause of morbidity and mortality, is increasingly predominant in the industrialized world, reflecting more prevalent risk factors such as obesity and age (Allender, Peto, Scarborough, Laur, & Rayner, 2008). In Europe, cardiovascular disease (CVD) accounts for 4 million deaths each year (Nichols, Townsend, Scarborough, & Rayner, 2013). The most common form of heart disease is myocardial infarction (MI), an acute ischemic injury resulting in widespread cardiac muscle cell death, biomechanical dysfunction, and eventually heart failure. The incidence is >100,000 per year in the UK (British Heart Foundation Health Promotion Research Group, 2012) and 7 million per year worldwide (White & Chew, 2008). The mortality of heart failure is particularly high, with 50% of patients who require hospitalization, not surviving beyond 3 years (Heidenreich et al., 2013). The resulting burden on healthcare services is massive (Luengo-Fernández Leal, Gray, Petersen, & Rayner, 2006) and is expected to increase globally despite the decline in CVD in developed countries due to improved primary and secondary treatments. Paradoxically, a marked rise is expected in the incidence of heart failure as the population's life expectancy increases, including chronic disability from heart failure in survivors of MI. Globally, CVD is expected to rise from 17.1 million in 2004 to 23.4 million in 2030 and heart failure to increase by 23% from 2012 to 2030 (Heidenreich et al., 2013). There is thus a compelling need for animal models capturing the key features of human heart failure, to understand key regulatory proteins, their genes, gene networks, and processes, to pinpoint potential therapeutic interventions, and improve their translation from laboratory-based research into the clinic.

Heart failure comprises organ-level dysfunction of the heart as a biomechanical pump. Heart failure can arise from innate defects in the mechanical properties of single cardiomyocytes (either inherited or acquired), abnormalities in the extracellular matrix surrounding the myocytes, or defects in cardiomyocyte survival. Cardiomyocyte loss is a defining feature of heart failure, not only in the context of antecedent infarction but also in chronic heart failure from diverse etiologies. Since replacement of lost cardiomyocytes is meager, maintenance of pump function relies primarily on remodeling the existing cardiomyocyte structure and environment, although it should be noted that minimal endogenous replacement does occur (Quaini et al., 1994; Kajstura et al., 1998; Bergmann et al., 2009). While the origin of new cardiomyocytes in humans is unclear (whether derived from stem or progenitor cells or from reentry of differentiated cardiomyocytes into the cell cycle), in adult mice at least it appears to be primarily due to stem cell replenishment (Mercola, Ruiz-Lozano, & Schneider, 2011; Ellison et al., 2013; Hsueh, Wu, Yu, Wu, & Hsieh, 2014). Despite the existence of some form of self-repair, this is insufficient to compensate for cardiomyocyte cell death. Heart failure can therefore be considered, in part, as a cardiomyocyte deficiency disease.

Here, we provide a comprehensive review of genetically engineered mouse models that exhibit explicit, instructive features of severe heart failure. For surgical models and pharmacological interventions that mimic aspects of, or induce heart failure, see Monnet (2005), and for additional information on clinical features, diagnosis, and current treatment paradigms see Hunt (2005), Mann (2011), and Houser et al. (2012). Germline mutations affecting the heart can result in embryonic lethality reflecting functions in a developmental context, such as morphogenesis or cardiomyocyte creation, rather than dysfunction in the terminally differentiated, mature environment in which adult human disease occurs. Consequently, this review will address only postnatal heart failure rather than perturbation of developmental cues in the immature heart. For a description of genes associated with or contributing to developmental heart malformation, see Preuss & Andelfinger (2013); Wessels & Willems (2010); Noseda, Peterkin, Simões, Patient, & Schneider (2011); Chu, Haghighi, & Kranias (2002); Fahed, Gelb, Seidman, & Seidman (2013).

A variety of different approaches have been explored to regenerate cardiomyocytes and restore pump function (Fig. 4.1). Many of these strategies aim to increase cardiac muscle cell number through reactivation of cell division or production of new myocytes from existing or *ex vivo* stem or

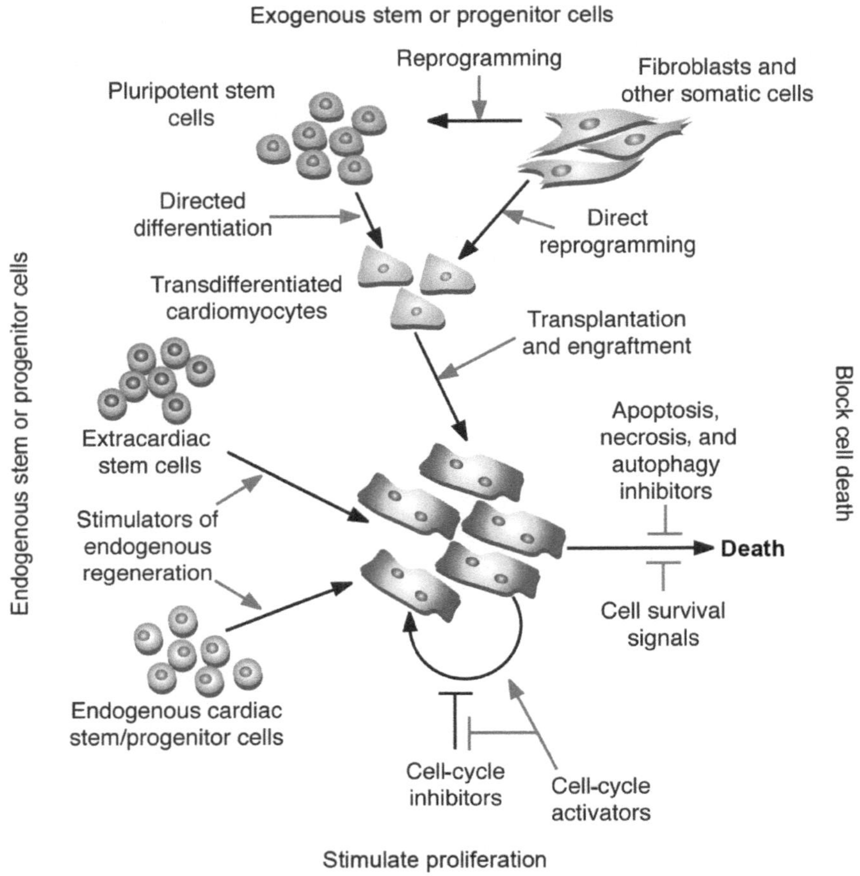

Figure 4.1 Strategies to increase cardiac muscle cell number as a therapeutic target. In principle, the limited ability of the heart to replace cardiomyocytes can be improved by reactivating cell division of preexisting cardiomyoctes and/or inhibiting cell death or augmenting survival. Alternatively, new myocytes can be produced from multipotent stem or progenitors that reside within niches in the myocardium, circulating stem cells with cardiac potency, or *ex vivo* cells transplanted into the injured heart. Challenges to regeneration include an endogenous restorative capacity that appears limited by an insufficient number of available stem or progenitor cells, and the need to develop efficient means to produce or deliver exogenous cells. Developmental signals are being investigated for use in enhancing therapeutic regeneration from endogenous and exogenous sources. *Reproduced from Mercola, Ruiz-Losano & Schneider, 2011.* (See the color plate.)

progenitor cells (Mercola, Ruiz-Losano & Schneider, 2011). An alternative strategy for improving pump function is to modify contractile responses and reduce pathological hypertrophy, arrhythmia, and sudden death by targeting the calcineurin–NFAT pathway, histone deacetylases (HDACs), and β-adrenergic receptor (β-AR) signaling to reduce hypertrophy. The use of animal models in this approach has been the subject of a recent review (van Berlo, Maillet, & Molkentin, 2013) and will not be covered here. However, if cardiomyocyte death could be prevented or mitigated, this would relieve or complement the present efforts to optimize the mechanical properties of the surviving muscle cells. In summary, cell death is the main focus of this review. Signaling components and pathways that have been implicated specifically in cell death and heart failure in both mouse and man may present auspicious and potential therapeutic targets.

2. PATHOBIOLOGICAL FEATURES

Heart failure is a culmination of cardiac muscle pathophysiology from diverse etiologies and can be defined as the inability of the heart to function in its capacity to meet the metabolic demands of the organism. Clinically, this presents as fatigue, breathlessness (dyspnea), exertional dyspnea, fluid retention, and reduced tissue perfusion, with death resulting from lethal arrhythmias or insufficient pump function (Hunt, 2005; Houser et al., 2012). At a cellular level, defects intrinsic to heart failure encompass those in cardiomyocyte contractile function (Boudoulas & Hatzopoulos, 2009), altered cardiomyocyte geometry/organization, and excessive myocyte loss unmatched by cell replacement, in addition to extrinsic defects such as interstitial fibrosis as a result of inflammatory infiltration (Dimmeler, Zeiher, & Schneider, 2005).

Heart "remodeling" is fundamentally the aggregate response to injury or stress that occurs in both pathological and physiological circumstances (with a number of key differences) to meet increased metabolic and contractile demands and/or to maintain structural integrity. Chronically, by itself, this can lead to cell death and pump dysfunction but on a background of existing mutations, structural defects or other diseases can be greatly exacerbated. Cardiomyopathies leading to heart failure are typically dilated (DCM) or hypertrophic (HCM), with restrictive cardiomyopathies and arrhythmogenic right ventricular cardiomyopathy occurring less frequently (Houser et al., 2012). Causes of such myopathies encompass damage from injury (e.g., viral or drug-induced), systemic conditions, or genetic mutations chiefly affecting the sarcomere, cytoskeleton, or myocardial metabolism. MI is the commonest form of CVD leading to heart failure and elicits both hypertrophy and dilatation.

The defining feature of DCM is ventricular dilatation, concurrent with normal or thinned ventricle walls. Increased wall stress and ongoing cell death contributes to the inability of the muscle to increase in thickness and perpetuates dilation. Cardiomyocyte hypertrophy occurs mainly by addition of new sarcomeres in series (eccentric hypertrophy), resulting in an elongated phenotype and chamber enlargement. Diffuse fibrosis is increasingly present due to progressive cell death of stressed cardiomyocytes, reducing myocardial compliance and impairing function (Francis & Tang, 2003; van Berlo et al., 2013).

Hypertrophic growth, without dilation, is the characteristic early response to excessive afterload, such as hypertension, obstruction to left ventricular (LV) outflow, or increased wall stress following infarction. Cardiomyocytes are stimulated to increase both DNA and protein content to generate further force and normalize wall stress. As a result, the cross-sectional thickness of the cells increases, adding sarcomeres in parallel, and causing concentric hypertrophy (Francis & Tang, 2003; van Berlo et al., 2013). Prolonged hypertrophy leads to impaired diastolic relaxation and can progress to dilatation, decompensation, and heart failure, one key component of that transition being deficient angiogenesis in supporting hypertrophic growth (Sano et al., 2007). Many of the human mutations associated with HCM are found in structural proteins and often contain missense residues or small deletions. As such, they are incorporated into sarcomeric or cytoskeletal structures as dominant-interfering proteins, perturb normal function, and impinge on mechanosensation directly (Morita Seidman, & Seidman, 2005) or indirectly (Seo, Rainer, Lee, et al., 2014; Seo, Rainer, Shalkey Hahn, et al., 2014). Whereas genetic etiology is, by definition, known in the case of identified mutations, the broader pathobiological question is how the mutations become coupled to transmission of aberrant signals for growth, fibrosis, and other components of the hypertrophic phenotype. From this perspective, understanding the downstream molecular consequences of these etiological mutations is formally similar to understanding the molecular consequences of altered loading conditions. Dissecting both cascades is essential in driving rational drug discovery.

Mice genetically engineered to carry gain- or loss-of-function mutations in putative signaling cascades, along with mice engineered to carry human mutations causing cardiomyopathies, together permit direct disease characterization and pathway dissection to identify key nodal points. Such studies can elucidate whether altered expression or activity of such genes and their products are causative, counter-regulatory adaptive responses, or merely epiphenomena in the heart following stress or injury. Distinguishing among these three interpretations of the "observational biochemistry" will better instruct therapeutic strategies, identify more worthwhile targets, and uncover common signaling pathways between very different etiologies.

3. CARDIAC MUSCLE CELL APOPTOSIS IN HEART FAILURE: DEATH KILLS

It had long been suggested that myocyte dropout/cell death leads to progressive deterioration in myocardial function, culminating in end-stage heart failure (Bing, 1994). This concept was supported by observations in end-stage, dilated human hearts where sporadic apoptotic cardiomyocytes (measured by the presence of DNA fragmentation) were documented (Narula et al., 1996; Olivetti et al., 1997). Analogously, DNA fragmentation and caspase-3 activity were more prevalent in cardiomyocytes from patients who died of heart failure than those with sudden cardiac arrest (Kytö et al., 2004). In addition to apoptosis, necrosis also features in heart failure (inferred on the basis of association with reparative fibrosis and a diffuse smearing pattern of DNA), suggesting the contributions of multiple mechanisms for cell death in end-stage heart failure (Olivetti et al., 1997). Furthermore, failing human hearts exhibit increased number of autophagosomes suggesting increased autophagy responses in failing heart (Hein et al., 2003; Kostin et al., 2003; Matsui et al., 2007; Tannous et al., 2008). In addition, studies on surgical rodent models indicate that dispersed cardiomyocyte death is equally, if not more so, deleterious to heart function than the immediate localized cell death associated with MI. Up to 50% of LV cardiomyocytes can be lost without inducing heart failure, while only up to 20% of sporadic myocyte loss/dropout is required in the case of pressure overload (Nadal-Ginard, Kajstura, Anversa, & Leri, 2003). Consequently, inhibition of cell death is a viable therapeutic strategy in heart failure arising from different origins. Furthermore, the extent of cell death in acute ischemia (i.e., infarct size) is a major determinant in the likelihood of survival (Miller et al., 1995) and earlier restoration of blood flow following infarct (by thrombolysis) results in a smaller infarct size, improved heart function, and prolonged survival (Simoons et al., 1986). Successful microcirculatory reperfusion in patients treated with percutaneous coronary intervention is also associated with smaller infarct size, improved function, and reduced mortality (Brener et al., 2013).

While apoptotic cell death had been described in failing human hearts, and has been suggested to be causative, a direct link was not clearly established until pioneering studies proved that chronic low levels of sporadic cardiac muscle cell death were sufficient to cause heart failure in mice, and that inhibition of caspase-dependent cell death in this model was protective (Wencker et al., 2003). In this model, administration of a small molecule (FK1012H2) causes dimerization of FKBP-caspase-8 fusion protein, inducing activation of caspase-8. Since expression of the chimeric caspase-8 protein is under the control of the cardiac α-myosin heavy chain (*Myh6*) promoter, the dimer triggers cardiomyocyte-specific caspase activity. Following treatment, all transgenic mice died in a dose-dependent manner, within 4–18 h concurrent with caspase-dependent cell death. Thus, acute activation provoked cell death and heart failure (Wencker et al., 2003). However, mice expressing high levels of the transgene spontaneously developed DCM, even in the absence of activating ligand, due to low-level caspase activation by spontaneous oligomerization, with death ensuing at 8 weeks of age. Slightly increased levels of cardiomyocyte cell death were seen and pharmacological caspase inhibition prevented the onset of cardiomyopathy (Wencker et al., 2003).

This cogent demonstration that "death kills" was pivotal in the field, as prior to this, it would have been plausible to contend that preventing the death and removal of damaged cells through apoptosis might have adverse effects, not beneficial ones. Interestingly, the observed levels of TUNEL-positive cardiomyocytes in these spontaneously failing hearts were less than those reported in failing human hearts, suggesting that even low levels of apoptosis are sufficient to induce lethal cardiomyopathies (Wencker et al., 2003). Diverse gain- or loss-of-function mutations and pharmacological interventions in murine models directly implicate cell-death signaling pathways as relevant therapeutic targets in reducing the inexorable progression to heart failure (summarized in Table 4.1).

In the setting of acute infarction, a pharmacological agent that inhibits cell death, salvages jeopardized myocardium and reduces infarct size would be expected to prove highly effective and provide additional benefit over current therapies that merely restore blood flow. While not extensive, a number of trials are utilizing pharmacological approaches to limit injury in patients undergoing urgent reperfusion therapy (summarized in Hausenloy & Yellon, 2013). While many such trials have not yielded the anticipated results, some promising candidates have emerged; ANP, exenatide, and cyclosporine A (Hausenloy & Yellon, 2013). ANP and exenatide both activate prosurvival kinase signaling pathways (Chang et al., 2013; Nishikimia, Maeda, & Matsuokaa, 2006). Cyclosporin A, a cyclic peptide, protects against cell death by inhibiting calcineurin and opening of the mitochondrial permeability transition pore (mPTP) (Hausenloy,

Table 4.1 Genetic models implicating cell-death regulators in cardiomyocyte death

Gene	Model	Phenotype	Reference
Casp1	*Casp1*$^{-/-}$	Reduced cardiomyocyte hypertrophy, increase survival, rescued decline of left ventricular function after MI	Merkle et al. (2007)
	YVAD inhibitor	Attenuation of ischemia-induced myocardial dysfunction (contractility improvement)	Pomerantz, Reznikov, Harken, & Dinarello (2001)
Casp3	*Myh6-Casp3*	Nuclear and myofibrillar damage, increased infarct size with MI and decreased survival	Condorelli et al. (2001)
Casp8	*Casp8*$^{-/-}$	Embryonic lethality (E11–E12.5), impaired cardiac muscle development and hematopoietic cell development, MEFs resistant to death-receptor-induced apoptosis	Varfolomeev et al. (1998)
	Myh6-FKBP-Casp8 (conditionally active)	Massive cardiomyocyte apoptosis and death; low levels of casp-8 activity in the absence of dimerizer led to sustained low apoptosis and lethal DCM; abrogated with a pan-caspase inhibitor	Wencker et al. (2003)
Ppif/CypD	CsA or SfA	Cell death inhibition (LDH release), improved left ventricular developed pressure (LVDP)	Clarke McStay, & Halestrap (2002)
	Ppif$^{-/-}$	Isolated mitochondria resistant to mPTP opening in response to Ca^{2+} and oxidative stress; mice resistant to I/R injury, but sensitive to apoptotic stimuli	Baines (2007), Baines et al. (2005), Nakayama et al. (2007)

Continued

Table 4.1 Genetic models implicating cell-death regulators in cardiomyocyte death—cont'd

Gene	Model	Phenotype	Reference
Vdac1/3	*Vdac1*$^{-/-}$ *Vdac3*$^{-/-}$ Double KO	Isolated cardiac mitochondria exhibit Ca^{2+}-induced swelling, no difference in cyt c release, caspase cleavage in response to Bax and Bid compared to WT controls	Baines, Kaiser, Sheiko, Craigen, & Molkentin (2007)
Tnf	*Myh6-Tnf*	Spontaneous death of weaning mice, diffuse cardiac inflammation, and interstitial edema (severe myocarditis)	Kubota, McTiernan, Frye, Demetris, & Feldman (1997)
	Myh6-Tnf	Higher heart weight-to-body weight ratio, reduced ejection fraction, dilatation, cardiac inflammation, cardiomyocyte death, increased ANF mRNA	Kubota, McTiernan, Frye, Slawson, et al. (1997), Sivasubramanian et al. (2001)
Nix/Bnip3L	Tet-off ER-targeted Nix	Hearts protected against cell death in *PPIF*$^{-/-}$ background; Bax/Bak-independent and mPTP-dependent cell death	Chen et al. (2010)
	Tet-off mitochondria-targeted Nix	Hearts not protected against cell death in *PPIF*$^{-/-}$ background; Bax/Bak- and caspase-dependent apoptosis	
	Myh6-sNix	Protection against peripartum cardiomyopathy in Gaq transgenic females	Yussman et al. (2002)
Bnip3	*Bnip3*$^{-/-}$	Diminished apoptosis in I/R, preserved LV systolic performance and diminished LV dilatation	Diwan et al. (2007)

Table 4.1 Genetic models implicating cell-death regulators in cardiomyocyte death—cont'd

Gene	Model	Phenotype	Reference
	Myh6-Bnip3	Increased apoptosis in unstressed mice leading to progressive LV dilatation; conditional overexpression exacerbates apoptosis, and infarct size	
Atf6	*Myh6-ATF6-MER; Myh6-caATF6*	Better functional recovery (increased LVDP); reduced cell death (TUNEL-positive cardiomyocytes)	Martindale et al. (2006), Toko et al. (2010)
	Myh6-dnATF6	LV dilatation, reduced fractional shortening, and heart failure	Toko et al. (2010)
Puma	*Puma*$^{-/-}$	Reduced pressure overload-induced apoptosis and fibrosis; preserved pump function; also rescued DCM in Mdm4 cardiac CKO	Mandl et al. (2011)
Fas/CD95/Tnfrsf6	*Fas*^{lpr}	Reduction in apoptosis and infarct size after MI	Lee et al. (2003)
Tnfrsf1a/b	*Tnfrsf1a*$^{-/-}$ *Tnfrsf1b*$^{-/-}$ Double KO	Increased infarct size in double KO compared to WT or single KO mice after MI with accelerated apoptosis	Kurrelmeyer et al. (2000)
Bax	*Bax*$^{-/-}$	Decreased creatine kinase (CK) release, caspase-3 activity, and infarct size after I/R	Hochhauser et al. (2003)
Omi/HtrA2	Small-molecule inhibitor	Reduction in infarct size and cardiac dysfunction in two rodent models following I/R; reduction in IAP loss; decrease in casp-9 and casp-3 activities	Bhuiyan & Fukunaga (2007), Liu et al. (2005)

Continued

Table 4.1 Genetic models implicating cell-death regulators in cardiomyocyte death—cont'd

Gene	Model	Phenotype	Reference
Fadd	*Fadd*$^{-/-}$	Embryonic lethality (E9–E12.5), impaired cardiomyogenesis and abdominal hemorrhage	Yeh et al. (1998)
Bcl2	*Myh6-Bcl2*	Increased functional recovery and decreased LDH release after I/R; decreased infarct size and apoptosis	Chen, Fujii, Zhang, Roberts, & Fu (2001), Chen, Chua, Ho, Hamdy, & Chua (2001a), Imahashi, Schneider, Steenbergen, & Murphy (2004)
Xiap	*Xiap*$^{-/-}$	Increased sensitivity of cardiomyocytes to cyt c-induced apoptosis	Potts, Singh, Knezek, Thompson, & Deshmukh (2003)
Cflar/Casper	*Casper*$^{-/-}$	Embryonic lethality, impaired heart development, increased FasL and TNF-induced apoptosis	Yeh et al. (2000)
Nol3/ARC	*Nol3*$^{-/-}$	Larger infarcts after I/R; mice developed DCM after TAC	Foo, Chan, Kitsis, & Bennett (2007)
Ripk1	Small-molecule inhibitor *Nec-1*	RIP1 kinase inhibition; reduction of infarct size following I/R	Lim et al. (2007)
L-type calcium channel/ LTCC/ Cav1.2; *Cacnab2 LTCC*	*Myh6-tTA*x Tet responder *Myh6-β2a*	Development of cardiac hypertrophy with interstitial fibrosis; increased cardiomyocyte death after Ca^{2+} overload	Nakayama et al. (2007)
	LTCC inhibitor	Ca^{2+} influx-induced cell death and cardiomyopathy were prevented	

Boston-Griffiths, & Yellon, 2012). These studies clearly support the logic of targeting the machinery or intracellular signals for cardiac cell death in future efforts.

The most promising candidates so far are peptides that stimulate cardioprotective signaling cascades or protect mitochondria from oxidative stress. However, an alternative perspective is to directly inhibit pro-death kinases rather than activate prosurvival kinases. In pharmacological approaches to cardiac cell death, small molecule inhibitors are particularly under-represented; suggesting an exciting avenue for further exploration, with potential targets unmasked in part through mouse models discussed in this review. In summary, pharmacological interventions designed to minimize cardiomyocyte death are relatively unexplored; however, a few clinical trials have begun to focus on this strategy, summarized in Table 4.2.

4. PATHWAYS EXECUTING CARDIAC MUSCLE CELL DEATH

While the lines are becoming increasingly blurred, and cell death is less easily compartmentalized, as being one idealized platonic type or another, it is still considered to comprise three main processes: apoptosis, necrosis, and autophagy. The latter is least well studied in its contribution to cell death and heart failure and is an intracellular recycling process initiated in response to cellular stress, nutrient limitation, organelle damage, and accumulation of protein aggregates (Pattingre et al., 2005; Rothermel & Hill, 2008). Conversely, the mitochondrial DNA that escapes autophagy causes toll-like receptor (TLR)-9-mediated inflammation and consequently cardiomyopathy (Oka et al., 2012). Both apoptosis and necrosis share convergent signaling pathways, and of late, the term "necroptosis" (a more regulated form of necrosis) has been introduced, indicating that these originally divergent types of cell death are more related than previously thought (Oerlemans et al., 2013). However classified, and whatever the mechanism, cell death contributes toward the pathophysiology of heart failure and dissecting its signaling pathways has increasing relevance for future therapeutic strategies (Scarabelli & Gottlieb, 2004; Whelan, Kaplinskiy, & Kitsis, 2010).

4.1. Apoptosis

Apoptosis is an evolutionary conserved mechanism of programmed cell death triggered by different stimuli and further classified into two central pathways: the extrinsic pathway, utilizing cell-surface receptors, and the

Table 4.2 Kinase targets for drug development in CVD

Target	Compound	Indication	Most advanced stage of testing
p38 MAPK	VX-702[a]	Acute coronary syndromes	Phase II
PI3K	TG100–115[b]	Acute myocardial infarction	Phase I/II
PKCδ	KAI-9803[c]	Myocardial infarction, adjunct to angioplasty	Phase II
ROCK	Fasudil	Angina	Phase II[d]
p38 c-raf	Semapimod	Congestive heart failure	Preclinical[e]
CaMKII[f]	KN-93	Arrhythmia/heart failure	Preclinical
Cdk9	EXE-8647[g]	Acute cardioprotection	Preclinical
Chk2	NA	Cardiovascular disease	Preclinical
Grk2[h]	M119	Heart failure	Preclinical
HDAC kinase[i]	Trichostatin A MPT0E014	Heart failure/fibrosis	Preclinical
HGK[j]	Compound 1 Compound 2	Cardiomyocyte survival	Preclinical
JAK2	AG490 (Tyrphostin)[k]	Heart failure	Preclinical[k]
JNK	JNK9395	Ischemia reperfusion injury	Preclinical
PKCε	Activator peptide (ψεRACK)[l]	Cardiovascular disease	Preclinical

PKC[m]	LY333531 (Ruboxistaurin) Ro-32-0432 Ro-31-8220	Heart failure	Preclinical[m]
PKCβ	LY379196	Heart failure	Preclinical[m]

[a]An oral p38 MAPK inhibitor (Vertex Pharmaceuticals Inc., Cambridge, MA).
[b]PI3K-selective inhibitor (TargeGen Inc., San Diego, CA).
[c]A PKCδ inhibitor (KAI Pharmaceuticals Inc., South San Francisco, CA).
[d]Approved for the treatment of cerebral vasospasm in Japan.
[e]Phase II for inflammatory disease.
[f]End-stage human heart failure (dilated and ischemic cardiomyopathy) where inhibition of CaMKII acutely improved contractility (Sossalla et al., 2010).
[g]Discovered by Exelixis Inc. in a high-throughput screen and is currently the most selective inhibitor against Cdk9 (Krystof et al., 2010).
[h]Utility as a biomarker in heart failure (Johnson & Johnson). M119 inhibits GRK2 binding to bG$\beta_1\gamma_2$ subunit (Bonacci et al., 2006) and halted HF progression in mice (Casey et al., 2010).
[i]Trichostatin A (TSA) attenuated cardiac hypertrophy by suppressing autophagy (Cao et al., 2011) and MPT0E014 had antifibrotic activity in rat hearts (Kao et al., 2013).
[j]Pfizer compounds (Guimarães et al., 2011) and others currently being investigated in Prof. Michael D. Schneider's lab.
[k]Selective inhibition used for cancer treatment; patent by Daniel L. Beckles. CEP-701 (lestaurtinib) is currently in Phase I/II trials for patients with acute monocytic leukemia.
[l]Activator peptide combined with δV1-1 (PKCδ inhibitor in isolated perfused hearts showed greater cardioprotection after I/R compared to each treatment alone) (Inagaki et al., 2003).
[m]Broad-activating PKC inhibitors (Liu et al., 2009). Table updated and adapted from Anderson et al. (2006).

intrinsic pathway entailing mitochondria and endoplasmic reticulum (ER), summarized in Fig. 4.2. Central to both pathways are the caspases, which in the absence of the stimulus exist primarily as inactive zymogens referred to as procaspases.

In the *extrinsic* (caspase-8/10-dependent) pathway, death ligands bind to cell-surface receptors and initiate a signaling cascade of programmed cell

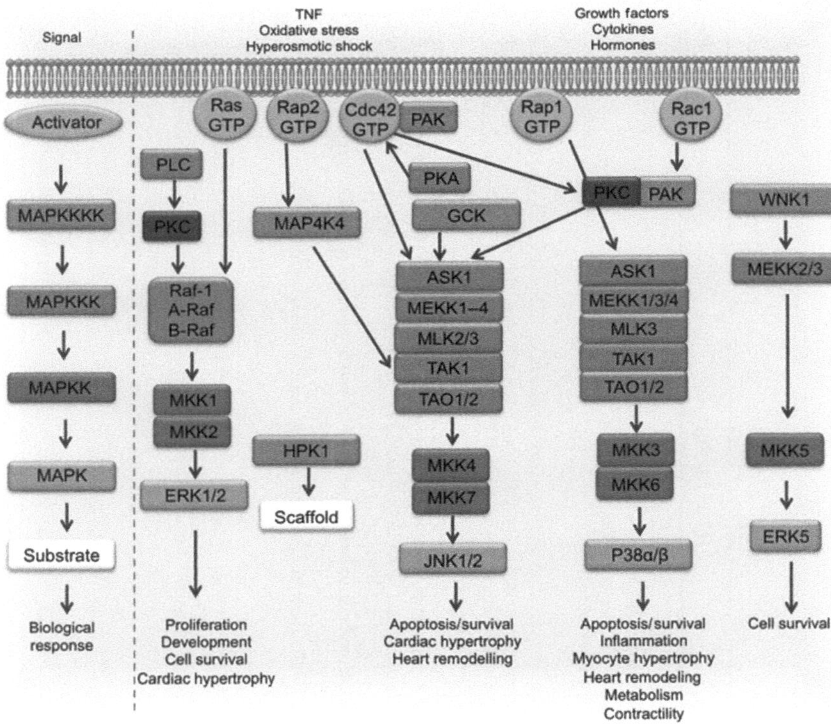

Figure 4.2 Schematic representation of signal transduction pathways in the heart by kinases. Signal transduction through a hierarchical MAPK pathway is elicited by stimuli that include cytokines, growth factors, and stress responses such as oxidative stress. Activation of the terminal MAPKs, ERK1/2, JNK1/2/3, p38α/β, or ERK5 leads to a biological response that can range from proliferation, apoptosis, cell motility, osmoregulation, hypertrophy, or metabolism. Thus, kinase signaling couples extracellular signals to cardiac hypertrophy, contractility, remodeling, cell death, and heart failure. In the case of MAP4K4, Rap2-GTP interacts with MAP4K4 through its C-terminal citron homology domain (CHD) and can enhance its activity (Machida et al., 2004). MAP4K4 has also been shown to couple to TAK1 and JNK1/2 in control of cell death. Cardiomyocyte, and non-myocyte, cell death is a complex process of interactive protein signaling cascades. *Adapted from Coulthard White, Jones, McDermott, & Burchill (2009).* (See the color plate.)

death. Binding of death ligands (e.g., FasL, tumor necrosis factor-α; TNF-α) leads to activation of caspase-8 by its recruitment to the activated receptors, leading ultimately to activation of caspase-3. Ligand binding induces the recruitment of adaptor proteins, such as Fas-associated via death domain (FADD), which interacts with procaspase-8 to form the death-inducing signaling complex (DISC). Dimerization of procaspase-8 in the complex results in its activation, which then proteolytically cleaves and activates effector procaspase-3.

The *intrinsic* (Bcl-2-sensitive, mitochondrial) pathway relies on an intermediary mitochondrial response through a variety of Bcl-2 family proteins, responding to a range of extracellular and intracellular stimuli (e.g., loss of survival factors, toxins, DNA damage, hypoxia, radiation, and oxidative stress). The Bcl-2 family of proapoptotic proteins (i.e., Bax, Bak; and BH3-domain only members: Bid, Bam, Bim, Bmf, Noxa, Puma, Bnip3, and Bnip3L) and antiapoptotic (i.e., Bcl-2, Bcl-x_L) proteins unites the extrinsic and intrinsic death-signaling pathways. A convergent signal transducer is BH3-interacting domain death agonist (Bid), which is cleaved by caspase-8/10, whereupon its C-terminal portion (tBid) translocates into the outer mitochondrial membrane (Wei et al., 2000). Bax, a prerequisite for the intrinsic signaling pathway together with Bak, is activated through a conformational change and also inserted into the outer mitochondrial membrane forming a conductance channel for release of cytochrome c (cyt c) and other apoptogens (Crow, Mani, Nam, & Kitsis, 2004; Suzuki, Youle, & Tjandra, 2000). The precise mechanism by which cyt c and other apoptogens are released is not completely understood, but may depend on interactions amongst the three proteins (Bim, Bad, and Bax) leading to the opening of the mPTP (Belzacq et al., 2003; Cheng, Sheiko, Fisher, Craigen, & Korsmeyer, 2003; Marzo et al., 1998; Shimizu, Narita, & Tsujimoto, 1999). Released cyt c binds to the apoptosis activating factor-1 (Apaf-1), along with ATP, recruiting homo-oligomers of procaspase-9 into a complex called the apoptosome, which subsequently activates downstream procaspase-3 (Acehan et al., 2002; Yu et al., 2005).

Apoptosis is held in check by "mitochondrial gatekeepers"; the antiapoptotic Bcl family proteins (i.e., Bcl-2 and Bcl-x_L), and endogenous inhibitors, the best-illustrated being the inhibitor-of-apoptosis (IAP) proteins. One postulated mechanism for the death antagonist activities of Bcl-2 and Bcl-x_L is via interactions with Bax and Bak, or indirectly by preventing BH3-only proteins binding to Bax and Bak (Crow et al., 2004; Kim, Pedram, Razandi, & Levin, 2006; Whelan et al., 2010). Bcl-2 is also suggested to bind to proapoptotic factors or by binding to proteins, such as the voltage-dependent anion channel (VDAC), that block the formation of the pore for cyt c release (Imahashi et al., 2004; Shimizu et al., 1999). Interestingly, however, Bcl-2 has a more general role in cardiomyocyte homeostasis and its survival effect is not limited to an antiapoptotic function. In support of this notion, cardiac-specific overexpression of Bcl-2 minimized ischemic reperfusion (I/R) injury due to reduced acidification and reduced rate of ATP consumption (Imahashi et al., 2004), indicating cardioprotection through modulation of metabolism. Lastly, Bcl-2 also plays a pivotal role in autophagy through its physical interaction with Beclin1 (He et al., 2012; Pattingre et al., 2005).

IAPs utilize their E3 ubiquitin/ligase activity to inhibit the catalytic activity of caspase-3/7. Upon direct binding of IAPs, caspases are targeted for proteasome inhibition, so preventing unregulated cyt c leakage. IAPs can bind procaspase-9 to directly inhibit apoptosome formation. To achieve sufficient activation of caspase-9, dissipation of mitochondrial membrane potential ($\Delta\Psi_m$) is required, leading to release of mitochondrial Smac/DIABLO and Omi/HtrA2 apoptogens that then bind to the multiple IAPs (c-IAP1, c-IAP2, XIAP), thereby neutralizing endogenous caspase inhibition (Du, Fang, Li, Li, & Wang, 2000). Cellular IAPs also play a critical role in the regulation of cell death that occurs by death-receptor activation (Whelan et al., 2010). Another potential inhibitor cFLIP (FLICE-inhibitory protein), with a bifunctional dose-dependent role (Peter, 2004), is enriched in striated muscle and binds to procaspase-8 to prevent DISC formation.

In contrast to inhibitors that exclusively block either the extrinsic or intrinsic pathway, apoptosis repressor protein with a CARD domain (ARC), expressed in cardiac and skeletal myocytes, exerts an inhibitory effect on both pathways (Nam et al., 2004). ARC engages with the death domains of Fas and FADD to inhibit Fas–FADD binding and DISC assembly, as well as antagonizing the intrinsic pathway by ARC–Bax complex formation to inhibit Bax activation and translocation to the mitochondria (Nam et al., 2004). Furthermore, ARC protein levels decrease in response to death stimuli in MI as a result of protein destabilization mediated by the ubiquitin–proteasomal pathway (Nam et al., 2007).

4.2. Necrosis

Necrosis is characterized by two branches: (i) the mitochondrial branch, which involves the opening of the mPTP and (ii) the regulated death-receptor signaling branch, which involves TNF ligand binding. Perhaps, the most obvious connection between necrosis and prolonged myocardial ischemia, with or without reperfusion injury, is the involvement of mPTP opening (Whelan et al., 2010). mPTP is a VDAC for small molecules (up to 1.5 kDa), and is regulated by adenine nucleotide (ADP/ATP) translocators (ANTs), members of the Bcl-2 protein family and cyclophilin D (Kokoszka et al., 2004).

In the context of I/R, high oxygen concentrations lead to increased oxygen uptake by the respiratory chain yielding high levels of reactive oxygen species (ROS). When accompanied by a drop in pH gradient, due to decreased ATP/ADP ratios, this leads to a collapse of the proton gradient, causing a collapse in the mitochondrial membrane potential and action potential destabilization leading to mPTP opening. CypD-deficient mice, whose mitochondria are more resistant to mPTP opening in response to calcium and oxidative stress stimuli, show a high resistance level to I/R cardiac injury (Baines, 2007; Baines et al., 2005; Nakagawa et al., 2005) compared to wild-type littermates, though they are still sensitive to "classic" apoptotic stimuli, such as TNF-α and staurosporine (Baines et al., 2005). Interestingly, cardiac mitochondria isolated from *Vdac1*-, *Vdac3*-, and *Vdac1–Vdac3*-null mice exhibited pronounced calcium-induced swelling compared to wild-type mice, which was still inhibited with cyclosporin A (Baines, 2007), suggesting that ANT is dispensable in the function of mPTP.

The second branch of necrosis involves the activation of death receptors via TNF signaling, which can alternatively result in survival, not solely cell death. The need for more refined counter-measures targeting this pathway is suggested by the discouraging results of general TNF-α antagonists in early clinical trials for heart failure (Feldman et al., 2000; Mann et al., 2004). In general, death receptor-dependent necrosis can be described as binding of TNF-α to TNFR1 to stimulate the formation of either of two complexes, with recruitment of various adaptor proteins and the serine/threonine kinase RIP1.

Complex I comprises the adaptor TRADD protein, RIP1 kinase, TNF receptor-associated factor 2 (TRAF2), and inhibitors of apoptosis c-IAP1/2. The latter utilize their ubiquitin ligase activity to stimulate the ubiquitination of RIP1 kinase and TRAF2. Polyubiquitinated RIP1 and TRAF2 recruit and activate transforming growth factor (TGF)-β-activated kinase 1 (TAK1) through TAK-1-binding protein (TAB2/3). The downstream effect is the activation of survival genes (Whelan et al., 2010), though TAK1 can drive cardiomyocyte death in other settings (Zhang et al., 2000). TAK1 phosphorylates the inhibitor of nuclear factor-kappa-B (NFκB), IκB, as part of the NFκB/IκB (IKK) complex to promote nuclear translocation of NFκB and stimulate expression of survival genes (Ea, Deng, Xia, Pineda, & Chen, 2006). Interestingly, polyubiquitinated RIP1 recruits IKK through binding to NEMO via its polyubiquitin chains and mutations of NEMO were shown to abolish IKK activation (Ea et al., 2006). Similarly, cells lacking TAK1 expression following TNF-α stimulation undergo rapid RIP1-mediated cell death (Arslan et al., 2011).

The transition of multiprotein complex I to *complex II* and cell-death signaling involves dissociation of complex I from TNFR1 and cytosolic internalizion. RIP1 is deubiquitinated by CYLD, a K63-specific deubiquitinase, and stimulates recruitment of the FADD–procaspase-8 complex (Chan et al., 2003; Wang, Du, & Wang, 2008). Unless procaspase-8 is inhibited, complex II will stimulate apoptosis. Caspase-8 targets RIP1 for C-terminal cleavage (Chan et al., 2003), obstructing RIP1 signaling toward necrosis or survival via NFκB. However, pharmacological or genetic inhibition of caspase-8 leads to recruitment of RIP3 into a complex with RIP1 (RIP1/RIP3), where both kinases undergo activating phosphorylation (He et al., 2009; Lin, Devin, Rodriguez, & Liu, 1999). The molecular interplay between apoptotic and necroptotic signals is reinforced by evidence from the RIP3–caspase-8 and RIP1–FADD double-knockout mice, where embryonic lethality caused by caspase-8 and FADD deficiency is rescued by loss of RIP3 or RIP1 (Feoktistova et al., 2011).

In summary, the kinase activities of both RIP1 and RIP3 are critical for necrosis (He et al., 2009; Holler et al., 2000), but downstream mechanisms of regulated necrosis remain unclear. Identification of these downstream mechanisms will be required to further define the connections to the mitochondrial pathway and to delineate the role of the RIP1–RIP3 axis in MI and heart failure.

4.3. Autophagy

Autophagy is initiated in response to a lack of nutrients, cellular stress, ROS production, protein aggregation, and damage to organelles, as a short-term mechanism by which to tolerate starvation, with the mammalian target of rapamycin (mTOR) being the core sensor of nutrient availability (Gottlieb et al., 2010). mTOR stimulates protein synthesis and suppresses degradation by inhibitory phosphorylation of autophagy-related (Atg) proteins (Whelan et al., 2010). Another key regulator of autophagy is AMP-activated protein kinase (AMPK), which can override suppression of mTOR. AMPK activation activates a gene regulatory cascade of multiple Atg proteins that leads to the formation of a phagophore that will attach and enclose its targets (i.e., a protein aggregate or a damaged organelle) by forming a double-membrane autophagosome (Gottlieb et al., 2010).

Briefly, following AMPK activation, the BH3-only protein Beclin1, acting through VPS34, a class III PI3-kinase, triggers activation of two ubiquitin-like signaling events involving Atg proteins. Phagophore formation is a prerequisite for the cleavage of Atg8 (also termed LC3) by a cysteine protease Atg4 (Gottlieb et al., 2010). Cleavage of LC3 exposes a terminal glycine residue now termed LC3-I. Another form of LC3, LC3-II, is a result of reaction with phospatidylethanolamine by Atg7 E2-like enzyme. This sequence of events gives rise to the autophagosome that when mature fuses with lysozomes for degradation (for more details, see Gottlieb et al., 2010).

Beclin1 is important for autophagic protein localization in the preautophagosome and depends on the interaction and regulation by multiple proteins (comprehensively reviewed by Kang, Zhe, Lotze, & Tang, 2011). Disruption of the Beclin1–Bcl-2 complex by mutations in the BH3-only domain results in the stimulation of autophagy. However, the relationship between Bcl-2 and Bcl-x_L is complicated since Beclin1 cannot neutralize Bcl-2's antiapoptotic action, whereas Bcl-2 and Bcl-x_L reduces the autophagic activity of Beclin1 (Maiuri et al., 2007; Pattingre et al., 2005). Interestingly, inhibition of autophagy can be achieved with ER-localized, but not mitochondrial-localized Bcl-2. Beclin1 interaction with these antiapoptotic proteins is inhibited by tBid, Bad, and BNIP3. Moreover, proapoptotic proteins such as BNIP3, Bad, Noxa, and PUMA all induce autophagy and may function as competitive inhibitors of Beclin1–Bcl-2/Bcl-x_L interactions (Sinha & Levine, 2008).

There has been considerable debate as to whether autophagy is an independent form of cell death (Whelan et al., 2010). However, autophagy is induced by coronary occlusion, I/R, in failing human hearts and in cardiomyopathies (Knaapen et al., 2001; Kostin et al., 2003; Matsui et al., 2007; Tannous et al., 2008). In addition, ischemia-induced autophagy is accompanied by AMPK activation, similar to the case of permanent coronary occlusion, and is inhibited by transgenic cardiac overexpression of dominant-negative AMPK (Matsui et al., 2007). Autophagy is further enhanced by heart reperfusion as measured by an increased LC3II/LC3I ratio. In contrast, autophagy during reperfusion is independent of AMPK activation and accompanied by Beclin1 upregulation, which was decreased in heterozygous Beclin-null mice. Furthermore, I/R injury in these mice showed a marked decrease in infarct size concurrent with decreased numbers of TUNEL-positive cells (Matsui et al., 2007). Further, heterozygous disruption of Beclin1 decreased cardiomyocyte autophagy and diminished pathological remodeling. Conversely, overexpression of Beclin1 heightened autophagic activity and exacerbated pathological remodeling (Zhu et al., 2007).

The mechanism by which inhibition of Beclin1 leads to decreased autophagy associated with diminished pathological remodeling and decreased infarcts in I/R remains largely unknown. A recent study suggests that I/R markedly impairs autophagosome clearance, and that this contributes to cell death that can be inhibited by cyclosporine A (CypD inhibitor) (Ma et al., 2012).

5. RECAPITULATING SPECIFIC MUTATIONS ASSOCIATED WITH HEART FAILURE IN MICE

5.1. Sarcomeric proteins

Instructive mouse models of human HCM and DCM mutations, which progress to heart failure to different degrees include missense mutations of cardiac myosin heavy chain gene (Geisterfer-Lowrance et al., 1996), and the thin filament protein cardiac α-actin (Song et al., 2011), cardiac Troponin I (cTnI) (James et al., 2000), and α-tropomyosin (Prabhakar et al., 2001).

The human HCM missense mutation, Arg403Gln, in the cardiac myosin heavy chain gene is one of the most comprehensively analyzed for pathobiological mechanisms among the many mouse models of hereditary heart disease (Berul et al., 1997; Gao, Perez, Seidman, Seidman, & Marban, 1999; Georgakopoulos et al., 1999). Among the major insights gleaned are the identification of abnormal calcium homeostasis as an instigator of hypertrophic growth and remodeling (Fatkin et al., 2000), proof for the causal role of TGF-β in cardiac fibrosis (Teekakirikul et al., 2010), and the identification of Fhl1 as a genetic modifier of cardiomyopathy, through transcriptomic analysis (5′RNA-Seq) of alternative start site usage (Christodoulou et al., 2014).

TnI is a component of the thin filament-associated troponin–tropomyosin complex involved in calcium-responsive regulation of muscle contraction. Multiple mutations in cTnI are associated with HCM, one of which was modeled directly in mice by cardiac-specific overexpression of either wild-type or mutant (Arg145Gly in human; 146Gly in mice) cTnI.

Mice overexpressing wild-type cTnI had no phenotype in contrast to cTnI(146Gly) mice, which exhibited cardiomyocyte disarray, interstitial fibrosis, heart failure, and premature death. The cardiac muscle had an increased sensitivity to calcium and, at the whole-organ level, hypercontractility and diastolic dysfunction was evident. In addition, sarcomeric disorganization and abnormal shortening was seen (James et al., 2000). In cTnI-null mice, which also died of heart failure, the major defect also lies in muscle responsiveness to calcium. In addition, mitochondria were more prevalent and enlarged in these hearts, indicating metabolic deficiencies (Huang et al., 1999).

α-Tropomyosin is another thin filament protein, which interacts with troponin and cardiac α-actin to regulate heart muscle contraction. A mouse model was generated based on a specific mutation in human HCM (Glu180Gly), which lies in a troponin T-binding site. Similarly to mice overexpressing wild-type cTnI, cardiac-specific wild-type protein overexpression does not induce any phenotype, however, in mutant mice, concentric hypertrophy, fibrosis, myocyte disorganization, and atrial enlargement are evident by 2 months of age, progressing to LV enlargement and heart failure. Hearts exhibit decreased diastolic performance and increased calcium ion sensitivity, presumed to be due to disruption of normal sarcomere function (Prabhakar et al., 2001).

Seen in diverse HCM mutant mice, altered myofilament calcium sensitivity and calcium homeostasis handling are potential unifying features, contributing to the secondary manifestations of disease involved in the progression to heart failure, such as fibrosis, remodeling, and cell death. Mutations of structural Z-disc proteins including titin, telethonin/Tcap, and muscle LIM protein (MLP) are also a major cause of DCM, with the titin/Tcap/MLP complex thought to serve as a mechanical stretch sensor. Rodent models of this spectrum of disorders include null mutations of MLP (Arber et al., 1997) and telethonin (Knöll et al., 2011) in mice, and a spontaneous mutation in rats affecting *RNA-binding motif protein 20* (RBM20), a gene for human DCM, which causes aberrant titin mRNA splicing (Guo et al., 2012). The development of heart failure after biomechanical stress in telethonin-deficient mice is ascribed to an essential role for telethonin in MDM2-mediated p53 degradation and, hence, enhanced p53 accumulation and p53-dependent apoptosis (Knöll et al., 2011).

Progressive dilation and heart failure in MLP-null mice is prevented by simultaneous deletion of the sarcoplasmic/endoplasmic reticulum calcium ATPase 2 (SERCA2a) inhibitor phospholamban (PLB) (Minamisawa et al.,

1999). PLB inhibits SERCA2a activity in its unphosphorylated form, while phosphorylated PLB dissociates resulting in augmented contractile performance in PLB-null mice. It was suggested that this indicated that defects in excitation–contraction coupling in DCM are related to the enhanced inhibition of SERCA2a by PLB, and inhibition of either PLB expression or PLB–SERCA2a interaction can prevent heart failure (Minamisawa et al., 1999). Similarly, the DCM and heart failure phenotype of MLP-null mice was rescued by crossing with PKCα-null mice that (like the PLB-null model), have increased cardiac performance and hypercontractile myocardium (Braz et al., 2004).

However, enhancing contractile function alone might not be sufficient to prevent heart failure. When crossed with the TNF1.6 model of heart failure, PLB-null mice were unable to restore survival, cardiac function, HCM or DCM, despite enhanced contractile function. The authors postulate that extracellular matrix deposition and downregulation of connexin 43 by chronic TNF-α stimulation cannot be overcome by PLB deletion; thus enhancing contractile function alone may not be sufficient to prevent heart failure (Janczewski et al., 2004).

5.2. Muscular dystrophies

Muscular dystrophies are hereditary disorders most commonly affecting the dystrophin complex of proteins shared by heart and skeletal muscle, that result in premature mortality as a consequence of respiratory or heart failure. The primary cardiac defect is muscle wasting with necrosis, fibrosis, and dilatation leading to impaired pump function and heart failure (Finsterer & Cripe, 2014; Shirokova & Niggli, 2013). Dystrophin is localized to the cytoplasmic leaflet of the plasma membrane and links the cytoskeleton to the transmembrane protein dystroglycan. Thus, loss of dystrophin or associated proteins like sarcoglycans perturbs signal transmission between ECM and the cytoskeleton, as well as conferring membrane fragility. The muscle fibers are easily damaged, resulting in extensive necrosis. Scrutiny of the dystrophin gene in patients exhibiting a more apparent cardiomyopathic phenotype has identified a number of mutations that may be related to heart failure (Muntoni et al., 1993; Ortiz-Lopez, Li, Su, Goytia, & Towbin, 1997). In particular, N-terminal deletions or mutations are associated with greater reduction in expression and more severe phenotypes (Beggs et al., 1991), and appear to cause cardiomyopathy more selectively (Towbin et al., 1993; Ortiz-Lopez, Lopez, Li, Su, Goytia, & Towbin, 1997). It has also been shown that compensation by a number of alternative isoforms occurs in skeletal muscle, but is very limited in the myocardium (Milasin et al., 1996; Muntoni et al., 1997; Neri et al., 2012).

Whereas humans with severe dystrophin defects die due to cardiac or respiratory failure (Moser, 1984; Mukoyama, Kondo, Hizawa, Nishitani, 1987), *mdx* mice (having a spontaneous, or chemically induced dystrophin mutation) exhibit only a mild skeletal muscle phenotype and do not develop heart failure (Bulfield et al., 1984; Chapman, Miller, Armstrong, & Caskey, 1989; Hoffman, Brown, & Kunkel, 1987; Im et al., 1996). Some explanation for this might lie in the observation that a proportion of muscle cells contain low levels of dystrophin, termed revertants (Danko, Chapman, & Wolff, 1992). It should be noted that with the advent of more sophisticated technical measurements of heart function, *mdx* mice may be more representative of early (subclinical) stages of human DMD cardiomyopathies than previously thought, although decompensation up to 6 months of age is followed by recovery (Li, Liu, Zhong, & Yu, 2009; Stuckey et al., 2012). One explanation for the milder phenotype is compensation by a homologous protein, utrophin, occurring more effectively in mice than in humans. Indeed, cardiomyopathy in double utrophin/dystrophin mice was more severe and mice did not survive beyond 14 weeks of age (Grady et al., 1997).

The laminin-binding integrin α7β1 was also suggested to play a compensatory role in dystrophin deficiency. Mice lacking both dystrophin and the α7 subunit displayed severe skeletal muscular dystrophy and died at 2–4 weeks of age due to respiratory failure (Rooney et al., 2006). However, a similar study suggested that heart failure was in fact the primary cause of death, despite the absence of hypertrophy or dilatation. Instead, severe dystrophy in respiratory tissues was thought to induce cardiac necrosis and myofibrillar disarray, along with accumulation of mitochondria (Guo et al., 2006). In accordance with this, in dystrophin/utrophin-deficient double-knockout mice, cardiac function and survival was rescued by restoring utrophin expression selectively in the diaphragm and respiratory muscles (Crisp et al., 2011). MyoD is a myogenic transcription factor expressed in skeletal muscle but not in heart. Bitransgenic *mdx*/MyoD-null mice unsurprisingly not only develop severe skeletal muscle dystrophy but also develop progressive LV DCM, concurrent with necrotic, fibrotic tissue in regions with hypertrophied myocytes (Megeney et al., 1999). Several lines of evidence indicate that skeletal myopathy contributes indirectly to cardiomyopathy, though in the case of γ-sarcoglycan, selectively rescuing the skeletal muscle compartment did not suppress focal necrosis in the heart (Zhu, Wheeler, Hadhazy, Lam, & McNally, 2002).

Dysfunctional coronary vasculature is also suggested to impact on heart failure. δ-Sarcoglycan, another component of the dystrophin-glycoprotein complex, is mutated and reduced in patients that present with DCM. Thus, loss and mutations of δ-sarcoglycan may be specifically associated with cardiac rather skeletal myopathies (Tsubata et al., 2000). In support of this, mice deficient for δ- but not α-sarcoglycan developed cardiomyopathy, necrosis, and fibrosis with increased mortality around 6 months of age (Coral-Vazquez et al., 1999). Absence of the sarcoglycan–sarcospan (SG–SSPN) complex in skeletal and cardiac tissue was observed in both animal models, however, this was additionally absent in vascular smooth muscle in δ-sarcoglycan-null mice. The resultant irregularities of the coronary vasculature were suggested to be responsible for myocardial necrosis (Coral-Vazquez et al., 1999).

In the search for a mouse model with a cardiomyopathy typical of patients, it was postulated that dystrophin-dependent defects might be unmasked in genetically aged mice. Telomeres are regions of G-rich repeats located at the ends of the chromosomes. The telomere shortens during each cell division and cumulative telomere shortening results in senescence and/or cell death. Telomeres are maintained in early life and in stem cells by telomerase, which consists of a reverse transcriptase, TERT, and its RNA template, *Terc*. In *Terc*-null mice, a premature aging phenotype becomes more apparent over the generations (Wong et al., 2009). *Mdx* mice were therefore crossed with mice lacking *Terc* at G1 or G2 (considered as being moderately aged mice). The severe progressive skeletal muscle weakness typical of DMD was recapitulated and fatalities observed from 7–8 months or 4–5 months of age at G1 and G2, respectively (Sacco et al., 2010). Of particular relevance, a follow-up study on the cardiac phenotype found similar features to patients with X-linked cardiomyopathies, such as contractile and conductance dysfunction in the LV, mitochondrial dysfunction, progressive DCM, fibrosis, and heart failure (Mourkioti et al., 2013). Heart failure in this model appears to be driven by oxidative damage, since LV function and overall survival was improved by treatment with antioxidants. Interestingly, when the hearts of four DMD patients who died below the age of 20 were examined, telomere lengths were reduced compared to three sex- and age-matched control hearts (Mourkioti et al., 2013). Although the number of studies and patient numbers are limited, telomere shortening has been shown to be a common feature in the heart tissue of patients with cardiomyopathies (Oh et al., 2003) and in circulating leukocytes from heart failure patients (Wong et al., 2009). The role of telomere shortening in CVD is discussed further in Fyhrquist, Saijonmaa, & Strandberg (2013).

The biological background of aging predisposes the heart to greater risk of cell death and dysfunction, indicating that this comorbidity should be considered when modeling heart failure in the mouse. Mice with accelerated aging phenotypes may be useful in heart failure studies, even if not fully recapitulating all aspects of chronological aging. Both the *mdx* and *Terc*-null single mutant mice have a very mild cardiomyopathic phenotype, which when combined, significantly drive heart failure. Hence, this bigenic model may be more relevant than those used previously to study dystrophic cardiomyopathies. Other genetic causes of muscular dystrophy with cardiac involvement, for which mouse models have been useful to dissect pathogenesis and test therapeutic interventions, include mutations in δ-sarcoglycan (Millay et al., 2008), myotonic dystrophy protein kinase (Wang et al., 2009), and the inner nuclear lamina protein lamin A/C (Azibani, Muchir, Vignier, Bonne, & Bertrand, 2014; Millay et al., 2008).

5.3. Desminopathies and crystallinopathies

Misfolded dysfunctional proteins are actively sequestered and removed by formation of inclusion bodies; essentially masses of aggregated, insoluble proteins rich in ubiquitin and proteasome pathway components. At its most severe abnormal protein accumulation in cardiac or skeletal muscle, causes heart or respiratory failure in the early twenties. In these diseases, cardiomyocytes typically contain both abnormal protein accumulation (aggresomes, typically containing Z-disc proteins and related chaperones) and precursor oligomer intermediates that are related to amyloid oligomers. As a result, myofibril disintegration and disorganization are common features, and desmin and α-Crystallin B (CryAB) are key disease-causing genes in this setting (Goebel & Bornemann, 1993; Sanbe, 2011).

Desmin is a protein component of desmosomes, membrane-associated structures that mediate cell–cell contacts. Desmin forms intermediate filaments that maintain mechanical integrity and transmit tension between the cell membrane, myofibrils, and the nuclear envelope. The most severe phenotypes are associated with mutations that cannot form normal filaments (Muñoz-Mármol et al., 1998; Wang, Osinska, Klevitsky, et al., 2001; Wang, Osinska, Dorn, et al., 2001). Disruption of desmin filament formation is predicted to result in myofibrillar misalignment and mechanical dysfunction.

CryAB is a small heat shock protein chaperone that directs proper folding of desmin, and colocalizes with its mutated form (Vicart et al., 1998). Through its chaperone activities, CryAB is considered to be cardioprotective by limiting dysfunction/damage of misfolded or denatured proteins triggered by stress stimuli, ischemic injury, and genetic mutations (Sanbe, 2011). Another desmosome component, desmoplankin, is also implicated in these myopathies (Norgett et al., 2000; Rampazzo et al., 2002), although has not been studied in as much detail.

In accordance with a role in maintaining mechanical integrity and fibril structure, disorganized myofibrils, abnormal mitochondria morphology, and organization were observed in desmin-null mice (Milner, Weitzer, Tran, Bradley, & Capetanaki, 1996). By 10 weeks of age, cardiomyocyte cell death was extensive with replacement fibrosis and calcification mainly localized to the LV. Despite seemingly severe cardiac degeneration, desmin-null mice do not display early mortality (Milner et al., 1996). In addition, cardiomyocyte expression of a human desmin mutation (DesD7) did not progress to heart failure, although this truncated protein was unable to form intermediate filament networks *in vitro*, although a disrupted desmin network with perturbed myofibril alignment was observed *in vivo* (Muñoz-Mármol et al., 1998; Wang, Osinska, Dorn, et al., 2001). It was suggested that the lack of heart failure in this model might be, in part, due to increased CryAB expression slowing desmin aggregation (Wang, Osinska, Dorn, et al., 2001).

A missense mutation (R120G) of CryAB was discovered in a family with desmin-related myopathy in which desmin was not mutated (Vicart et al., 1998). The consequences of this mutation have been directly investigated by generating mice overexpressing cardiac-specific mutant CryAB, and this model has proved invaluable in understanding the pathology and signaling mechanisms underlying desminopathies, and in contributing to development of new therapeutic strategies.

Indicatative that overexpression *per se* does not cause cardiomyopathy, desmin, overexpression of cardiomyocyte-specific wild-type CryAB did not cause cardiomyopathy or heart failure. However, mutant $CryAB^{R120G}$ induced formation of CryAB-positive aggresomes and death occurred from DCM/heart failure at 5–7 months. At the molecular level, expression of $CryAB^{R120G}$ triggered a hypertrophic response, though initially compensatory, by 3 months deficits in relaxation were evident indicating decompensation prior to dilatation and heart failure (Wang, Osinska, Klevitsky, et al., 2001). Similarly, adenovirus-mediated overexpression of wild-type or mutant CryAB in rat neonatal cardiomyocytes elicited formation of aggregates (Sanbe et al., 2004).

It was suggested that the lack of heart failure in the DesD7 model might be, in part, due to increased CryAB expression (Wang, Osinska, Dorn, et al., 2001). When crossed into a low copy cardiomyocyte-specific CryAB^{R120G} line (viable up to 18 months), bitransgenic mice had increased levels of desmin and more extensive desmin aggregates compared to single transgenics. This was associated with a severely hypertrophic response, with heart failure by 7 weeks. Thus, CryAB was suggested to limit abnormal desmin aggregation, promote degradation, and thus serve a cardioprotective role (Wang et al., 2003).

Prior to cell death and organ dysfunction, disruptions to the cytoskeletal/sarcomeric network in desminopathies become apparent, as a consequence of which mitochondrial localization becomes disorganized. This altered localization is postulated to perturb mitochondrial morphology, resulting in dysfunctional metabolism and cell death. However, the mechanistic proof for this is not in hand from human tissue studies, but may be uncovered by studying the role of these proteins and their mutations in engineered mice.

Indeed, disorganization of the desmin network precedes mitochondrial mislocalization in the CryAB^{R120G} mouse, followed by dysfunction and cell death, prior to impaired heart function (Maloyan et al., 2005). The desmin network is disrupted by 3 months and at this point mitochondria were unaligned with sarcomeres and showed evidence of altered morphology, with dysfunctional isolated mitochondria. By 6 months of age, where heart failure is apparent in transgenic mice, mitochondrial release of cyt c into the cytosol was detected, concurrent with activation of caspase-3 and the appearance of TUNEL-positive cells (Maloyan et al., 2005). Interestingly, VDAC was found to interact with mutant CryAB^{R120G} but not with the wild-type protein (Maloyan et al., 2005). In further support for a mitochondrial-dependent cell-death pathway being relevant to these diseases, *in vitro* studies indicated that mutant CryAB protein can mediate loss of mitochondrial membrane potential and mPTP opening (Maloyan et al., 2005).

Abnormal desmin may interact with CryAB, which acts as a chaperone, promoting degradation and preventing formation of disruptive aggregates. Indeed, when CryAB function is lost or compromised, abnormal desmin protein forms aberrant aggregates and disrupts the integrity of the desmin network with subsequent pathological consequences. Conceivably though, soluble intermediates rather than the aggregates themselves might be the damaging entity, and aggresomes might work as a survival mechanism to sequester preamyloid oligomers (PAO) and reduce cell death.

In support of this interpretation, PAO, but not aggresomes, correlated with decreased cardiomyocyte viability and heart failure in CryAB^{R120G} mice (Sanbe et al., 2005). In mutant CryAB^{R120G} mice, exercise promoted survival and decreased PAO levels (Maloyan, Gulick, Glabe, Kayed, & Robbins, 2007). Adding to the notion that PAOs can elicit cell death and thus heart failure, cardiomyocyte-specific expression of the PAO mutant PQ83 induced cell death, fibrosis, and hypertrophy from 3 months and heart failure by 8 months, while a control, nonamyloid forming peptide did not (Pattison et al., 2008). Interestingly, the mechanism of cell death in this instance was caspase-independent and occurred in the absence of DNA fragmentation. Ultrastructural examination revealed evidence of autophagy plus necrotic cell death indicated by increased inflammatory infiltration and membrane permeability (Evans blue uptake; Pattison et al., 2008).

In other experiments, tet-off mice were generated in which CryAB^{R120G} expression is suppressed by tetracycline (this system is described in detail in Sanbe et al., 2003). As shown previously, cardiomyocyte-specific expression of CryAB^{R120G} induced heart failure at 3.5 months. However, when mutant protein expression was terminated at 3 months, normal heart function and survival were restored. Interestingly, levels of CryAB-positive aggresomes were unaffected after mutant protein expression was suppressed, but PAO levels were reduced (Sanbe et al., 2005). The authors proposed that aggresome formation represents a cytoprotective mechanism, by which levels of soluble, toxic, PAO are reduced by incorporation into aggresomes, thereby reducing cell death (Sanbe et al., 2005). Thus, mouse models of aggresomopathy have proven essential in dissecting the toxic species, showing that targeting PAO would be beneficial and that aggresome formation is, in fact, a cardioprotective mechanism not a toxic one.

As proof that suppressing cell death would ameliorate the cardiac phenotype of CryAB^{R120G} mice, cardiomyocyte-specific overexpression of Bcl-2 was able to prolong survival by 20%, though did not prevent premature mortality. Bcl-2 overexpression was associated with decreased mitochondrial abnormalities, decreased caspase-3 activity, restoration of cardiac function, prevention of cardiac hypertrophy, and reduced aggresome formation (Maloyan, Sayegh, Osinska, Chua, & Robbins, 2010). Similarly, cardiac-specific Bcl-2 overexpression rescued defective mitochondrial function and improved cardiac function in desmin-null mice (Weisleder, Taffet, & Capetanaki, 2004).

However, suppression of Bcl-2-sensitive cell death resulted in increased levels of autophagy and necrosis (Maloyan et al., 2010) as Bcl-2 inhibits autophagy though Beclin1 (Pattingre et al., 2005). That this marker was unaffected by Bcl-2 overexpression in mutant CryAB mice indicates that Bcl-2-independent mechanisms of autophagy exist (Maloyan et al., 2010). Interestingly, a recent study demonstrated for the first time that two different organelles—mitochondria and ER—have predominant roles in mediating cardiomyocyte death signaling during hypertrophy and MI, respectively (Mitra et al., 2013). Of note, CryAB was implicated directly in mediating cell death-dependence on either mitochondria or ER, and suggested to act as a "molecular switch" in bypassing mitochondrial-dependent cell death in MI by binding to VDAC1 and limiting release of cyt c (Mitra et al., 2013). These data would indicate that CryAB is an important target to consider in modulating cell-death responses.

6. MITOGEN-ACTIVATED PROTEIN KINASES IN HEART FAILURE

Mitogen-activated protein kinases (MAPKs) are implicated in cardiomyocyte cell death, cardiac hypertrophy, and heart failure (Marber et al., 2011a; Marber 2011b; Rose et al., 2010). Indeed, MAPK signaling is a prominent feature in human heart failure (Rose et al., 2010). Despite this, there are currently no drugs in the clinic that target MAPKs in this setting (van Berlo et al., 2013). Many protein kinases may act as central control points for regulating cell-death decisions in cardiomyocytes. An overlapping role for diverse protein kinases in the development of heart failure following DCM, HCM, and MI is evident from many studies. PKC isozymes (PKCα, β, ε, λ, and ζ), AMPK and protein kinase D (PKD) have a more central role in hypertrophic growth and heart failure, some (CaMKII, ROCK1, TAK1) have an integral role in regulating cardiac cell death/survival decisions, and others (MAP4K4, PKCδ, PKCθ) are pro-death kinases. A comprehensive mechanistic understanding will uncover potential nodal control points for inhibiting cardiac cell death.

6.1. MAPK signaling cascades

Cardiomyocyte cell death is a complex process of intricately interconnected and interacting protein signal transduction cascades that serve multiple purposes in differentiation, growth, cell function, and pathobiology.

Proteins kinases couple extracellular inputs (from cell-surface molecules including receptor tyrosine kinases, receptor serine/threonine kinases, and G protein-coupled receptors (GPCRs)) as well as intracellular inputs to diverse effector pathways for cell death. Signal transduction is achieved by autophosphorylation, phosphorylation, and dephosphorylation events in a hierarchy that encompasses a recurring four-tier kinase cascade: MAP4Ks → MAP3Ks → MAP2Ks → MAPKs. The MAPK family consists of the familiar "terminal" MAPKs JNK, ERK1/2, and p38 that phosphorylate nonkinase substrates, such as transcription factors and scaffolding proteins, but also phosphorylate other kinases, the MAP kinase-activating protein kinases. MAPKs and upstream protein kinases are widely expressed with their specific cardiac roles being an intense focus for more than a decade, with recognizable functions in cardiac development and heart disease (Kyriakis et al., 2012; Rose et al., 2010; Wang, Whelan, Kaplinskiy, & Kitsis, 2007; summarized in Fig. 4.3).

6.1.1 MAP4 kinases

MAP4K4 (mitogen-activated protein kinase kinase kinase kinase-4, hepatocyte progenitor kinase-like/germinal center kinase-like kinase (HGK), Nck-interacting kinase (NIK)) is a serine/threonine protein kinase related to *S. cerevisiae* Sterile 20 (STE20). A summary of human STE20-related kinases is shown in Fig. 4.4.

Orthologs exist in *Drosophila* and *C. elegans*, known as Msn or mig-15, respectively, that also resemble the two most closely related STE20 kinases in mammals, TNIK (Traf2 and Nck-interacting kinase) and MINK (Misshapen/NIK-related kinase). Defined as members of the germinal center kinase (GCK) group, these all possess a characteristic N-terminal kinase domain and C-terminal citron homology (CNH) domain. MAP4K4 will be discussed in more detail in conjunction with TAK1 (MAP3K7), following other members of the MAPK superfamily.

A second subgroup of MAP4Ks comprises the p21-activated protein kinases. Deletion of Pak1 in mouse myocardium unmasked an anti-hypertrophic role, the conditional knockout mice developing increased hypertrophy in response either to pressure overload or angiotensin II infusion (Liu et al., 2011). Apoptosis induced by aortic banding was increased fivefold, in the absence of Pak1. By 5 weeks of aortic constriction, the Pak1-deficient hearts developed severely impaired systolic function, ventricular dilation, and marked fibrosis. Pak1 was activated by several of the

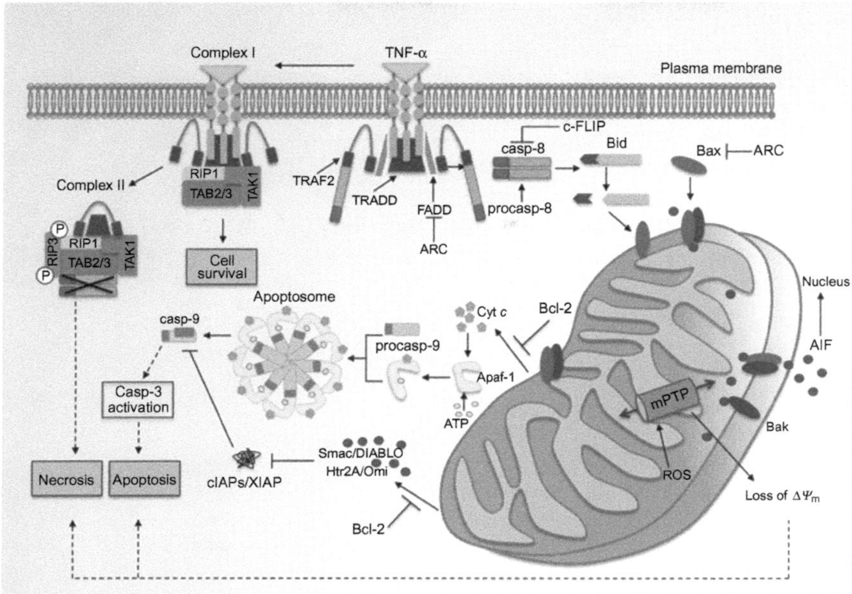

Figure 4.3 Illustration of the different branches of apoptosis and necrosis cell-death signaling. Ligand binding (TNF-α) induces recruitment of adaptor proteins and subsequent activation of procaspase-8. Caspase-8 will induce Bid cleavage to tBid that will lead to Bax translocation on the mitochondrial membrane and the formation of mitochondrial pores consisting of Bax/Bak duet. Apoptogens (Smac/DIABLO, Htr2A/Omi, and cyt c) will be released for the inhibition of XIAP/cIAPs and the formation of the apoptosome. Cyt c will account for procaspase-9 activation by stimulating its binding on Apaf-1, which along with ATP, ends in the formation of the apoptosome. Caspase-3 is activated via either caspase-8 directly (extrinsic pathway) or via caspase-9 (intrinsic pathway), which will subsequently target cytoskeletal proteins and other enzymes for apoptosis. AIF presumably translocates in the nucleus from the mitochondria to induce large-scale DNA fragmentation. ARC protein antagonizes both pathways and Bcl-2 antagonizes the release of apoptogens from the mitochondria. During ischemia, depletion of ATP and lower pH sensitizes mPTP opening, an event that occurs upon reperfusion. Mitochondria become overloaded with Ca^{2+} and ROS leading to events such as dissipation of mitochondrial membrane potential ($\Delta\Psi_m$), mitochondrial swelling, disruption of the outer mitochondrial membrane, and release of intermembrane apoptogens that mark irreversible cell death. Cyclophilin D (CypD or PPIF) has an essential role in protein folding and is involved in the regulation of mPTP (mPTP-dependent necrosis). Alternatively, necrosis can be initiated at the cell surface, when multiprotein complex I-TNFR1 is dissociated to form complex II, to further mediate cell death. Deubiquitination of RIP1 by CYLD (not shown) will lead to recruitment of FADD–procaspase-8 via DD interactions. If procaspase-8 is activated, RIP1 is cleaved by caspase-8 and rendered inactive to signal necrosis. In the case of procaspase-8 inhibition, RIP1 recruits RIP3 instead ensuing a series of phosphorylation events leading to cell death (death receptor-dependent necrosis). In unstimulated cells, complex I comprises TNFR1, TRADD, RIP1, TRAF2, and cIAP1/2, the latter of which polyubiquitinates RIP1 and TRAF2 so that they recruit TAK1 kinase via TAB2/3-binding proteins, which in turn will turn on survival genes via NFκB activation or IKK inhibition (for more details, see main body of review). (See the color plate.)

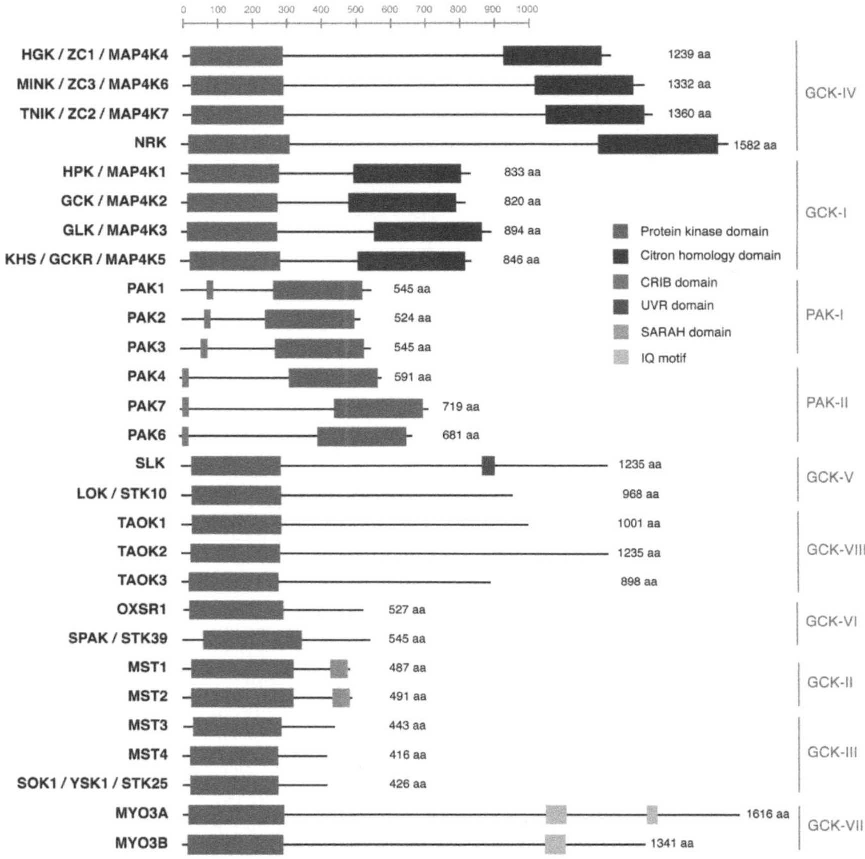

Figure 4.4 Human STE20-related kinases. Members of the GCK group possess a characteristic N-terminal kinase domain and C-terminal citron homology (CNH) domain and are divided into eight subgroups on the basis of sequence similarities (I–VIII). A second subgroup of MAP4Ks comprises the p21-activated protein kinases (PAKs) that consist of a C-terminal kinase domain and N-terminal CRIB domain. PAKs are divided into subclasses I and II. (See the color plate.)

same pathophysiological and pharmacological cues as the several pro-death kinases, but thus has a contrasting essential cardioprotective role.

6.1.2 MAP3 kinases

Apoptosis signal-regulating kinase 1 (ASK1) is a member of the MAP3K family that mediates growth and death decisions in cardiac myocytes. It is activated by a variety of stress stimuli (e.g., TNF-α, ROS, and Fas), and phosphorylates both MKK4/7 and MKK3/6, the activators of JNK and p38, respectively. TRAFs are also important in the regulation of ASK1 activity; among the TRAF family members only TRAF2, 5, and 6 increase its activity, despite direct binding to multiple other members. Isolated cells from TRAF2- and TRAF6-deficient mice had a dramatic attenuation of both JNK and p38 activities when ASK1 was activated by stimulation of hydrogen peroxide (Noguchi et al., 2005). This suggests that the receptor-mediated pathway of cell death plays an essential role in ROS-dependent activation of ASK1.

Inducible cardiac-specific overexpression of ASK1 in the heart leads to no pathology of hypertrophy at 3 and 12 months of age, that under stimulation with acute pressure overload or isoproterenol infusion is no different to control mice. However, upon extended pressure overload or MI, the mice exhibited increased cell death leading to the development of cardiomyopathy. Prominently activated downstream signaling components were MKK4/4 and JNK1/2 (in the absence of p38 and ERK1/2). The cardiomyopathy phenotype after 8 weeks of pressure overload was significantly reduced in ASK1 transgenic mice in a calcineurin Aβ-null background, suggesting that the cardiomyopathy propensity is, in part, mediated via a calcineurin-dependent mechanism (Liu et al., 2009). In a different model, when ASK1 was deleted in *Raf-1* null mice, this led to a rescue phenotype of heart dysfunction, dilation, and fibrosis, most likely due to a decrease in TUNEL-positive cardiomyocytes by 4 and 10 weeks (Yamaguchi et al., 2004). Indeed raf-1 and ASK1 interact both *in vitro* and *in vivo* in a MEK–ERK1/2-independent manner to inhibit apoptosis (Chen, Fujii, Zhang, Roberts, & Fu, 2001). During myoblast differentiation to myotubes, ASK1 and TAK1 also interacted with a promyogenic cell-surface molecule (Cdo) and a scaffold protein (JLP) (Tran et al., 2012).

6.1.3 MAP kinases

The Ras–Raf–MEK–ERK1/2 pathway is the founding or prototypic MAPK cascade, in which Ras acts as a molecular switch between receptors and downstream kinases (Kyriakis et al., 2012). Ras is a small GTP-binding protein that induces activation of downstream Raf (a MAP3K), which translocates from the membrane to the cytosol to activate MEK1 (MAP2K1) and ERK1/2. The pathway can be activated by mitogenic stimuli growth factors like EGF and PDGF, insulin, or cytokines like interleukin-1β (IL-1β) (Wan, Chi, Xie, Schneider, & Flavell, 2006; Yu et al., 2008) and TGF-β (Ieda et al., 2009; Matsumoto-Ida et al., 2006). ERK1/2 can be activated via tyrosine kinase receptors and GPCRs and plays a significant role in cardiac hypertrophy and protection against cell death. In general, the pathway is regarded as prohypertrophic and prosurvival, but not a necessary component for the development of cardiomyocyte hypertrophy.

JNK is stimulated by MAP3Ks (i.e., MEKK1, MEKK2 and MEKK3), as well as by mixed lineage kinase 2 and 3 (MLK2, MLK3) (Kyriakis et al., 2012). MAP2Ks (i.e., MKK4 and MKK7) act downstream of MAP3Ks, to activate JNK. Robust activation of JNK occurs as a result of different stimuli, such as inflammatory cytokines, heat shock, hyperosmolarity, I/R, UV, oxidative and ER stress, DNA damage, and to a lesser extent by growth factors. In the heart, JNK is activated by mechanical overload or I/R injury (Bogoyevitch et al., 1996; Ramirez et al., 1997; Wang et al., 1998) and has a role related to hypertrophy, heart remodeling, and cardiac muscle cell death.

p38 is the third major MAPK and comprises four different isoforms; from which the first two, the prototypic p38α (referred to often as just p38) and p38β are ubiquitously expressed and directly relevant to cardiomyocyte fate. The canonical pathway for p38 activation is formally similar to the JNK and ERK1/2 pathways: a number of upstream MAP3Ks including MEKK1–4, TAK1, and ASK1 activate MKK3, MKK6, and possibly MKK4 at the MAP2K level. So-called "noncanonical" p38 activation occurs via the TAK1-binding protein TAB1 and the necrosis-mediated death-receptor pathway (De Nicola et al., 2013). While gain- and loss-of-function studies overule the hypothesis that p38 activity is sufficient to promote cardiac hypertrophy *in vivo*, upstream activators of p38 (i.e., MAP2Ks) are involved in the hypertrophic process. In general, p38 reportedly regulates a cardiac gene program for the development of myocyte hypertrophy, heart remodeling, metabolism, contractility, proliferation, and cell death, though with significant controversies as to how p38-mediated signaling leads to cardiac pathologies (Wang et al., 2007). p38 induction is more closely related to pathological hypertrophy rather than physiological or compensatory hypertrophy, with chronic levels of p38 activity in the injured myocardium contributing toward maladaptive cardiac remodeling (i.e., contractility and ECM deposition). However, the role of p38 in ischemia is clearer since inhibition by various pharmacological and genetic models is cardioprotective, and a p38α/β inhibitor is currently being tested in Phase I/II clinical trials for respiratory and CVDs (Denise Martin, De Nicola, & Marber, 2012).

Abnormal MAPK activation occurs not only in acquired heart failure but also in hereditary cardiomyopathies, and cardiac dysfunction due to mutations of *Lmna* in mice is improved by pharmacological inhibitors of the terminal MAPKs (ERK1/2: selumetinib; JNK: SP600125; p38: ARRY797) or by germline deletion of *Erk1* (Azibani et al., 2014).

The final MAPK, and least studied among the four, is ERK5, activated by MEK5 (MAP2K) following activation by upstream MAP3Ks (MEKK2 and MEKK3). A variety of growth factors (EGF, NGF, VEGF, and FGF-2), serum, hyperosmotic stress, oxidative stress, and UV stimulate ERK5 activation (Wang et al., 2007). In the heart, ERK5 is thought to play anti-hypertrophic and antiapoptotic roles.

6.2. The MAP4K4–TAK1 module in cell death

An activating TAK1 mutation in mouse myocardium, which mimics the sustained increase in activity after mechanical load, was sufficient to induce hypertrophy, fibrosis, severe myocardial dysfunction, cell death, and heart failure, associated primarily with p38 activation (Zhang et al., 2000). TAK1 is implicated in a number of signaling pathways that might contribute to this role. It is an important component of multiple cell death-associated membrane receptor complexes, such as TNFR1 and TLRs, resulting in altered NFκB signaling. Further, TGF-β is upregulated during cardiac hypertrophy in both rodents and humans (Hein et al., 2003; Teekakirikul et al., 2010; Zhang, 2000), and is upregulated in cardiomyocytes by TAK1 activity (Zhang et al., 2000). In addition, we and others (Chen, Tu, Wu, & Bahl, 2000) have seen activation of TAK1 in cultured cardiomyocytes in response to hydrogen peroxide and C2-ceramide, cell death in this acute setting being ascribed to JNK and the activation of TAK1 ascribed to MAP4K4 (Xie et al., 2007). In support of the MAP4K4–TAK1–JNK pathway being involved in heart failure, tamoxifen-inducible activation of JNK in the adult heart caused progressive lethal cardiomyopathy, with ECM remodeling and abnormal gap junction signaling due to loss of connexin 43 (Petrich, Molkentin, & Wang, 2003; Petrich et al., 2004; Ursitti et al., 2007).

The Rho effector Rho-associated coiled-coil protein kinase 1 (ROCK-1) is cleaved by caspase-3 in human heart failure and is an activator of cell death associated with cardiac load (Chang et al., 2006). Cardiac-specific over-expression of wild-type MAP4K4 does not induce cell death or heart failure, but when crossed with the *Myh6*-Gaq line (which has an underlying, mild myopathy), bitransgenic mice die of apoptotic heart failure within 3 months

(Xie et al., 2007). The bigenic mice also had significantly increased ROCK-1 cleavage and caspase-3 activity compared to single transgenics (Chang et al., 2006). These data show that MAP4K4 activity sensitizes the heart to otherwise sublethal death stimuli and could act as a nodal point to mediate the switch from adaptive to maladaptive hypertrophy.

Another MAP4K, and member of the STE 20-like kinase family is mammalian sterile 20-like kinase 1 (Mst-1), which also mediates cardiomyocyte cell death and heart failure, albeit in an unexpected manner. Cardiac-specific overexpression of Mst1 results in increased cell death, DCM, fibrosis, and heart failure from 15 days of age (Yamamoto et al., 2003). Unexpectedly, cardiomyocytes did not appear elongated as expected, but rather were reduced in volume and in length. In this model, side-to-side slippage rather than elongation of cardiomyocytes maybe responsible for dilatation. In addition, cardiac-specific expression of dominant-negative Mst1 was protective against cell death induced by I/R and reduced infarct size. Mst1 is proapoptotic, but in addition, acts as an inhibitor of hypertrophy (Yamamoto et al., 2003). In contrast, TAK1 induced both apoptosis and hypertrophy *in vivo* (Zhang et al., 2000).

7. NEUROHORMONAL RESPONSES: G PROTEINS AND G PROTEIN-COUPLED RECEPTORS

GPCRs regulate chronotropy, inotropy, and cardiac growth and are important in the pathophysiology of cardiac hypertrophy and heart failure. They are transmembrane receptors that upon extracellular ligand binding undergo a conformational change and couple the intracellular domain with a heterotrimeric G protein consisting of α, β, and γ subunits, the former of which can be of type Gs, Gi, or Gq. There are three families of regulatory molecules involved in the desensitization (time-dependent attenuation) of the GPCR response: the second-messenger regulated kinases (cAMP-dependent protein kinase, PKA; calcium-activated phospholipid-dependent protein kinase, PKC), G protein-coupled receptor protein kinase (GRKs) and arrestins, which serve as scaffold molecules for receptor internalization.

β-AR are a subclass of GPCRs that can elicit a number of downstream signaling events by coupling to different G proteins, including activation of adenylyl cyclase that elevates cAMP to activate PKA, when coupled to Gs. Signaling through Gq activates phospholipase C, which activates PKC and calcium/calmodulin-dependent protein kinase (CaMK), whereas coupling to Gi inhibits adenyl cyclase and hence PKA. Consequently, other downstream effectors include PKD which is activated by PKC (Harrison et al., 2006; Haworth, 2000), MLP (or cysteine-rich protein 3), and PLB, an inhibitor of the SERCA2, which is phosphorylated and inhibited by PKA and CaMKs, relieving its inhibition of calcium reuptake and improving cardiac function (Bhupathy et al., 2007; Traaseth et al., 2008). Other GPCRs relevant to cardiac hypertrophy, remodeling, and heart failure include receptors for angiotensin II, endothelin, and the relaxin receptors RFXP1/2.

7.1. The β-adrenergic receptor

Mice overexpressing the β2-ARs develop progressive fibrotic DCM and heart failure in a dose-dependent manner (Liggett et al., 2000). At 10–15 weeks, the highest expressing line exhibited hypertrophy, impaired LV function, and increased chamber volume, whereas lower expressing lines showed hypertrophy and enhanced LV function. Thus, expression of β_2-ARs at levels that enhance the response to agonists, but do not cause ligand-independent signaling, may improve cardiac function without deleterious effects (Liggett et al., 2000).

Cardiomyocyte-specific overexpression of β1-AR also induces heart failure and increases mortality (Engelhardt, Hein, Wiesmann, & Lohse, 1999). Younger mice exhibited increased contractility that progressed to hypertrophy, fibrosis, and cell death indicating that, like β2-AR, stimulation of β1-AR may improve cardiac function, while prolonged stimulation leads to cell death and heart failure (Engelhardt et al., 1999). Though it has been proposed that the β1-AR is the "cardiotoxic subtype" and β2-AR the "cardioprotective subtype," deletion of β2-AR prevented heart failure induced either by aortic constriction or by deletion of MLP (Fajardo et al., 2013).

β-Adrenergic signaling results in phosphorylation and inhibition of PLB relieving inhibition of calcium uptake into the sarcoplasmic reticulum and improving cardiac function. Overexpressing PLB in cardiomyocytes resulted in hypertrophy that progressed from adaptive to maladaptive, with dilatation and diffuse fibrosis, resulting in declining LV function and heart failure. Initially, PLB was highly phosphorylated but with aging, phosphorylation decreased where β-adrenergic signaling was suppressed, due in part to downregulation of norepinephrine, epinephrine, and dopamine in transgenic hearts (Dash et al., 2001).

7.2. Receptor-coupled G proteins

Cardiac-specific overexpression of G proteins results in necrosis, replacement fibrosis, hypertrophy, DCM, and arrhythmias (Iwase et al., 1997). Conditional cardiac-specific expression of a synthetic Gi-coupled receptor was utilized to assess long-term effects of a precisely modifiable signal: chronic expression of Ro1 (receptor activated solely by a synthetic ligand, opioid 1) induced arrhythmia, dilatation, disarray, fibrosis, and cardiac dysfunction, resulting in heart failure (Redfern et al., 2000). In accordance with chronic β-AR signaling being detrimental, so too is activation of their coupled G proteins.

Cardiac-specific overexpression of Gαq has no effect at lower levels, while intermediate levels (Gαq-25 copy line) evoke hypertrophy, fetal gene expression, and impaired contractile function, though without necrosis and fibrosis. With higher levels of Gαq, further decompensation and dilation occurred, with development of fatal heart failure (D'Angelo et al., 1997). Similarly, mice expressing cardiomyocyte-specific Gαq at higher levels died of heart failure around 3 months of age, with excessive apoptosis. In cultured cardiomyocytes overexpressing Gαq, as in transgenic mice expressing Gαq, the increase in JNK and p38 activity was minimal but increased significantly both in cells expressing an active mutant, Q209L and in failing Gαq-overexpressing hearts (Adams et al., 1998). The milder myopathy in Gαq-25 mice is significantly exacerbated by stresses including pregnancy. Infusion of a pan-caspase inhibitor significantly reduces the prevalence of TUNEL-positive cells, reduces dysfunction and abrogates mortality (Hayakawa et al., 2003). These data strongly support the conclusion that myocyte cell death is a causal feature in heart failure and that high levels of Gαq-mediated signaling activate caspase-dependent cell death. Intriguingly, in Gαq hearts, a BH3-like protein Nix/Bnip3L (BCL2/adenovirus E1B 19 kDa interacting protein 3-like) is upregulated, which when overexpressed suffices to cause extensive cardiomyocyte death and heart failure. Conversely, a splice variant of Nix/Bnip3L (sNix) is protective against peripartum cardiomyopathy in Gαq transgenic females, diminishing cell death and improving cardiac function and survival (Yussman et al., 2002).

7.3. Desensitization: βARK1/GRK2

GRKs directly phosphorylate GPCRs, leading to desensitization, internalization, and inactivation. Thus, βARK1/GRK2 decreases stimulation and contraction, and the impaired β-AR signaling in heart failure is due at least in part, to desensitization by βAR kinase activity. Some effects of GRK2, such as reducing infarct size, in turn depend on endothelial nitric oxide synthase (abolished by *Nos3* knockout) and on S-nitrosylation of the kinase (largely abolished by a C340S knockin mutation of its principal nitrosylation site; Huang et al., 2013).

βARKct is a peptide inhibitor of βARK, comprising the last 195 amino acids of βARK1. This portion contains the binding site for Gβγ thus competes with full-length protein for binding to Gβγ. In MLP-null mice that develop heart failure (Arber et al., 1997), cardiac-specific expression of βARKct rescued the heart failure phenotype (Rockman et al., 1998). Inhibition of βARK1 activity therefore restored normal function, although basal LV contractility was not completely restored, and levels of fibrosis were not altered. In addition, in MLP-null mice, βARK1 expression and activity was increased (Rockman et al., 1998) and aspects of the MLP-null DCM phenotype that were not restored were independent of βARK1/β-AR signaling. As fibrosis was still evident, this indicated some level of cardiomyocyte dropout with improved function of remaining cardiomyocytes as a compensatory mechanism, but one that was not able to completely prevent cell death.

Three genetic models with increased cardiac contractile performance (PLB-null and cardiac-specific overexpression of β2-AR or βARKct) were each crossed into a background of HCM, each resulting in a different phenotype (Freeman, Lerman, et al., 2001). The HCM model expresses a mutated MHC with diminished actin-binding properties due to a R403Q missense mutation and deletion of amino acids 468–527. Male mice show signs of impaired function and heart failure by 8 months of age, initially with hypertrophy that progress to extensive dilation, concurrent with cardiac dysfunction (Freeman, Colon-Rivera, et al., 2001; Vikstrom, Factor, & Leinwand, 1996). Absence of PLB rescued cardiac dysfunction and improved fibrosis, indicating less of a pathological response, though hypertrophic growth increased. In contrast, β2-AR overexpression caused more rapid dilation and heart failure. βARKct rescued cardiac dysfunction, hypertrophy, and fibrotic deposition. Consequently, one conclusion from this combinatorial comparison is that the maximal stimulation provoked by β2-AR is detrimental, while the submaximal increases elicited by βARKct or PLB deletion allow the myocardium to respond to normal β-AR stimulation, an imperative functional property (Freeman, Lerman, et al., 2001). The pathological features of hypertrophy were at least partially PLB independent, and loss of PLB may therefore stimulate adaptive hypertrophy that can rescue cardiac function even in the presence of maladaptive hypertrophy. This study indicates that inhibition of PLB could pose a particularly attractive target to rescue cardiac function and retain the ability of the myocardium to generate non-pathological hypertrophic responses in order to preserve and improve heart function.

Inhibition of βARK1 by βARKct was also tested for its ability to rescue heart failure in mice overexpressing the sarcoplasmic reticulum Ca^{2+}-binding protein, calsequestrin (CSQ; Harding, Jones, Lefkowitz, Koch, & Rockman, 2001). CSQ is a Ca^{2+}-binding protein in SR that interacts with the ryanodine receptor (RyR). Transgenic mice with cardiac-specific CSQ overexpression showed suppressed calcium release, hypertrophy, and premature death by 16 weeks of age. At 7 weeks, mild LV enlargement, decreased LV function, and hypertrophy was observed, that by 14 weeks, had progressed to more marked impairment, concurrent with impaired β-AR responsiveness and activity. The authors suggest that alterations in β-AR signaling precede the development of heart failure in this model (Cho et al., 1999; Jones et al., 1998). βARK1 activity is increased in human heart failure, and similarly so in CSQ mice, but restored to normal levels in CSQ/βARKct mice; and as predicted, CSQ/βARKct mice had improved cardiac function, reduced hypertrophy and dilation, and increased survival (Harding et al., 2001). Further, survival could be further enhanced by adding a selective β1-blocker/antagonist, metoprolol (a current treatment for heart failure), compared to metoprolol or βARKct alone. These results indicate that inhibition of βARK1 could work successfully alongside current treatments in combating heart failure (Harding et al., 2001). In addition, PLB deletion in CSQ mice was also able to rescue the contractile and hypertrophic response (Sato et al., 2001).

7.4. Intracellular effectors

7.4.1 PKA

β-ARs activate Gs that in turn activate PKA by stimulating adenylyl cyclase and increasing cAMP formation. Conversely, receptor-mediated activation of Gi inhibits PKA (Dorn & Mochly-Rosen, 2002). PKA phosphorylates PLB and the sarcoplasmic reticulum calcium release channel RyR2, improving myocyte calcium handling and thus stimulating contractility. In accordance with chronic stimulation of this pathway being detrimental, constitutive activation of cardiac PKA results in DCM, reduced contractility, arrhythmias, and death (Antos et al., 2001). In younger animals, RyR2 and PLB phosphorylation was evident, with increased calcium signaling. By 8 weeks, this resulted in dilation, reduced function, and susceptibility to ventricular arrhythmias. From 10 weeks, hypertrophy occurred and by 13 weeks, DCM, fibrosis, and heart failure ensued, with no survival beyond 20 weeks (Antos et al., 2001). Many of these models recapitulate signaling seen in failing human hearts and represent useful models to test whether short-term inhibition of downstream components of the β-AR pathway might prevent progression from adaptive to maladaptive responses that occur with prolonged and chronic activation. This may provide a means to inhibit selective, deleterious aspects of signaling rather than blocking signaling at the level of the receptor itself, which would suppress multiple pathways including those that may be beneficial in protecting myocardium from degeneration.

CREB is a transcription factor activated by β-AR–PKA signaling that is downregulated in response to chronic activation. A dominant-negative form of CREB, transgenically expressed in myocardium, induces severe, lethal DCM and heart failure. Despite extensive cardiomyocyte disarray and fibrosis, TUNEL-positive cells were absent, and DNA laddering was not detected, indicating that apoptotic cell death was likely not present. Interestingly however, myocytes were abnormally vacuolated (Frenzke, Korcarz, Lang, & Leiden, 1998), a possible indication of autophagy, indicating that multiple mechanisms of cell death are activated by chronic β-AR activation.

7.4.2 PKC

The PKC family comprises calcium and/or lipid-activated Ser/Thr kinases that serve as effector molecules for seven GPCRs coupled to the Gq class of heterotrimeric G proteins. The 10 PKC isozymes/isoforms are classified by their activation characteristics. The Ca^{2+}- and lipid-activated isozymes are PKCα, PKCβI, PKCβII, and PKCγ; novel isozymes (requiring diacylglycerol but not calcium) are PKCε, PKCθ, PKCη, and PKCδ; and the atypical enzymes, PKCζ and PKCλ are calcium-independent and depend on alternative lipids for their activation. Some PKC isozymes (PKCα, β, ε, λ, and ζ) have a more central role in the development of cardiac hypertrophy, while others (PKCδ and θ) are proapoptotic.

PKCα, the least studied of the cardiac PKCs is not regulated in acute myocardial ischemia, but is associated with hypertrophy and DCM after ischemic injury and is upregulated in failing hearts (Bowling et al., 1999; Dorn & Force, 2005). PKCα-null mice have increased cardiac contractile function that is protective against heart failure induced by pressure overload or by deletion of MLP (Braz et al., 2004). Mechanistically, changes in PKCα activity correlate with PLB phosphorylation; increased in the case of PKCα deficiency and decreased with overexpression.

PKCβ is upregulated in failed hearts compared to nonfailed left ventricles (Bowling et al., 1999). Sustained activation of PKCβ in the heart induces concentric cardiac hypertrophy associated with impaired diastolic relaxation, whereas expression in newborns caused sudden death due to calcium transient relaxation defects (Bowman et al., 1997). In support of a role for PKCβ2 in heart failure, cardiomyocyte-restricted overexpression cause hypertrophy from 3 weeks of age, progressing to dilatation, necrosis, diffuse and replacement fibrosis, and decreased LV function (Wakasaki et al., 1997).

PKC translocation and activation involve coupling to isozyme-specific proteins, called receptors for activated C kinases (RACKs). Lesser expression of constitutively active PKCε was cardioprotective against ischemia, while higher expression caused cardiac hypertrophy, fibrillar disarray, diffuse fibrosis, impaired contractile function, and failure by 13 weeks of age (Pass, Zheng, et al., 2001; Pass, Gao, et al., 2001). Differential patterns of PKCε-RACK interactions may mediate these different cardiac responses (Pass, Zheng, et al., 2001) and the level of PKCε activity may in turn mediate progression from compensated to decompensated hypertrophy and failure by upregulating RACK1, and inducing interaction with both itself and PKCβ2 (Pass, Gao, et al., 2001).

PKC peptides derived from PKCε-RACK binding or pseudo-RACK sites can selectively act as PKCε translocation inhibitors (εV1) and activators (ψεRACK), respectively. When expression was driven in cardiomyocytes by the *Myh6* promoter, εV1 (the inhibitory peptide) induced hypertrophy and slightly reduced LV function at low levels, while higher levels resulted in dilation and heart failure, though without obvious cardiomyocyte dropout or fibrosis. In contrast, activation of PKCε by ψεRACK selectively increased β-MHC expression and resulted in cardiac enlargement, though unexpectedly, decreased cardiomyocyte size was observed, and hearts maintained normal function (Mochly-Rosen et al., 2000). The pattern of fetal gene expression in these hearts is similar to that in Gαq/β2-AR overexpressing mice: increased expression of βMHC but not ANF or α-skeletal actin. Heart enlargement was therefore presumed to be physiological, and suggested to reflect enhanced developmental or postnatal hyperplasia, rather than cardiomyocyte enlargement.

Conversely, myocardial Gαq overexpression results in PKCε activation (D'Angelo et al., 1997). Cardiac-specific εV1 or ψεRACK were coexpressed with Gαq to inhibit and exacerbate PKCε activity, respectively. Unexpectedly, further stimulation of PKCε by ψεRACK improved cardiac function and altered hypertrophy from an eccentric to a concentric type, indicating that activation of PKCε in Gαq mice is a compensatory event. This leads to the inference that further activation might be beneficial in heart failure, while inhibition would be detrimental, since reciprocally, inhibition by εV1 exacerbated the Gαq nonfailing hypertrophic phenotype, becoming instead a lethal dilated cardiomyopathy (Wu, Toyokawa, Hahn, & Dorn, 2000). Consequently, the net effect of a signaling component in heart failure models cannot be assumed to be deleterious, and tailored investigations should be carried out to distinguish those that speed the progression to heart failure, and thus should be inhibited, from compensatory, counter-regulatory responses that should be maintained or even enhanced.

Using a similar approach to test the involvement of PKCα in Gαq signaling, mice were established with cardiomyocyte-specific expression of peptides that inhibit or activate PKCα. While a minimal phenotype was observed in PKCα-activated mice, further activation in the Gαq background caused interstitial fibrosis, ventricular stiffness, and heart failure, while, conversely, suppression of PKCα activity in Gαq hearts improved function (Hahn et al., 2003). On the basis of such studies, PKCα would seem a better target for inhibition and PKCε for activation. It would be interesting to see the consequence of these interventions in concert.

Notably, given the large number of conventional and novel PKC isozymes, achieving stringent selectivity has been difficult with typical small molecular approaches, whereas the innovative studies of PKC using translocation peptides in mice has led to clinical investigations targeting the isozyme-specific protein–protein interactions (Mochly-Rosen, Das, & Grimes, 2012). Together these studies therefore indicate a method by which activity of individual PKC isoforms can be selectively modified to support physiological rather than pathophysiological responses and prevent progression to heart failure.

7.4.3 Calcium, calmodulin, CaMKs, and calcineurin

Calcineurin is arguably the most thoroughly studied calcium-dependent pathway in heart failure (Molkentin et al., 1998). Cardiac-specific overexpression of a constitutively active catalytic subunit of calcineurin or constitutively active NFAT3 elicits concentric hypertrophy and cardiomyocyte disarray, interstitial fibrosis, progression to dilation, and heart failure. Moreover, several hypertrophic or heart failure models can be rescued by administration of cyclosporine A, a calcineurin inhibitor (Molkentin et al., 1998), or deletion of calcineurin. Notably, a number of genes that suppress pathological cardiac remodeling work by inhibiting calcineurin–NFAT signaling in cardiomyocyte including *interferon regulatory factor 8* (Jiang et al., 2014) and *Trim63*, encoding the E3 ligase muscle-specific RING finger protein-1 (Maejima et al., 2014). In control of calcium signaling, transient receptor potential channels serve as key initiators of the calcium-dependent pathways for pathological remodeling, as evidenced most conclusively by genetic ablation of *Trpc1* or combined deletion of *Trpc3* and *Trpc6* (Eder & Molkentin, 2011; Seo, Rainer, Lee, et al., 2014; Seo, Rainer, Shalkey Hahn, et al., 2014).

CaMK is a major activator for a number of calcium-handling proteins and transcription factors. It shares many substrates with PKA central to the regulation of calcium (e.g., PLB, RyR, and calcineurin) and the development of cardiac hypertrophy (e.g., NFAT, MEF2). Mechanistically, CaMKs induce cardiac hypertrophy through activation of MEF2, acting in parallel and synergistically with the dephosphorylation of NFAT by calcineurin, a calcium-dependent phosphatase (Passier et al., 2000).

A solid body of evidence has emerged around CaMKII confirming the following: (i) that CaMKII activity is increased in cardiac hypertrophy and in failing myocardium of mice and patients, (ii) CaMKII overexpression exacerbates cell death and induces myocardial hypertrophy and arrhythmias (Sag et al., 2009) leading to heart failure, and (iii) inhibition of CaMKII by drugs, inhibitory peptides and, most conclusively, gene ablation improves myocardial hypertrophy, protects the myocardium against apoptosis, reduces infarct, and preserves contractile function (Ling et al., 2009; Vila-Petroff et al., 2007; Yang et al., 2006). Transgenic mice that overexpress the δ_C isoform of CaMKII in the heart develop hypertrophy, progressive DCM, impaired contractile function, and heart failure, associated with increased phosphorylation of RyR2 and PLB. Interestingly, RyR2 phosphorylation preceded development of hypertrophy and heart failure, indicating that this isoform could represent a relevant therapeutic target (Zhang et al., 2003). Further, albeit with the limitation of gain-of-function studies, this model showed a role for the isoform in regulating calcium handling and excitation–contraction coupling (Maier et al., 2003). Analogously, cardiac-specific CaMKIV transgenic mice develop progressive HCM (Passier et al., 2000). At 6 months, cardiac wall thickening was frequently accompanied by ventricular dilation, suggesting progression from concentric hypertrophy to a dilated hypertrophic phenotype, concurrent with LV dysfunction.

7.4.4 PKD

The diacylglycerol and PKC effector PKD is linked to cardiac hypertrophy and its inhibition by PKD-selective drugs is a promising mode of treating hypertrophic heart disease (Fu & Rubin, 2011). PKD interacts with, and phosphorylates cTnI myofilaments to regulate contraction (Haworth, 2000; Haworth et al., 2004) and regulates function of HDAC5 (Harrison et al., 2006). Phosphorylation by PKD stimulates nuclear export of HDAC5 following hypertrophic stimulation, thus relieving inhibition of prohypertrophic genes by MEF2 (Harrison et al., 2006; Vega et al., 2004). Deletion of PLCε protects mouse cardiac myocytes from stress-induced cardiac hypertrophy that is concomitant with decreased levels of nuclear PKD. Furthermore, PKD was found to be part of a multicomponent complex containing PLCε, Epac, PKCε, and RyR2 required for PKD activation (Zhang et al., 2013). Cardiac-specific overexpression of a constitutively active PKD transgene led to pathological cardiac hypertrophy in mice with ventricular chamber dilatation, contractile dysfunction, and wall thinning (Harrison et al., 2006), whereas conditional deletion of PKD1 diminished fibrosis and improved cardiac function after pressure overload or adrenergic stimulation (Fielitz et al., 2008). In conclusion, PKD regulates fundamental processes in the heart, such as contraction and hypertrophy, and is an emerging translational target in heart failure.

7.4.5 Cyclin-dependent kinase-9

Hypophosphorylated RNA polymerase II (RNAPIIa) is recruited to promoters and initiates productive transcript elongation once phosphorylated (IIo) in its C-terminal domain (CTD). Cdk7 and Cdk9 phosphorylate the CTD to mediate transition from initiation to elongation, with Cdk9 activity being particularly critical. Showing the relevance to heart disease, Cdk9 activity is increased in failing human hearts and RNAPII was hyperphosphorylated (Sano et al., 2004). Further, knocking down the non-coding 7SK RNA, an essential component of the endogenous Cdk9 inhibitory complex, was sufficient for spontaneous hypertrophy in culture.

Overexpressing *CycT1* the cyclin partner of Cdk9, persistently at the normal embryonic level increased cardiac Cdk9 activity and CTD phosphorylation. While the baseline phenotype was nonpathological hypertrophy, rapid DCM, apoptosis, and fibrosis resulted from simultaneous expression of cardiac-specific low-copy number *Gnaq*, a mild prohypertrophic stimulus. Cdk9 activity was further enhanced in bigenic mice, and heart failure concurrent with cardiomyocyte death and enhanced caspase-3 activity was evident by 4 weeks of age. In addition, Cdk9 activation suppressed mitochondrial biogenesis and function by suppressing *Ppargc1a* transcription, culminating in mitochondrial defects, predisposition to cell death and heart failure, and mortality. The pathophysiological levels of Cdk9 activity therefore render myocardium susceptible to cell death and heart failure, indicating that Cdk9 is a potential target for suppressing cardiomyocyte death and mitigating the transition from hypertrophy to dilatation (Sano et al., 2004; Sano & Schneider, 2004).

8. NEUROHUMORAL RESPONSES: TNF-α

TNF-α is proinflammatory cytokine that plays a convincing role in the pathogenesis of heart failure. Increased mortality in heart failure patients is associated with high levels of circulating TNF-α with an effect on myocyte contractility suggesting a direct role in the progression of disease (Müller-Ehmsen & Schwinger 2004). In animal models, administration of TNF-α at pathophysiologically relevant concentrations results in myocyte death LV dysfunction, and dilation (Bozkurt et al., 1998). These effects were partially reversed upon discontinuation of infusion or inhibition with anti-TNF-α antibodies, indicating TNF-α signaling to be a viable target in heart failure (Bozkurt et al., 1998).

Interestingly, TNF-α can elicit either survival or cell death, suggesting it may be pivotal in the cardiomyocyte decision to undergo one or the other. TNF-α signals through both of the TNF-α receptors TNFR1 and TNFR2, with activation of the former being associated with apoptosis and necrosis (Liu et al., 2005). Binding of TNF-α to TNFR1 stimulates the formation of either of two complexes (complex I or complex II) with the recruitment of various adaptor proteins and the serine/threonine kinase RIP1/RIP3. In general, complex I regulates cell survival and a transition to complex II induces cell death (Whelan et al., 2010). Treatment of cells with TNF-α alone does not promote cell death, since both survival and death signaling cascades are activated. However, when survival mechanisms are inhibited, cell death ensues, and when caspase activity is inhibited, necrosis (Whelan et al., 2010). These dual effects no doubt underlie the discouraging results of TNF-α antagonists in early clinical trials (Feldman et al., 2000; Mann et al., 2004). With a greater understanding of the specific signals elicited by TNF-α that are responsible for cell death rather than survival, efforts are being directed toward design of selective inhibitors against TNFR1 (Mann et al., 2004).

Cardiac-specific overexpression of TNF-α in mice was first shown to induce cardiac inflammation and heart failure 2 weeks after birth (Kubota McTiernan, Frye, Slawson, et al., 1997). Overexpression attenuated by modifying the 3′-UTR, however, produced a less severe phenotype more relevant to typical heart failure (Kubota, McTiernan, Frye, Demetris, & Feldman, 1997). The inflammatory response was milder, and cardiomyocyte death was evident, along with DCM and reduced heart function (Kubota McTiernan, Frye, Demetris, et al., 1997). In these mice, inhibition of TNF-α by adenovirus-mediated expression of a TNF receptor–IgG fusion protein abrogated inflammation, many molecular markers of hypertrophy, and LV dilation, though not increased wall thickness (Kubota et al., 2000). These data indicate that inhibition of TNF-α signaling suppresses dilatation and cell death, although is permissive for hypertrophic growth. Subsequent studies in mice with cardiac-restricted overexpression of TNF-α offered important insights into the role of fibrosis and cell death in heart failure (Engel et al., 2004; Li et al., 2000; Sivasubramanian et al., 2001). The prevalence of cardiomyocyte death increased from 4 to 12 weeks, and treatment with a pan-caspase inhibitor attenuated cell death and dilation (Engel et al., 2004). In short, cardiac-specific overexpression of TNF-α induces cardiomyocyte death and ECM remodeling, both of which contribute to progressive dilation, impaired LV dysfunction, and heart failure.

TNF-α is secreted from the cell-surface membrane and enzymatically cleaved by the TNF-α-converting enzyme (TACE), and it is the secreted form that has been most extensively characterized in the work above. Conversely, in mice overexpressing a cardiac-specific, noncleavable transmembrane form of TNF-α, concentric hypertrophy develops, along with increased heart function in the absence of dilation (Diwan et al., 2004). Indeed, in accordance with this disparity, mice expressing the secreted form of TNF-α, but treated with a TACE inhibitor did not develop DCM (Dibbs et al., 2003). It is the secreted form of TNF-α that induces cell death and heart failure, and its absence or inhibition can support adaptive mechanisms that improve heart function.

The prevalence of cardiomyocyte apoptosis increases in a time-dependent manner in cardiac-restricted soluble TNF-α expressing mice and this is accompanied by a progressive loss of myocardial Bcl-2 and release of cyt c from mitochondria (Haudek et al., 2007). A causal role of cell death in heart failure arising from pathological levels of TNF-α is strongly supported by simultaneous overexpression of cardiac-restricted Bcl-2 along with soluble TNF-α, which reduced cardiomyocyte death and progression to dilatation (Haudek et al., 2007). Bcl-2 did not prevent hypertrophy, but inhibited LV dilation and rescued contractile dysfunction, ascribed at least in part to suppressing cardiomyocyte dropout. Bcl-2 was unable to completely rescue cell death, however, due to concurrent activation of the Bcl-2-independent, extrinsic pathway (Haudek et al., 2007).

9. DISCUSSION: PROTECTING THE MYOCARDIUM

9.1. Where did the investment go?

Given that heart disease and heart failure specifically are among the largest contributors to mortality and burden on healthcare services, it might seem surprising that few novel pharmacological interventions have emerged in recent decades. In an environment of financial instability, investment from the pharmaceutical industry is ever more selective and "high risk" targets less likely to be deemed worthwhile. The risks in heart failure include patients' fragility, the intricate weave of disease etiologies and phenotypes, the necessity for sustained intervention, and a past history of therapeutic disappointments. Following a flurry of investment in previous decades, many pharmacological interventions were met with failure at the stage of progressing into clinical trials, at which point significant investment had already been made and lost. Preclinical testing did not translate into the clinic as expected, and this remains an ongoing problem (Hausenloy & Yellon, 2013; van Berlo et al., 2013).

Translational breakthroughs do exist, most notably cyclosporine A infusion for reducing reperfusion injury and SERCA2a gene therapy for restoring pump function in the failing heart. Cyclosporin A reduced reperfusion injury and infarct size in human trials (Hausenloy & Yellon, 2013; Hausenloy et al., 2014; Mewton et al., 2010; Piot et al., 2008) and patient outcome is currently being assessed in an ongoing trial (CIRCUS, National Clinical Trial (NCT) number 01502774). In the case of SERCA2a, current trials show promising results in improving heart function and decreasing recurrent cardiovascular events (Jaski et al., 2009; Jessup et al., 2011; Shareef et al., 2013; Zsebo et al., 2013), with a further trial currently planned for completion in 2016 (AGENT-HF, NCT01966887). It is worthwhile noting that each of these targets had been extensively studied in engineered mice, standard animal models, and *in vitro* systems, and that the signaling pathways have been extensively dissected. This approach—highly robust target identification—has no doubt contributed to success in the clinic, and as such, information gained from mouse models has proved particularly instructive in these cases.

In the following sections, we highlight the limitations of mouse models, how these may be overcome, and promising future directions to improve translation into the clinic.

9.2. Mouse models: Limitations, liabilities, and lessons

Certain limits inherently exist in using mice to model human heart failure. Simply put, a mouse cannot faithfully and completely replicate human physiology. However, such models exist for many cogent reasons. Not least, the mouse is a genetically tractable model organism that can be utilized readily to dissect the relevance of a gene or mutation of interest. Consequently, the mouse presently serves as a unique bridge between larger mammal models of heart disease (more faithful to human but less tractable genetically), and other organisms in which heart failure now can be modeled including zebrafish and even flies (more tractable, less faithful) (Becker et al., 2012; Knöll et al., 2007; Neely et al., 2010; Ocorr, Crawley, Gibson, & Bodmer, 2007).

In moving forward, it is essential to consider why in some instances, these models have not been sufficiently predictive. Failure to translate in many cases can come down to biological, interpretive, or technical reasons. As noted by others it is imperative that the caveats of a model system are not overlooked (Cook, Clerk, & Sugden, 2009; Molkentin & Robbins, 2009). Further, the study of individual genes has been suggested to have little relevance to a complex end-stage disorder like heart failure (Cook et al., 2009), for which a more quantitative systems or network approach might be most useful. This viewpoint has been challenged to some degree, as discrepancies may reflect improperly conducted or interpreted experiments (Molkentin & Robbins, 2009).

From a technical point of view, one issue is the increasing recognition of Cre-mediated toxicity. The cardiac-specific tamoxifen-inducible deleter, *Myh6-MerCreMer* (Sohal et al., 2001), is potent, tightly regulated, and deservedly adopted by many laboratories (Moga, Nakamura, & Robbins, 2008). However, Cre translocation to the nucleus can elicit promiscuous recombination at the genome's infrequent cryptic loxP sites, resulting in cell death, inflammation, and replacement fibrosis (Bersell et al., 2013; Lexow, Poggioli, Sarathchandra, Santini, & Rosenthal, 2013). Furthermore, the requisite ligands have acute perturbing effects on cardiomyocyte contractility and calcium handing, unrelated to Cre (Asp, Martindale, & Metzger, 2013). For both reasons, controlling tamoxifen dose and duration is critical. A single injection of tamoxifen at 40 mg/kg can induce efficient recombination in some cases, with negligible detrimental effects (Hall, Smith, Hall, & Stec, 2011; Hougen et al., 2010; Lexow et al., 2013), however, further tamoxifen was required for recombination of other "floxed" genes (Bersell et al., 2013; Koitabashi et al., 2009). Hence, individual lines should be thoroughly assessed to optimize treatment and limit toxicity. Most importantly, tamoxifen-treated *Myh6-MerCreMer* littermate controls must be utilized to avoid misinterpretation of mutant phenotypes (Bersell et al., 2013; Hall et al., 2011; Hougen et al., 2010; Koitabashi et al., 2009; Lexow et al., 2013). Overcoming these issues is important to increase confidence in this sophisticated system. Far from making such approaches prohibitive, these issues simply drive the continued evolution of mouse models to improve their predictive power.

Another key factor is that comorbidities or backgrounds commonly occurring in human heart failure have rarely been considered in reductionist mouse models. These cannot be ignored as being contributory, and perhaps it is the simpler "clean" background in mouse models that can result in different responses between mouse and man to a given drug. Conversely, however, it could be argued that this is a particular strength, where signaling mechanisms can be explored in a reductionist environment, in which the plethora of deranged signaling pathways found in comorbidities are not present. It might be expected that interpretation of results in this case may be rather more difficult.

With the myriad pathways activated in heart failure, the concept of a single gene or mutation as causative *per se* seems implausible. However, when considered together, reductionist models in which a single gene is modified can provide key information on convergence of several inputs on downstream signaling pathways or components. Further, many human cardiomyopathies result from a single mutation. Reductionist studies in mice can therefore probe the molecular mechanisms involved and uncover key nodal points common to cardiomyopathies that result in heart failure. When combined, this knowledge should ideally be used to drive a more quantitative, integrative, systems biology approach (MacLellan, Wang, & Lusis, 2012). Such networks can make the best convergent points more apparent, potentially allowing identification of targets that can be inhibited without blocking compensatory responses or eliciting other confounding effects.

In terms of the underlying mouse substrate, increased life expectancy and comorbid obesity are particularly important variables that can be more easily incorporated than previously. While naturally aging mice might be impracticable due to cost, genetically aged mice (e.g., *Terc*-null mice) provide an alternative that has already been successfully used and found to model more closely the heart failure that occurs in muscular dystrophy patients (Mourkioti et al., 2013). Furthermore, many models of obese or diabetic mice currently exist that could also be utilized.

A singular genetic model is unlikely to predict success in the clinic, especially for complex end-stage disease, and this can be substantially improved by using combinations of the genetic models at the field's disposal, in an effort to assess the generality of effect. In understanding the function of a specific gene of interest, a good starting point is to use loss-of-function models, such as cardiac-specific null mutation, or inducible deletion in adult myocardium to obviate long-term compensatory circuits. Overexpression, though at risk for promiscuous effects, should not be altogether dismissed, as this can be valuable both as a clue and as a replica of disease, where the human and animal pathobiology agree a signal is increased in expression or activity. Following this, specific mutations that activate or inhibit the gene of interest can be explored, and of most relevance, those that have been identified in the clinic provide an unequivocal tool for more closely modeling the function of a specific gene in heart failure. Once an appropriate target has been identified, it would be advantageous to assess efficacy of pharmacological intervention in both genetic and microsurgical models of heart failure, and in particular, to incorporate backgrounds such as age and obesity where possible.

Notably, all of the potential platforms for mutagenesis in mice can serve as a test bed for potential counter-measures, illustrated by the suppression of apoptosis, increased PLB phosphorylation, and improved cardiac pump function in aged *mdx* mice given a cardiotropic adeno-associated virus encoding PDZ domain-deleted nNOS (Lai, Zhao, Yue, Wasala, & Duan, 2014).

Mouse models can be used to investigate the relevance of a gene of interest, dissect its mechanism of action, and identify shared targets for intervention. The efficacy of inhibiting the target of interest should then be tested, though, in models more representative of human heart failure.

9.3. Inhibiting cell death in heart failure: From general considerations to the TNF-α-MAP4K4 axis

As discussed, cell death plays a causative role in heart failure, and many studies indicate that inhibition provides an important therapeutic strategy. However, pharmacological interventions against cell death are relatively unexplored. In particular, current examples exist (summarized in Table 4.1) that have been identified though use of mouse models, which might otherwise be unknown, or which could have led to disappointing results in the clinic due to lack of basic mechanistic understanding.

The emergence of a greater appreciation for the complexity of cell-death signaling, whereby apoptosis, necrosis, and autophagy are no longer discrete categories, will shape future therapeutic strategies. Key signaling components that mediate the decision of the cell to undergo cell death or survival might provide the best targets, such as TNFR1. Those that direct cell death toward mitochondria-dependent or -independent types of cell death, such as CryAB, might also provide especially suitable targets. Interruption of a kinase's localization, for example, disrupting membrane targeting, might block only a chosen subset of its functionst, and remains a novel approach to achieve specificity in kinase targeting (Churchill, Murriel, Chen, Mochly-Rosen, & Szweda, 2005; Dorn & Mochly-Rosen, 2002).

Given its proven role in cell death and heart failure, TNF-α signaling would seem a particularly viable therapeutic target, provided cell-death signals are selectively suppressed and survival signals maintained. Rather than blocking signaling at the level of the cytokine itself, or its receptor, a more productive strategy lies in pursuing a downstream effector that couples TNF-α selectively to cell death. However, many of the terminal MAPKs activated by TNF-α, in particular, p38 and JNK, are convergence points for multiple inputs, inhibition of which might be deleterious. An alternative strategy would therefore be to inhibit an effector that couples TNF-α more selectively to cell death.

In this regard, MAP4K4 might prove a worthwhile target. Increased activity of MAP4K4 is associated with TNF-α-dependent disorders in addition to heart failure, like obesity and diabetes (Elbein et al., 2009; Isakson et al., 2009; Sartorius et al., 2012), and many consequences of TNF-α signaling, though importantly, not all, are dependent on MAP4K4 (Bouzakri et al., 2009; Guilherme et al., 2008; Tang et al., 2006; Tesz et al., 2007; Wang et al., 2013). Similarly, in the context of skeletal muscle differentiation, TNF-α is inhibitory, but silencing MAP4K4 only marginally rescued differentiation (Wang et al., 2013). Thus, MAP4K4 selectively mediates a subset of TNF-α effects.

In considering downstream kinases, MAP4K4 knockdown selectively blocks activation of JNK by TNF-α, but not p38 or NFκB in pancreatic beta cells (Bouzraki et al., 2009), and in HEK293 cells, TNF-α activates JNK through MAP4K4–TAK1–MKK4/MKK7 (Yao et al., 1999). In contrast, loss of MAP4K4 did not prevent phosphorylation of JNK by TNF-α in adipocytes or skeletal muscle (Tang et al., 2006; Wang et al., 2013), indicating MAP4K4-independent mechanisms for activation of JNK exist in certain cell types. In cardiomyocytes, we have found that MAP4K4–TAK1–JNK mediates cell death, and that TNF-α, of all inputs tested, induces the greatest level of activation of this pathway *in vivo* (Xie et al., 2007).

An activating mutant of TAK1 resulted in heart failure, indicating a pathological role for this kinase (Zhang et al., 2000). However, cardiac-specific expression of a kinase inactive mutant of TAK1 also results in early mortality and highlights an important role for TAK1 in the metabolic AMPK pathway (Xie et al., 2006). This would indicate that TAK1 is required for proper maintenance of cardiac viability, hypertrophy, and function, but that it can be coupled to heart failure in certain environments. Although MAP4K4 does appear to mediate at least some of these responses, uncoupling of MAP4K4–TAK1 might be relevant in the context of cell death, while maintaining activation of TAK1 by other inputs.

In support of a strategy in which inhibition of MAP4K4 might be pertinent to protecting the myocardium, knockdown in primary β-cells promoted survival (Bouzakri et al., 2011), and knockdown in macrophages was protective and decreased mortality in an inflammatory model of LPS-induced lethality (Aouadi et al., 2009). Kenpaullone, a pharmacological inhibitor of MAP4K4 (though also GSK-3) confers a survival benefit to mouse and human stem cell-derived motor neurons, suggesting a potential therapeutic strategy in amyotrophic lateral sclerosis (Yang et al., 2013). Furthermore, in cardiomyocytes, MAP4K4 mediates both intrinsic and extrinsic cell death, and its suppression by RNA interference or a dominant-negative mutation is protective (Xie et al., 2007). That MAP4K4 mediates multiple cell-death pathways indicates that it might offer an improved approach over inhibiting just one aspect. Of note, activation of MAP4K4 by TNF-α in beta cells suppresses multiple survival pathways including Akt, IRS-2, and ERK (Bouzakri et al., 2009). Plausibly, inhibiting MAP4K4 might enhance diverse survival signals to protect the myocardium.

Knockout and knockdown strategies, while giving insight into whether an inhibitory approach is warranted, might elicit rather different effects than pharmacological inhibition, for reasons including the innate potential for loss of scaffold and docking functions unrelated to enzymatic activity when gene expression is disrupted. In this regard, we have identified two novel compounds that inhibit MAP4K4 activity, and protect against hydrogen peroxide and C2-ceramide mediated cell death in cardiomyocytes (Fiedler et al., 2014), with potency and selectivity at least equal to the best reported (Guimarães et al., 2011), and lack of toxicity to cardiomyocytes.

9.4. New horizons

Some of the caveats of current murine systems, where technical, will no doubt be overcome in the future through use of more sophisticated procedures. Relatively blunt whole-body-null mutations and overexpression have already given way to tissue-specific modifications and inducible systems that may better represent the diseased human heart. Recent advances will allow even more subtle and representative manipulations to be carried out, although the value and contributions of previous and current approaches should not be undermined, and should be considered together to gain the greatest understanding. Future directions might be less about complete deletion or overexpression, but more about introducing mutations identified in the clinic (an under-utilized strategy), titrating expression more finely, and by manipulating specific binding sites or activities of the proteins of interest to more subtly modify signaling pathways. In addition, the study of miRNA is a fast-growing field, and cardiac-specific, inducible systems for manipulating these would no doubt prove fruitful.

New technologies that are likely to emerge in the future include nuclease-mediated gene targeting, exploiting transcription activator-like effector nucleases, or zinc-finger nucleases for very specific genetic modifications (Doestschman & Azhar, 2012; Gaj, Gersbach, & Barbas, 2013). These future directions will likely encompass another emerging technology, clustered regulatory interspaced short palindromic repeat/Cas-based RNA-guided DNA endonucleases (Gaj et al., 2013). In addition, the issue of mouse strain-specific effects resulting in different phenotypes can be better understood following the availability of the mouse genome; recent efforts have focussed on characterizing genotype-phenotype interactions between different strains in order to understand these different responses (Keane et al., 2011; Simon et al., 2013). These new tools for rapid and efficient genome editing will have further applicability in the rat (Ponce de Leon, Merillat, Tesson, Anegon, & Hummler, 2014), opening the transformative insights from rat genomics (Atanur et al., 2013) to direct functional testing.

Another avenue of increasing interest that directly addresses the problem of a mouse not being able to fully replicate human physiology or signaling responses is the use of human pluripotent stem cell-derived cardiomyocytes (Liang et al., 2013; Mordwinkin, Burridge, & Wu, 2013). Genetic manipulations in these, alongside those in mice can further validate a target with the expectation of better translation into the clinic. Patient-specific-induced pluripotent stem cells carrying disease-causing genetic mutations provide a useful model and direct means of dissecting signaling pathways relevant to heart failure, as well as providing a relevant test bed for more tailored therapeutic interventions (Sun et al., 2012). In short, a target that is validated both in mouse models (providing information from the adult, intact heart) and human cardiomyocytes (providing information from a human platform, even if stem cell-derived) could be considered to carry a lower risk for investment, and provide better-posed targets with significantly improved translational potential, than programs based on either criterion alone.

Many previous failures and the perceived high risk involved in current and future investment have been detrimental to progress in bringing heart failure therapeutics forward to the clinic. However, with the advent of more sophisticated mouse and human models, improved mechanistic knowledge, the identification of common targets in diverse etiologies—in short, a more rational approach to target discovery and validation—a new wave of interest and opportunity may drive drug discovery for heart failure into a new era.

REFERENCES

Acehan, D., Jiang, X. J., Morgan, D. G., Heuser, J. E., Wang, X. D., & Akey, C. W. (2002). Three-dimensional structure of the apoptosome: Implications for assembly, procaspase-9 binding, and activation. *Molecular Cell*, *9*(2), 423–432.

Adams, J. W., Sakata, Y., Davis, M. G., Sah, V. P., Wang, Y., Liggett, S. B., et al. (1998). Enhanced Galphaq signalling: A common pathway mediates cardiac hypertrophy and apoptotic heart failure. *Proceedings of the National Academy of Sciences of the United States of America*, *95*(17), 10140–10145.

Allender, S., Peto, V., Scarborough, P., Laur, A., & Rayner, M. (2008). *Coronary heart disease statistics* (2008 ed.). British Heart Foundation Statistics Database.

Anderson, M. E., Zeng, H., Higgins, L. S., Park, J. W., Schulman, H., Guo, M., et al. (2006). Disease mechanisms and emerging therapies: Protein kinases and their inhibitors in myocardial disease. *Nature Clinical Practice Cardiovascular Medicine*, *3*, 437–445.

Antos, C. L., Frey, N., Marx, S. O., Reiken, S., Gaburjakova, M., Richardson, J. A., et al. (2001). Dilated cardiomyopathy and sudden death resulting from constitutive activation of protein kinase A. *Circulation Research*, *89*(11), 997–1004.

Aouadi, M., Tesz, G. J., Nicoloro, S. M., Wang, M., Chouinard, M., Soto, E., et al. (2009). Orally delivered siRNA targeting macrophage Map4k4 suppresses systemic inflammation. *Nature*, *458*(7242), 1180–1184.

Arber, S., Hunter, J. J., Ross, J., Jr., Hongo, M., Sansig, G., Borg, J., et al. (1997). MLP-deficient mice exhibit a disruption of cardiac cytoarchitectural organization, dilated cardiomyopathy, and heart failure. *Cell*, *88*(3), 393–403.

Arslan, S. C., Yuasa, S., Scheidereit, C., Itabashi, Y., Koshimizu, U., Tanaka, T., et al. (2011). The prevalence of TNF-α-induced necrosis over apoptosis is determined by TAK1-RIP1 interplay. *PLoS One*, *6*(10), e26069.

Asp, M. L., Martindale, J. J., & Metzger, J. M. (2013). Direct, differential effects of tamoxifen, 4-hydroxytamoxifen, and raloxifene on cardiac myocyte contractility and calcium handling. *PLoS One*, *8*(10), e78768.

Atanur, S. S., Diaz, A. G., Maratou, K., Sarkis, A., Rotival, M., Game, L., et al. (2013). Genome sequencing reveals loci under artificial selection that underlie disease phenotypes in the laboratory rat. *Cell*, *154*(3), 691–703.

Azibani, F., Muchir, A., Vignier, N., Bonne, G., & Bertrand, A. T. (2014). Striated muscle laminopathies. *Seminars in Cell & Developmental Biology*. Jan 15 [E-pub ahead of print].

Baines, C. P. (2007). The mitochondrial permeability transition pore as a target of cardioprotective signalling. *American Journal of Physiology. Heart and Circulatory Physiology*, *293*(2), H903–H904.

Baines, C. P., Kaiser, R. A., Purcell, N. H., Blair, N. S., Osinska, H., Hambleton, M. A., et al. (2005). Loss of cyclophilin D reveals a critical role for mitochondrial permeability transition in cell death. *Nature*, *434*(7033), 658–662.

Baines, C. P., Kaiser, R. A., Sheiko, T., Craigen, W. J., & Molkentin, J. D. (2007). Voltage-dependent anion channels are dispensable for mitochondrial-dependent cell death. *Nature Cell Biology*, *9*(5), 550–555.

Becker, J. R., Robinson, T. Y., Sachidanandan, C., Kelly, A. E., Coy, S., Peterson, R. T., et al. (2012). In vivo natriuretic peptide reporter assay identifies chemical modifiers of hypertrophic cardiomyopathy signalling. *Cardiovascular Research*, *93*(3), 463–470.

Beggs, A. H., Hoffman, E. P., Snyder, J. R., Arahata, K., Specht, L., Shapiro, F., et al. (1991). Exploring the molecular basis for variability among patients with Becker muscular dystrophy: Dystrophin gene and protein studies. *Journal of Human Genetics*, *49*(1), 54–67.

Belzacq, A. S., Vieira, H., Verrier, F., Vandecasteele, G., Cohen, I., Prévost, M. C., et al. (2003). Bcl-2 and bax modulate adenine nucleotide translocase activity. *Cancer Research*, *63*(2), 541–546.

Bergmann, O., Bhardwaj, R. D., Bernard, S., Zdunek, S., Barnabé-Heider, F., Walsh, S., et al. (2009). Evidence for cardiomyocyte renewal in humans. *Science*, *324*(5923), 98–102.

Bersell, K., Choudhruy, S., Mollova, M., Polizzotti, B. D., Ganapathy, B., Walsh, S., et al. (2013). Moderate and high amounts of tamoxifen in αMHC-MerCreMer mice induce a DNA damage response, leading to heart failure and death. *Disease Models & Mechanisms*, *6*(6), 1459–1469.

Berul, C. I., Christe, M. E., Aronovitz, M. J., Seidman, C. E., Seidman, J. G., & Mendelsohn, M. E. (1997). Electrophysiological abnormalities and arrhythmias in alpha MHC mutant familial hypertrophic cardiomyopathy mice. *The Journal of Clinical Investigation*, *99*(4), 570–576.

Bhuiyan, M. S., & Fukunaga, K. (2007). Inhibition of HtrA2/Omi ameliorates heart dysfunction following ischemia/reperfusion injury in rat heart in vivo. *European Journal of Pharmacology*, *557*(2–3), 168–177.

Bhupathy, P., Babu, G. J., & Periasamy, M. (2007). Sarcolipin and phospholamban as regulators of cardiac sarcoplasmic reticulum Ca2+ ATPase. *Journal of Molecular and Cellular Cardiology*, *42*(5), 903–911.

Bing, O. H. L. (1994). Hypothesis: Apoptosis may be a mechanism for the transition to heart failure with chronic pressure overload. *Journal of Molecular and Cellular Cardiology*, *26*(8), 943–948.

Bogoyevitch, M. A., GillespieBrown, J., Ketterman, A. J., Fuller, S. J., BenLevy, R., Ashworth, A., et al. (1996). Stimulation of the stress-activated mitogen-activated protein kinase subfamilies in perfused heart—p38/RK mitogen-activated protein kinases and c-Jun N-terminal kinases are activated by ischemia/reperfusion. *Circulation Research*, *79*(2), 162–173.

Bonacci, T. M., Mathews, J. L., Yuan, C., Lehmann, D. M., Malik, S., Wu, D., et al. (2006). Differential targeting of Gbetagamma-subunit signaling with small molecules. *Science*, *312*(5772), 443–446.

Boudoulas, K. D., & Hatzopoulos, A. K. (2009). Cardiac repair and regeneration: The Rubik's cube of cell therapy for heart disease. *Disease Models & Mechanisms*, *2*(7–8), 344–358.

Bouzakri, K., Ribaux, P., & Halban, P. A. (2009). Silencing mitogen-activated protein 4 kinase 4 (MAP4K4) protects beta cells from tumor necrosis factor-alpha-induced decrease of IRS-2 and inhibition of glucose-stimulated insulin secretion. *The Journal of Biological Chemistry*, *284*(41), 27892–27898.

Bowling, N., Walsh, R. A., Song, G., Estridge, T., Sandusky, G. E., Fouts, R. L., et al. (1999). Increased protein kinase C activity and expression of Ca2+-sensitive isoforms in the failing human heart. *Circulation*, *99*(3), 384–391.

Bowman, J. C., Steinberg, S. F., Jiang, T., Geenen, D. L., Fishman, G. I., & Buttrick, P. M. (1997). Expression of protein kinase C beta in the heart causes hypertrophy in adult mice and sudden death in neonates. *The Journal of Clinical Investigation*, *100*(9), 2189–2195.

Bozkurt, B., Kribbs, S. B., Clubb, F. J., Jr., Michael, L. H., Didenko, V. V., Hornsby, P. J., et al. (1998). Pathophysiologically relevant concentrations of tumor necrosis factor-alpha promote progressive left ventricular dysfunction and remodeling in rats. *Circulation*, *97*(14), 1382–1391.

Braz, J. C., Gregory, K., Pathak, A., Zhao, W., Sahin, B., Klevitsky, R., et al. (2004). PKCalpha regulates cardiac contractility and propensity towards heart failure. *Nature Medicine*, *10*(3), 248–254.

Brener, S. J., Maehara, A., Dizon, J. M., Fahy, M., Witzenbichler, B., Parise, H., et al. (2013). Relationship between myocardial reperfusion, infarct size, and mortality: The INFUSE-AMI (Intracoronary Abciximab and aspiration thrombectomy in patients with large anterior myocardial Infarction) trial. *JACC: Cardiovascular Interventions*, *6*(7), 718–724.

British Heart Foundation Health Promotion Research Group. (2012). *Coronary heart disease statistics: A compendium of health statistics*. London: British Heart Foundation.

Bulfield, G., Siller, W. G., Wight, P. A., & Moore, K. J. (1984). X chromosome-linked muscular dystrophy (mdx) in the mouse. *Proceedings of the National Academy of Sciences of the United States of America*, *81*(4), 1189–1192.

Cao, D. J., Wang, Z. V., Battiprolu, P. K., Jiang, N., Morales, C. R., Kong, Y., et al. (2011). Histone deacetylase (HDAC) inhibitors attenuate cardiac hypertrophy by suppressing autophagy. *Proceedings of the National Academy of Sciences of the United States of America*, *108*(10), 4123–4128.

Casey, L. M., Pistner, A. R., Belmonte, S. L., Migdalovich, D., Stolpnik, O., Nwakanma, F. E., et al. (2010). Small molecule disruption of G beta gamma signaling inhibits the progression of heart failure. *Circulation Research*, *107*(4), 532–539.

Chan, F., Shisler, J., Bixby, J. G., Felices, M., Zheng, L. X., Appel, M., et al. (2003). A role for tumor necrosis factor receptor-2 and receptor-interacting protein in programmed necrosis and antiviral responses. *The Journal of Biological Chemistry*, *278*(51), 51613–51621.

Chang, J., Xie, M., Shah, V. R., Schneider, M. D., Entman, M. L., Wei, L., et al. (2006). Activation of Rho-associated coiled-coil protein kinase 1 (ROCK-1) by caspase-3 cleavage plays an essential role in cardiac myocyte apoptosis. *Proceedings of the National Academy of Sciences of the United States of America*, *103*(39), 14495–14500.

Chang, G., Zhang, D., Yu, H., Zhang, P., Wang, Y., Zheng, A., et al. (2013). Cardioprotective effects of exenatide against oxidative stress-induced injury. *International Journal of Molecular Medicine*, *32*(5), 1011–1020.

Chapman, V. M., Miller, D. R., Armstrong, D., & Caskey, C. T. (1989). Recovery of induced mutations for X chromosome-linked muscular dystrophy in mice. *Proceedings of the National Academy of Sciences of the United States of America*, *86*(4), 1292–1296.

Chen, Z., Chua, C. C., Ho, Y.-S., Hamdy, R. C., & Chua, B. H. L. (2001a). Overexpression of Bcl-2 attenuates apoptosis and protects against myocardial I/R injury in transgenic mice. *American Journal of Physiology. Heart and Circulatory Physiology*, *280*(5), H2313–H2320.

Chen, J., Fujii, K., Zhang, L., Roberts, T., & Fu, H. (2001b). Raf-1 promotes cell survival by antagonizing apoptosis signal-regulating kinase 1 through a MEK-ERK independent mechanism. *Proceedings of the National Academy of Sciences of the United States of America*, *98*(14), 7783–7788.

Chen, Y., Lewis, W., Diwan, A., Cheng, E. H.-Y., Matkovich, S. J., & Dorn, G. W. (2010). Dual autonomous mitochondrial cell death pathways are activated by Nix/BNip3L and induce cardiomyopathy. *Proceedings of the National Academy of Sciences of the United States of America*, *107*(20), 9035–9042.

Chen, Q. M., Tu, V. C., Wu, Y., & Bahl, J. J. (2000). Hydrogen peroxide dose dependent induction of cell death or hypertrophy in cardiomyocytes. *Archives of Biochemistry and Biophysics*, *373*(1), 242–248.

Cheng, E., Sheiko, T. V., Fisher, J. K., Craigen, W. J., & Korsmeyer, S. J. (2003). VDAC2 inhibits BAK activation and mitochondrial apoptosis. *Science*, *301*(5632), 513–517.

Cho, M. C., Rapacciuolo, A., Koch, W. J., Kobayashi, Y., Jones, L. R., & Rockman, H. A. (1999). Defective β-Adrenergic receptor signalling precedes the development of dilated cardiomyopathy in transgenic mice with calsequestrin overexpression. *The Journal of Biological Chemistry*, *274*(32), 22251–22256.

Christodoulou, D. C., Wakimoto, H., Onoue, K., Eminaga, S., Gorham, J. M., Depalma, S. R., et al. (2014). 5'RNA-Seq identifies Fhl1 as a genetic modifier in cardiomyopathy. *The Journal of Clinical Investigation*. Mar 3 [E-pub ahead of print].

Chu, G., Haghighi, K., & Kranias, E. G. (2002). From mouse to man: Understanding heart failure through genetically altered mouse models. *Journal of Cardiac Failure*, *8*(Suppl. 6), S432–S449.

Churchill, E. N., Murriel, C. L., Chen, C.-H., Mochly-Rosen, D., & Szweda, L. I. (2005). Reperfusion-induced translocation of deltaPKC to cardiac mitochondria prevents pyruvate dehydrogenase reactivation. *Circulation Research*, *97*(1), 78–85.

Clarke, S. J., McStay, G. P., & Halestrap, A. P. (2002). Sanglifehrin A acts as a potent inhibitor of the mitochondrial permeability transition and reperfusion injury of the heart by binding to cyclophilin-D at a different site from cyclosporin A. *The Journal of Biological Chemistry*, *277*(38), 34793–34799.

Condorelli, G., Roncarati, R., Ross, J., Pisani, A., Stassi, G., Todaro, M., et al. (2001). Heart-targeted overexpression of caspase3 in mice increases infarct size and depresses cardiac function. *Proceedings of the National Academy of Sciences of the United States of America*, *98*(17), 9977–9982.

Cook, S. A., Clerk, A., & Sugden, P. H. (2009). Are transgenic mice the 'alkahest' to understanding myocardial hypertrophy and failure? *Journal of Molecular and Cellular Cardiology*, *6*(2), 118–129.

Coral-Vazquez, R., Cohn, R. D., Moore, S. A., Hill, J. A., Weiss, R. M., Davisson, R. L., et al. (1999). Disruption of the sarcoglycan-sarcospan complex in vascular smooth muscle: A novel mechanism for cardiomyopathy and muscular dystrophy. *Cell*, *98*(4), 465–474.

Coulthard, L. R., White, D. E., Jones, D. L., McDermott, M. F., & Burchill, S. A. (2009). p38(MAPK): Stress responses from molecular mechanisms to therapeutics. *Trends in Molecular Medicine*, *15*(8), 369–379.
Crisp, A., Yin, H., Goyenvalle, A., Betts, C., Moulton, H. M., Seow, Y., et al. (2011). Diaphragm rescue alone prevents heart dysfunction in dystrophic mice. *Human Molecular Genetics*, *20*(3), 413–421.
Crow, M. T., Mani, K., Nam, Y.-J., & Kitsis, R. N. (2004). The mitochondrial death pathway and cardiac myocyte apoptosis. *Circulation Research*, *95*(10), 957–970.
D'Angelo, D. D., Sakata, Y., Lorenz, J. N., Boivin, G. P., Walsh, R. A., Liggett, S. B., et al. (1997). Transgenic Galphaq overexpression induces cardiac contractile failure in mice. *Proceedings of the National Academy of Sciences of the United States of America*, *94*(15), 8121–8126.
Danko, I., Chapman, V., & Wolff, J. A. (1992). The frequency of revertants in mdx mouse genetic models for duchenne muscular dystrophy. *Pediatric Research*, *32*(1), 128–131.
Dash, R., Kadambi, V., Schmidt, A. G., Tepe, N. M., Biniakiewicz, D., Gerst, M. J., et al. (2001). Interactions between phospholamban and beta-adrenergic drive may lead to cardiomyopathy and early mortality. *Circulation*, *103*(6), 889–896.
De Nicola, G. F., Martin, E. D., Chaikuad, A., Bassi, R., Clark, J., Martino, L., et al. (2013). Mechanism and consequence of the autoactivation of p38α mitogen-activated protein kinase promoted by TAB1. *Nature Structural & Molecular Biology*, *20*(10), 1182–1190.
Denise Martin, E., De Nicola, G. F., & Marber, M. S. (2012). New therapeutic targets in cardiology: p38 alpha mitogen-activated protein kinase for ischemic heart disease. *Circulation*, *126*(3), 357–368.
Dibbs, Z. I., Diwan, A., Nemoto, S., DeFreitas, G., Abdellatif, M., Carabello, B. A., et al. (2003). Targeted overexpression of transmembrane tumor necrosis factor provokes a concentric cardiac hypertrophic phenotype. *Circulation*, *108*(8), 1002–1008.
Dimmeler, S., Zeiher, A. M., & Schneider, M. D. (2005). Unchain my heart: The scientific foundations of cardiac repair. *The Journal of Clinical Investigation*, *115*(3), 572–583.
Diwan, A., Dibbs, Z., Nemoto, S., DeFreitas, G., Carabello, B. A., Sivasubramanian, N., et al. (2004). Targeted overexpression of noncleavable and secreted forms of tumor necrosis factor provokes disparate cardiac phenotypes. *Circulation*, *109*(2), 262–268.
Diwan, A., Krenz, M., Syed, F. M., Wansapura, J., Ren, X., Koesters, A. G., et al. (2007). Inhibition of ischemic cardiomyocyte apoptosis through targeted ablation of Bnip3 restrains postinfarction remodeling in mice. *The Journal of Clinical Investigation*, *117*(10), 2825–2833.
Doestschman, T., & Azhar, M. (2012). Cardiac-specific inducible and conditional gene targeting in mice. *Circulation Research*, *110*(11), 1498–1512.
Dorn, G. W., & Mochly-Rosen, D. (2002). Intracellular transport mechanisms of signal transducers. *Annual Review of Physiology*, *64*, 407–429.
Dorn, G. W., & Force, T. (2005). Protein kinase cascades in the regulation of cardiac hypertrophy. *The Journal of Clinical Investigation*, *115*(3), 527–537.
Du, C., Fang, M., Li, Y., Li, L., & Wang, X. (2000). Smac, a mitochondrial protein that promotes cytochrome c-dependent caspase activation by eliminating IAP inhibition. *Cell*, *102*(1), 33–42.
Ea, C.-K., Deng, L., Xia, Z.-P., Pineda, G., & Chen, Z. J. (2006). Activation of IKK by TNFα requires site-specific ubiquitination of RIP1 and polyubiquitin binding by NEMO. *Molecular Cell*, *22*(2), 245–257.
Eder, P., & Molkentin, J. D. (2011). TRPC channels as effectors of cardiac hypertrophy. *Circulation Research*, *108*(2), 265–272.
Elbein, S. C., Das, S. K., Hallman, D. M., Hanis, C. L., & Hasstedt, S. J. (2009). Genome-wide linkage and admixture mapping of type 2 diabetes in African American families

from the American Diabetes Association GENNID (Genetics of NIDDM) Study Cohort. *Diabetes, 58*(1), 268–274.

Ellison, G. M., Vicinanza, C., Smith, A. J., Aquila, I., Leone, A., Waring, C. D., et al. (2013). Adult c-kit(pos) cardiac stem cells are necessary and sufficient for functional cardiac regeneration and repair. *Cell, 154*(4), 827–842.

Engel, D., Peshock, R., Armstong, R. C., Sivasubramanian, N., & Mann, D. L. (2004). Cardiac myocyte apoptosis provokes adverse cardiac remodeling in transgenic mice with targeted TNF overexpression. *The Americal Journal of Physiology. Heart Circulation Physiology, 287*(3), H1303–H1311.

Engelhardt, S., Hein, L., Wiesmann, F., & Lohse, M. J. (1999). Progressive hypertrophy and heart failure in beta1-adrenergic receptor transgenic mice. *Proceedings of the National Academy of Sciences of the United States of America, 96*(12), 7059–7064.

Fahed, A. C., Gelb, B. D., Seidman, J. G., & Seidman, C. E. (2013). Genetics of congenital heart disease: The glass half empty. *Circulation Research, 112*(4), 707–720.

Fajardo, G., Zhao, M., Urashima, T., Farahani, S., Hu, D. Q., Reddy, S., et al. (2013). Deletion of the beta2-adrenergic receptor prevents the development of cardiomyopathy in mice. *Journal of Molecular and Cellular Cardiology, 63*, 155–164.

Fatkin, D., McConnell, B. K., Mudd, J. O., Semsarian, C., Moskowitz, I. G., Schoen, F. J., et al. (2000). An abnormal Ca(2+) response in mutant sarcomere protein-mediated familial hypertrophic cardiomyopathy. *The Journal of Clinical Investigation, 106*(11), 1351–1359.

Feldman, A. M., Combes, A., Wagner, D., Kadakomi, T., Kubota, T., Li, Y. Y., et al. (2000). The role of tumor necrosis factor in the pathophysiology of heart failure. *Journal of the American College of Cardiology, 35*(3), 537–544.

Feoktistova, M., Fujita, J., Luo, X., Stewart, S. A., Geserick, P., Crane, A. M., et al. (2011). cIAPs block ripoptosome formation, a RIP1/caspase-8 containing intracellular cell death complex differentially regulated by cFLIP isoforms. *Molecular Cell, 43*(3), 449–463.

Fiedler, L. R., Chapman, K., Guo, W., Kanaganayagam, S., Low, C., Traulau-Stewart, C., et al. (2014). MAP4K4 as a therapeutic target in cardiomyocyte death: Novel inhibitors identified through pharmacophore modelling and virtual screening. *Heart, 100*(Suppl. 1), A5.

Fielitz, J., Kim, M.-S., Shelton, J. M., Qi, X., Hill, J. A., Richardson, J. A., et al. (2008). Requirement of protein kinase D1 for pathological cardiac remodeling. *Proceedings of the National Academy of Sciences of the United States of America, 105*(8), 3059–3063.

Finsterer, J., & Cripe, L. (2014). Treatment of dystrophin cardiomyopathies. *Nature Reviews Cardiology, 11*(3), 168–179.

Foo, R. S.-Y., Chan, L. K., Kitsis, R. N., & Bennett, M. R. (2007). Ubiquitination and degradation of the anti-apoptotic protein ARC by MDM2. *The Journal of Biological Chemistry, 282*(8), 5529–5535.

Francis, G. S., & Tang, W. H. (2003). Pathophysiology of congestive heart failure. *Reviews in Cardiovascular Medicine, 4*(Suppl. 2), S14–S20.

Freeman, K., Colon-Rivera, C., Olsson, M. C., Moore, R. L., Weinberger, H. D., Grupp, I. L., et al. (2001). Progression from hypertrophic to dilated cardiomyopathy in mice that express a mutant myosin transgene. *The American Journal of Physiology, 280*(1), H151–H159.

Freeman, K., Lerman, I., Kranias, E. G., Bohlmeyer, T., Bristow, M. R., Lefkowitz, R. J., et al. (2001). Alterations in cardiac adrenergic signalling and calcium cycling differentially affect the progression of cardiomyopathy. *The Journal of Clinical Investigation, 107*(8), 967–974.

Frenzke, R., Korcarz, C., Lang, R., & Leiden, J. (1998). Dilated cardiomyopathy in transgenic mice expressing a dominant-negative CREB transcription factor in the heart. *The Journal of Clinical Investigation, 101*(11), 2415–2426.

Fu, Y., & Rubin, C. S. (2011). Protein kinase D: Coupling extracellular stimuli to the regulation of cell physiology. *EMBO Reports*, *12*(8), 785–796.
Fyhrquist, F., Saijonmaa, O., & Strandberg, T. (2013). The roles of senescence and telomere shortening in cardiovascular disease. *Nature Reviews Cardiology*, *10*(5), 274–283.
Gaj, T., Gersbach, C. A., & Barbas, C. F., 3rd. (2013). ZFN, TALEN, and CRISPR/Cas-based methods for genome engineering. *Trends in Biotechnology*, *31*(7), 397–405.
Gao, W. D., Perez, N. G., Seidman, C. E., Seidman, J. G., & Marban, E. (1999). Altered cardiac excitation-contraction coupling in mutant mice with familial hypertrophic cardiomyopathy. *The Journal of Clinical Investigation*, *103*(5), 661–666.
Geisterfer-Lowrance, A. A., Christe, M., Conner, D. A., Ingwall, J. S., Schoen, F. J., Seidman, C. E., et al. (1996). A mouse model of familial hypertrophic cardiomyopathy. *Science*, *272*(5262), 731–734.
Georgakopoulos, D., Christe, M. E., Giewat, M., Seidman, C. M., Seidman, J. G., & Kass, D. A. (1999). The pathogenesis of familial hypertrophic cardiomyopathy: Early and evolving effects from an alpha-cardiac myosin heavy chain missense mutation. *Nature Medicine*, *5*(3), 327–330.
Goebel, H. H., & Bornemann, A. (1993). Desmin pathology in neuromuscular diseases. *Virchows Archiv. B, Cell Pathology Including Molecular Pathology*, *64*(3), 127–315.
Gottlieb, R. A., Echeverri, C. J., Liu, J., Mentzer, R. M., Perrimon, N., Lieu, D. K., et al. (2010). Autophagy during cardiac stress: Joys and frustrations of autophagy. *Annual Review of Physiology*, *72*, 45–59.
Grady, R. M., Teng, H., Nichol, M. C., Cunningham, J. C., Wilkinson, R. S., & Sanes, J. R. (1997). Skeletal and cardiac myopathies in mice lacking utrophin and dystrophin: A model for Duchenne muscular dystrophy. *Cell*, *90*(4), 729–738.
Guilherme, A., Virbasius, J. V., Puri, V., & Czech, M. P. (2008). Adipocyte dysfunctions linking obesity to insulin resistance and type 2 diabetes. *Nature Reviews in Molecular Cell Biology*, *9*(5), 367–377.
Guimarães, C. R., Rai, B. K., Munchhof, M. J., Liu, S., Wang, J., Bhattacharya, S. K., et al. (2011). Understanding the impact of the P-loop conformation on kinase selectivity. *Journal of Chemical Information and Modeling*, *51*(6), 1199–1204.
Guo, W., Schafer, S., Greaser, M. L., Radke, M. H., Liss, M., Govindarajan, T., et al. (2012). RBM20, a gene for hereditary cardiomyopathy, regulates titin splicing. *Nature Medicine*, *18*(5), 766–773.
Guo, C., Willem, M., Werner, A., Raivich, G., Emerson, M., Neyses, L., et al. (2006). Absence of alpha 7 integrin in dystrophin-deficient mice causes a myopathy similar to Duchenne muscular dystrophy. *Human Molecular Genetics*, *15*(6), 989–998.
Hahn, H. S., Marreez, Y., Odley, A., Sterbling, A., Yussman, M. G., Hilty, K. C., et al. (2003). Protein kinase C alpha negatively regulates systolic and diastolic function in pathological hypertrophy. *Circulation Research*, *93*(11), 1111–1119.
Hall, M. E., Smith, G., Hall, J. E., & Stec, D. E. (2011). Systolic dysfunction in cardiac-specific ligand-inducible MerCreMer transgenic mice. *American Journal of Physiology. Heart and Circulatory Physiology*, *301*(1), H253–H260.
Harding, V. B., Jones, L. R., Lefkowitz, R. J., Koch, W. J., & Rockman, H. A. (2001). Cardiac beta ARK1 inhibition prolongs survival and augments beta blocker therapy in a mouse model of severe heart failure. *Proceedings of the National Academy of Sciences of the United States of America*, *98*(10), 5809–5814.
Harrison, B. C., Kim, M.-S., van Rooij, E., Plato, C. F., Papst, P. J., Vega, R. B., et al. (2006). Regulation of cardiac stress signalling by protein kinase D1. *Molecular and Cellular Biology*, *26*(10), 3875–3888.
Haudek, S. B., Taffet, G. E., Schneider, M. D., & Mann, D. L. (2007). TNF provokes cardiomyocyte apoptosis and cardiac remodeling through activation of multiple cell death pathways. *The Journal of Clinical Investigation*, *117*(9), 2692–2701.

Hausenloy, D. J., Boston-Griffiths, E. A., & Yellon, D. M. (2012). Cyclosporin A and cardioprotection: From investigative tool to therapeutic agent. *British Journal of Pharmacology, 165*(5), 1235–1245.
Hausenloy, D. J., Kunst, G., Boston-Griffiths, E., Kolvekar, S., Chaubey, S., John, L., et al. (2014). The effect of cyclosporin-A on peri-operative myocardial injury in adult patients undergoing coronary artery bypass graft surgery: A randomised controlled clinical trial. *Heart, 100*(7), 544–549.
Hausenloy, D. J., & Yellon, D. M. (2013). Myocardial ischemia-reperfusion injury: A neglected therapeutic target. *The Journal of Clinical Investigation, 123*(1), 92–100.
Haworth, R. (2000). Expression and activity of protein kinase D/protein kinase C μ in myocardium: Evidence for α1 -adrenergic receptor- and protein kinase C-mediated regulation. *Journal of Molecular and Cellular Cardiology, 32*(6), 1013–1023.
Haworth, R. S., Cuello, F., Herron, T. J., Franzen, G., Kentish, J. C., Gautel, M., et al. (2004). Protein kinase D is a novel mediator of cardiac troponin I phosphorylation and regulates myofilament function. *Circulation Research, 95*(11), 1091–1099.
Hayakawa, Y., Chandra, M., Miao, W., Shirani, J., Brown, J. H., Dorn, G. W., 2nd, et al. (2003). Inhibition of cardiac myocyte apoptosis improves cardiac function and abolishes mortality in the peripartum cardiomyopathy of Galpha(q) transgenic mice. *Circulation, 108*(24), 3036–3041.
He, C., Bassik, M. C., Moresi, V., Sun, K., Wei, Y., Zou, Z., et al. (2012). Exercise-induced BCL2-regulated autophagy is required for muscle glucose homeostasis. *Nature, 481*(7382), 511–515.
He, S., Wang, L., Miao, L., Wang, T., Du, F., Zhao, L., et al. (2009). Receptor interacting protein kinase-3 determines cellular necrotic response to TNF-α. *Cell, 137*(6), 1100–1111.
Heidenreich, P. A., Albert, N. M., Allen, L. A., Bluemke, D. A., Butler, J., Fonarow, G. C., et al. (2013). Forecasting the impact of heart failure in the United States: A policy statement from the American Heart Association. *Circulation. Heart Failure, 6*(3), 606–619.
Hein, S., Arnon, E., Kostin, S., Schönburg, M., Elsässer, A., Polyakova, V., et al. (2003). Progression from compensated hypertrophy to failure in the pressure-overloaded human heart: Structural deterioration and compensatory mechanisms. *Circulation, 107*(7), 984–991.
Hochhauser, E., Kivity, S., Offen, D., Maulik, N., Otani, H., Pannet, H., et al. (2003). Bax ablation protects against myocardial ischemia-reperfusion injury in transgenic mice. *American Journal of Physiology. Heart and Circulatory Physiology, 284*(6), H2351–H2359.
Hoffman, E. P., Brown, R. H., Jr., & Kunkel, L. M. (1987). Dystrophin: The protein product of the Duchenne muscular dystrophy locus. *Cell, 51*(6), 919–928.
Holler, N., Zaru, R., Micheau, O., Thome, M., Attinger, A., Valitutti, S., et al. (2000). Fas triggers an alternative, caspase-8-independent cell death pathway using the kinase RIP as effector molecule. *Nature Immunology, 1*(6), 489–495.
Hougen, K., Aronsen, J. M., Stokke, M. K., Enger, U., Nygard, S., Andersson, K. B., et al. (2010). Cre-loxP DNA recombination is possible with only minimal unspecific transcriptional changes and without cardiomyopathy in Tg(alphaMHC-MerCreMer) mice. *American Journal of Physiology. Heart and Circulatory Physiology, 299*(5), H1671–H1678.
Houser, S. R., Margulies, K. B., Murphy, A. M., Spinale, F. G., Francis, G. S., Prabhu, S. D., et al. (2012). Animal models of heart failure: A scientific statement from the American Heart Association. *Circulation Research, 111*(1), 131–150.
Hsueh, Y. C., Wu, J. M., Yu, C. K., Wu, K. K., & Hsieh, P. C. (2014). Prostaglandin E promotes post-infarction cardiomyocyte replenishment by endogenous stem cells. *EMBO Molecular Medicine, 64*(4), 496–503.
Huang, Z. M., Gao, E., Fonseca, F. V., Hayashi, H., Shang, X., Hoffman, N. E., et al. (2013). Convergence of G protein-coupled receptor and S-nitrosylation signalling determines the outcome to cardiac ischemic injury. *Science Signaling, 6*(299), ra95.

Huang, X., Pi, Y., Lee, K. J., Henkel, A. S., Gregg, R. G., Powers, P. A., et al. (1999). Cardiac troponin I gene knockout: A mouse model of myocardial troponin I deficiency. *Circulation Research*, *84*(1), 1–8.

Hunt, S. A.; American College of Cardiology; American Heart Association Task Force on Practice Guidelines (Writing Committee to Update the 2001 Guidelines for the Evaluation and Management of Heart Failure). (2005). ACC/AHA 2005 guideline update for the diagnosis and management of chronic heart failure in the adult: A report of the American College of Cardiology/American Heart Association Task Force on Practice Guidelines (Writing Committee to Update the 2001 Guidelines for the Evaluation and Management of Heart Failure). *Journal of the American College of Cardiology*, *46*(6), e1–e82.

Ieda, M., Umezawa, Y., Tsuchihashi, T., Ivey, K. N., Ross, R. S., Hong, T.-T., et al. (2009). Cardiac fibroblasts regulate myocardial proliferation through beta 1 integrin signalling. *Developmental Cell*, *16*(2), 233–244.

Im, W. B., Phelps, S. F., Copen, E. H., Adams, E. G., Slightom, J. L., & Chamberlain, J. S. (1996). Differential expression of dystrophin isoforms in strains of mdx mice with different mutations. *Human Molecular Genetics*, *5*(8), 1149–1153.

Imahashi, K., Schneider, M. D., Steenbergen, C., & Murphy, E. (2004). Transgenic expression of Bcl-2 modulates energy metabolism, prevents cytosolic acidification during ischemia, and reduces ischemia/reperfusion injury. *Circulation Research*, *95*(7), 734–741.

Inagaki, K., Hahn, H. S., Dorn, G. W., 2nd., & Mochly-Rosen, D. (2003). Additive protection of the ischemic heart ex vivo by combined treatment with delta-protein kinase C inhibitor and epsilon-protein kinase C activator. *Circulation*, *108*(7), 869–875.

Isakson, P., Hammarstedt, A., Gustafson, B., & Smith, U. (2009). Impaired preadipocyte differentiation in human abdominal obesity: Role of Wnt, tumor necrosis factor-alpha, and inflammation. *Diabetes*, *58*(7), 1550–1557.

Iwase, M., Uechi, M., Vatner, D. E., Asai, K., Shannon, R. P., Kudej, R. K., et al. (1997). Cardiomyopathy induced by cardiac Gs alpha overexpression. *The American Journal of Physiology*, *272*(1 Pt. 2), 585–589.

James, J., Zhang, Y., Osinska, H., Sanbe, A., Klevitsky, R., Hewett, T. E., et al. (2000). Transgenic modeling of a cardiac troponin I mutation linked to familial hypertrophic cardiomyopathy. *Circulation Research*, *87*(9), 805–811.

Janczewski, A. M., Zahid, M., Lemster, B. H., Frye, C. S., Gibson, G., Higuchi, Y., et al. (2004). Phospholamban gene ablation improves calcium transients but not cardiac function in a heart failure model. *Cardiovascular Research*, *62*(3), 468–480.

Jaski, B. E., Jessup, M. L., Mancini, D. M., Cappola, T. P., Pauly, D. F., Greenberg, B., et al. (2009). Calcium Up-regulation by percutaneous administration of gene therapy in cardiac disease (CUPID) trial Investigators. Calcium upregulation by percutaneous administration of gene therapy in cardiac disease (CUPID Trial), a first-in-human phase 1/2 clinical trial. *Journal of Cardiac Failure*, *15*(3), 171–181.

Jessup, M., Greenberg, B., Mancini, D., Cappola, T., Pauly, D. F., Jaski, B., et al. (2011). Calcium upregulation by percutaneous administration of gene therapy in cardiac disease (CUPID). A phase 2 trial of intracoronary gene therapy of sarcoplasmic reticulum Ca2+-ATPase in patients with advanced heart failure. *Circulation*, *124*(3), 304–313.

Jiang, D. S., Wei, X., Zhang, X. F., Liu, Y., Zhang, Y., Chen, K., et al. (2014). IRF8 suppresses pathological cardiac remodelling by inhibiting calcineurin signalling. *Nature Communications*, *5*, 3303.

Jones, L. R., Suzuki, Y. J., Wang, W., Kobayashi, Y. M., Ramesh, V., Franzini-Armstrong,-C., et al. (1998). Regulation of Ca2+ signalling in transgenic mouse cardiac myocytes overexpressing calsequestrin. *The Journal of Clinical Investigation*, *101*(7), 1385–1393.

Kajstura, J., Leri, A., Finato, N., Di Loreto, C., Beltrami, C. A., & Anversa, P. (1998). Myocyte proliferation in end-stage cardiac failure in humans. *Proceedings of the National Academy of Sciences of the United States of America*, *95*(15), 8801–8805.

Kang, R., Zeh, H. J., Lotze, M. T., & Tang, D. (2011). The Beclin1 network regulates autophagy and apoptosis. *Cell Death and Differentiation, 18*(4), 571–580.
Kao, Y. H., Liou, J. P., Chung, C. C., Lien, G. S., Kuo, C. C., Chen, S. A., et al. (2013). Histone deacetylase inhibition improved cardiac functions with direct antifibrotic activity in heart failure. *The International Journal of Cardiology, 168*(4), 4178–4183.
Keane, T. M., Goodstadt, L., Danecek, P., White, M. A., Wong, K., Yalcin, B., et al. (2011). Mouse genomic variation and its effect on phenotypes and gene regulation. *Nature, 477*(7364), 289–294.
Kim, J. K., Pedram, A., Razandi, M., & Levin, E. R. (2006). Estrogen prevents cardiomyocyte apoptosis through inhibition of reactive oxygen species and differential regulation of p38 kinase isoforms. *The Journal of Biological Chemistry, 281*(10), 6760–6767.
Knaapen, M. W., Davies, M. J., De Bie, M., Haven, A. J., Martinet, W., & Kockx, M. M. (2001). Apoptotic versus autophagic cell death in heart failure. *Cardiovascular Research, 51*(2), 304–312.
Knöll, R., Linke, W. A., Zou, P., Miocic, S., Kostin, S., Buyandelger, B., et al. (2011). Telethonin deficiency is associated with maladaptation to biomechanical stress in the mammalian heart. *Circulation Research, 109*(7), 758–769.
Koitabashi, N., Bedja, D., Zaiman, A. L., Pinto, Y. M., Zhang, M., Gabrielson, K. L., et al. (2009). Avoidance of transient cardiomyopathy in cardiomyocyte-targeted tamoxifen-induced MerCreMer gene deletion models. *Circulation Research, 105*(1), 12–15.
Kokoszka, J. E., Waymire, K. G., Levy, S. E., Sligh, J. E., Cai, J., Jones, D. P., et al. (2004). The ADP/ATP translocator is not essential for the mitochondrial permeability transition pore. *Nature, 427*(6973), 461–465.
Kostin, S., Pool, L., Elsässer, A., Hein, S., Drexler, H. C. A., Arnon, E., et al. (2003). Myocytes die by multiple mechanisms in failing human hearts. *Circulation Research, 92*(7), 715–724.
Krystof, V., Chamrád, I., Jorda, R., & Kohoutek, J. (2010). Pharmacological targeting of CDK9 in cardiac hypertrophy. *Medical Research Reviews, 30*(4), 646–666.
Kubota, T., Bounoutas, G. S., Miyagishima, M., Kadokami, T., Sanders, V. J., Bruton, C., et al. (2000). Soluble tumor necrosis factor receptor abrogates myocardial inflammation but not hypertrophy in cytokine-induced cardiomyopathy. *Circulation, 101*(21), 2518–2525.
Kubota, T., McTiernan, C. F., Frye, C. S., Demetris, A. J., & Feldman, A. M. (1997a). Cardiac-specific overexpression of tumor necrosis factor-alpha causes lethal myocarditis in transgenic mice. *Journal of Cardiac Failure, 3*(2), 117–124.
Kubota, T., McTiernan, C. F., Frye, C. S., Slawson, S. E., Lemster, B. H., Koretsky, A. P., et al. (1997b). Dilated cardiomyopathy in transgenic mice with cardiac-specific overexpression of tumor necrosis factor-alpha. *Circulation Research, 81*(4), 627–635.
Kurrelmeyer, K. M., Michael, L. H., Baumgarten, G., Taffet, G. E., Peschon, J. J., Sivasubramanian, N., et al. (2000). Endogenous tumor necrosis factor protects the adult cardiac myocyte against ischemic-induced apoptosis in a murine model of acute myocardial infarction. *Proceedings of the National Academy of Sciences of the United States of America, 97*(10), 5456–5461.
Kyriakis, J. M., Carpenter, M. K., Avruch, J., Rosler, E. S., Fisk, G. J., Brandenberger, R., et al. (2012). Mammalian MAPK signal transduction pathways activated by stress and inflammation: A 10-year update. *Physiological Reviews, 92*(2), 689–737.
Kytö, V., Saraste, A., Saukko, P., Henn, V., Pulkki, K., Vuorinen, T., et al. (2004). Apoptotic cardiomyocyte death in fatal myocarditis. *The American Journal of Cardiology, 94*(6), 746–750.
Lai, Y., Zhao, J., Yue, Y., Wasala, N. B., & Duan, D. (2014). Partial restoration of cardiac function with DeltaPDZ nNOS in aged mdx model of Duchenne cardiomyopathy. *Human Molecular Genetics*.

Lee, P., Sata, M., Lefer, D. J., Factor, S. M., Walsh, K., & Kitsis, R. N. (2003). Fas pathway is a critical mediator of cardiac myocyte death and MI during ischemia-reperfusion in vivo. *American Journal of Physiology. Heart and Circulatory Physiology*, *284*(2), H456–H463.

Lexow, J., Poggioli, T., Sarathchandra, P., Santini, M. P., & Rosenthal, N. (2013). Cardiac fibrosis in mice expressing an inducible myocardial-specific Cre driver. *Disease Models & Mechanisms*, *6*(6), 1470–1476.

Li, Y. Y., Feng, Y. Q., Kadokami, T., McTiernan, C. F., Draviam, R., Watkins, S. C., et al. (2000). Myocardial extracellular matrix remodeling in transgenic mice overexpressing tumor necrosis factor alpha can be modulated by anti-tumor necrosis factor alpha therapy. *Proceedings of the National Academy of Sciences of the United States of America*, *97*(23), 12746–12751.

Li, W., Liu, W., Zhong, J., & Yu, X. (2009). Early manifestation of alteration in cardiac function in dystrophin deficient mdx mouse using 3D CMR tagging. *Journal of Cardiovascular Magnetic Resonance*, *11*, 40.

Liang, P., Lan, F., Lee, A. S., Gong, T., Sanchez-Freire, V., Wang, Y., et al. (2013). Drug screening using a library of human induced pluripotent stem cell-derived cardiomyocytes reveals disease-specific patterns of cardiotoxicity. *Circulation*, *127*(16), 1677–1691.

Liggett, S. B., Tepe, N. M., Lorenz, J. N., Canning, A. M., Jantz, T. D., Mitarai, S., et al. (2000). Early and delayed consequences of beta(2)-adrenergic receptor overexpression in mouse hearts: Critical role for expression level. *Circulation*, *101*(14), 1707–1714.

Lim, S. Y., Cao, F., Davidson, S. M., Wagner, R. A., Mocanu, M. M., Wilson, K. D., et al. (2007). The cardioprotective effect of necrostatin requires the cyclophilin-D component of the mitochondrial permeability transition pore. *Cardiovascular Drugs and Therapy*, *21*(6), 467–469.

Lin, Y., Devin, A., Rodriguez, Y., & Liu, Z. G. (1999). Cleavage of the death domain kinase RIP by Caspase-8 prompts TNF-induced apoptosis. *Genes & Development*, *13*(19), 2514–2526.

Ling, H., Zhang, T., Pereira, L., Means, C. K., Cheng, H., Gu, Y., et al. (2009). Requirement for Ca2+/calmodulin–dependent kinase II in the transition from pressure overload–induced cardiac hypertrophy to heart failure in mice. *The Journal of Clinical Investigation*, *119*(5), 1230–1240.

Liu, H.-R., Gao, E., Hu, A., Tao, L., Qu, Y., Most, P., et al. (2005). Role of Omi/HtrA2 in apoptotic cell death after myocardial ischemia and reperfusion. *Circulation*, *111*(1), 90–96.

Liu, Q., Silva, J. P., Sargent, M. A., Sardão, V. A., York, A. J., Coutinho, O. P., et al. (2009). ASK1 regulates cardiomyocyte death but not hypertrophy in transgenic mice. *Circulation Research*, *105*(11), 1110–1117.

Liu, W., Zi, M., Naumann, R., Ulm, S., Jin, J., Taglieri, D. M., et al. (2011). Pak1 as a novel therapeutic target for antihypertrophic treatment in the heart. *Circulation*, *124*(24), 2702–2715.

Luengo-Fernández, R., Leal, J., Gray, A., Petersen, S., & Rayner, M. (2006). Cost of cardiovascular diseases in the United Kingdom. *Heart*, *92*(10), 1384–1389.

Ma, X., Liu, H., Foyil, S. R., Godar, R. J., Weinheimer, C. J., Hill, J. A., et al. (2012). Impaired autophagosome clearance contributes to cardiomyocyte death in ischemia/reperfusion injury. *Circulation*, *125*(25), 3170–3181.

Machida, N., Umikawa, M., Takei, K., Sakima, N., Myagmar, B. E., Taira, K., et al. (2004). Mitogen-activated protein kinase kinase kinase kinase 4 as a putative effector of Rap2 to activate the c-Jun N-terminal kinase. *The Journal of Biological Chemistry*, *279*(16), 15711–15714.

MacLellan, W. R., Wang, Y., & Lusis, A. J. (2012). Systems-based approaches to cardiovascular disease. *Nature Reviews Cardiology*, *9*(3), 172–184.

Maejima, Y., Usui, S., Zhai, P., Takamura, M., Kaneko, S., Zablocki, D., et al. (2014). MuRF1 negatively regulates pathological cardiac hypertrophy through downregulation of calcineurin A. *Circulation. Heart Failure.* Feb 13 [E-pub ahead of print].

Maier, L. S., Zhang, T., Chen, L., DeSantiago, J., Brown, J. H., & Bers, D. M. (2003). Transgenic CaMKIIdeltaC overexpression uniquely alters cardiac myocyte Ca2+ handling: Reduced SR Ca2+ load and activated SR Ca2+ release. *Circulation Research, 92*(8), 904–911.

Maiuri, M. C., Le Toumelin, G., Criollo, A., Rain, J.-C., Gautier, F., Juin, P., et al. (2007). Functional and physical interaction between Bcl-XL and a BH3-like domain in Beclin1. *The EMBO Journal, 26*(10), 2527–2539.

Maloyan, A., Gulick, J., Glabe, C. G., Kayed, R., & Robbins, J. (2007). Exercise reverses preamyloid oligomer and prolongs survival in alphaB-crystallin-based desmin-related cardiomyopathy. *Proceedings of the National Academy of Sciences of the United States of America, 104*(14), 5995–6000.

Maloyan, A., Sanbe, A., Osinska, H., Westfall, M., Robinson, D., Imahashi, K., et al. (2005). Mitochondrial dysfunction and apoptosis underlie the pathogenic process in alpha-B-crystallin desmin-related cardiomyopathy. *Circulation, 112*(22), 3451–3461.

Maloyan, A., Sayegh, J., Osinska, H., Chua, B. H., & Robbins, J. (2010). Manipulation of death pathways in desmin-related cardiomyopathy. *Circulation Research, 106*(9), 1524–1532.

Mandl, A., Huong Pham, L., Toth, K., Zambetti, G., & Erhardt, P. (2011). Puma deletion delays cardiac dysfunction in murine heart failure models through attenuation of apoptosis. *Circulation, 124*(1), 31–39.

Mann, D. L. (2011). *Heart failure: A companion to Braunwald's heart disease* (2nd ed.). Elsevier, Saunders, St. Louis, Missouri, USA.

Mann, D. L., McMurray, J. J. V., Packer, M., Swedberg, K., Borer, J. S., Colucci, W. S., et al. (2004). Targeted anticytokine therapy in patients with chronic heart failure: Results of the Randomized Etanercept Worldwide Evaluation (RENEWAL). *Circulation, 109*(13), 1594–1602.

Marber, M. S., Molkentin, J. D., & Force, T. (2011). Developing small molecules to inhibit kinases unkind to the heart: p38 MAPK as a case in point. *Drug Discovery Today Disease Mechanisms*, 7(2), e123–e127.

Marber, M. S., Nostro, M. C., Rose, B., Cheng, X., Wang, Y., Keller, G. M., et al. (2011). The p38 mitogen-activated protein kinase pathway—A potential target for intervention in infarction, hypertrophy, and heart failure. *Journal of Molecular and Cellular Cardiology, 51*(4), 485–490.

Martindale, J. J., Fernandez, R., Thuerauf, D., Whittaker, R., Gude, N., Sussman, M. A., et al. (2006). Endoplasmic reticulum stress gene induction and protection from ischemia/reperfusion injury in the hearts of transgenic mice with a tamoxifen-regulated form of ATF6. *Circulation Research, 98*(9), 1186–1193.

Marzo, I., Brenner, C., Zamzami, N., Jürgensmeier, J. M., Brozena, S., Vieira, H. L., et al. (1998). Bax and adenine nucleotide translocator cooperate in the mitochondrial control of apoptosis. *Science, 281*(5385), 2027–2031.

Matsui, Y., Takagi, H., Qu, X., Abdellatif, M., Sakoda, H., Asano, T., et al. (2007). Distinct roles of autophagy in the heart during ischemia and reperfusion. *Circulation, 100*(6), 914–922.

Matsumoto-Ida, M., Takimoto, Y., Aoyama, T., Akao, M., Takeda, T., & Kita, T. (2006). Activation of TGF-beta1-TAK1-p38 MAPK pathway in spared cardiomyocytes is involved in left ventricular remodeling after myocardial infarction in rats. *American Journal of Physiology. Heart and Circulatory Physiology, 290*(2), H709–H715.

Megeney, L. A., Kablar, B., Perry, R. L., Ying, C., May, L., & Rudnicki, M. A. (1999). Severe cardiomyopathy in mice lacking dystrophin and MyoD. *Proceedings of the National Academy of Sciences of the United States of America, 96*(1), 220–225.

Mercola, M., Ruiz-Lozano, P., & Schneider, M. D. (2011). Cardiac muscle regeneration: Lessons from development. *Genes & Development, 25*(4), 299–309.
Merkle, S., Frantz, S., Schön, M. P., Bauersachs, J., Buitrago, M., Frost, R. J. A., et al. (2007). A role for caspase-1 in heart failure. *Circulation Research, 100*(5), 645–653.
Mewton, N., Croisille, P., Gahide, G., Rioufol, G., Bonnefoy, E., Sanchez, I., et al. (2010). Effect of cyclosporine on left ventricular remodeling after reperfused myocardial infarction. *Journal of the American College of Cardiology, 55*(12), 1200–1205.
Milasin, J., Muntoni, F., Severini, G. M., Bartoloni, L., Vatta, M., Krajinovic, M., et al. (1996). A point mutation in the 5′ splice site of the dystrophin gene first intron responsible for X-linked dilated cardiomyopathy. *Human Molecular Genetics, 5*(1), 73–79.
Millay, D. P., Sargent, M. A., Osinska, H., Baines, C. P., Barton, E. R., Vuagniaux, G., et al. (2008). Genetic and pharmacologic inhibition of mitochondrial-dependent necrosis attenuates muscular dystrophy. *Nature Medicine, 14*(4), 442–447.
Miller, T. D., Christian, T. F., Hopfenspirger, R. N., Hodge, D. O., Gersh, B. J., & Gibbons, R. J. (1995). Infarct size after acute myocardial infarction measured by quantitative tomographic 99mTc sestamibi imaging predicts subsequent mortality. *Circulation, 92*(3), 334–341.
Milner, D. J., Weitzer, G., Tran, D., Bradley, A., & Capetanaki, Y. (1996). Disruption of muscle architecture and myocardial degeneration in mice lacking desmin. *The Journal of Cell Biology, 134*(5), 1255–1270.
Minamisawa, S., Hoshijima, M., Chu, G., Ward, C. A., Frank, K., Gu, Y., et al. (1999). Chronic phospholamban-sarcoplasmic reticulum calcium ATPase interaction is the critical calcium cycling defect in dilated cardiomyopathy. *Cell, 99*(3), 313–322.
Mitra, A., Basak, T., Datta, K., Naskar, S., Sengupta, S., & Sarkar, S. (2013). Role of α-crystallin B as a regulatory switch in modulating cardiomyocyte apoptosis by mitochondria or endoplasmic reticulum during cardiac hypertrophy and myocardial infarction. *Cell Death & Disease, 4*, e582.
Mochly-Rosen, D., Das, K., & Grimes, K. V. (2012). Protein kinase C, an elusive therapeutic target? *Nature Reviews Drug Discovery, 11*(12), 937–957.
Mochly-Rosen, D., Wu, G., Hahn, H., Osinska, H., Liron, T., Lorenz, J. N., et al. (2000). Cardiotrophic effects of protein kinase C epsilon: Analysis by in vivo modulation of PKCepsilon translocation. *Circulation Research, 86*(11), 1173–1179.
Moga, M.-A., Nakamura, T., & Robbins, J. (2008). Genetic approaches for changing the heart and dissecting complex syndromes. *Journal of Molecular and Cellular Cardiology, 45*(2), 148–155.
Molkentin, J. D., Lu, J. R., Antos, C. L., Markham, B., Richardson, J., Robbins, J., et al. (1998). A calcineurin-dependent transcriptional pathway for cardiac hypertrophy. *Cell, 93*(2), 215–228.
Molkentin, J. D., & Robbins, J. (2009). With great power comes great responsibility: Using mouse genetics to study cardiac hypertrophy and failure. *Journal of Molecular and Cellular Cardiology, 46*(2), 130–136.
Monnet, E. (2005). Animal models of heart failure: What is new? *The Annals of Thoracic Surgery, 79*(4), 1445–1453.
Mordwinkin, N. M., Burridge, P. W., & Wu, J. C. (2013). A review of human pluripotent stem cell-derived cardiomyocytes for high-throughput drug discovery, cardiotoxicity screening, and publication standards. *Journal of Cardiovascular Translational Research, 6*(1), 22–30.
Morita, H., Seidman, J., & Seidman, C. E. (2005). Genetic causes of human heart failure. *The Journal of Clinical Investigation, 115*(3), 518–526.
Moser, H. (1984). Duchenne muscular dystrophy: Pathogenetic aspects and genetic prevention. *Human Genetics, 66*(1), 17–40.

Mourkioti, F., Kustan, J., Kraft, P., Day, J. W., Zhao, M. M., Kost-Alimova, M., et al. (2013). Role of telomere dysfunction in cardiac failure in Duchenne muscular dystrophy. *Nature Cell Biology*, *15*(8), 895–904.

Mukoyama, M., Kondo, K., Hizawa, K., & Nishitani, H. (1987). Life spans of Duchenne muscular dystrophy patients in the hospital care program in Japan. *The Journal of Neurological Science*, *81*(2–3), 155–158.

Müller-Ehmsen, J., & Schwinger, R. H. (2004). TNF and congestive heart failure: Therapeutic possibilities. *Expert Opinion on Therapeutic Targets*, *8*(3), 203–209.

Muñoz-Mármol, A. M., Strasser, G., Isamat, M., Coulombe, P. A., Yang, Y., Roca, X., et al. (1998). A dysfunctional desmin mutation in a patient with severe generalized myopathy. *Proceedings of the National Academy of Sciences of the United States of America*, *95*(19), 11312–11317.

Muntoni, F., Cau, M., Ganau, A., Congiu, R., Arvedi, G., Mateddu, A., et al. (1993). Brief report: Deletion of the dystrophin muscle-promoter region associated with X-linked dilated cardiomyopathy. *The New England Journal of Medicine*, *329*(13), 921–925.

Muntoni, F., Di Lenarda, A., Porcu, M., Sinagra, G., Mateddu, A., Marrosu, G., et al. (1997). Dystrophin gene abnormalities in two patients with idiopathic dilated cardiomyopathy. *Heart*, *78*(6), 608–612.

Nadal-Ginard, B., Kajstura, J., Anversa, P., & Leri, A. (2003). A matter of life and death: Cardiac myocyte apoptosis and regeneration. *The Journal of Clinical Investigation*, *111*(10), 1457–1459.

Nakagawa, T., Shimizu, S., Watanabe, T., Yamaguchi, O., Otsu, K., Yamagata, H., et al. (2005). Cyclophilin D-dependent mitochondrial permeability transition regulates some necrotic but not apoptotic cell death. *Nature*, *434*(7033), 652–658.

Nakayama, H., Chen, X., Baines, C. P., Klevitsky, R., Zhang, X., Zhang, H., et al. (2007). Ca2+- and mitochondrial-dependent cardiomyocyte necrosis as a primary mediator of heart failure. *The Journal of Clinical Investigation*, *117*(9), 2431–2444.

Nam, Y.-J., Mani, K., Ashton, A. W., Peng, C.-F., Krishnamurthy, B., Hayakawa, Y., et al. (2004). Inhibition of both the extrinsic and intrinsic death pathways through nonhomotypic death-fold interactions. *Molecular Cell*, *15*(6), 901–912.

Nam, Y.-J., Mani, K., Wu, L., Peng, C.-F., Calvert, J. W., Foo, R. S.-Y., et al. (2007). The apoptosis inhibitor ARC undergoes Ubiquitin-Proteasomal-mediated degradation in response to death stimuli identification of a degradation-resistant mutant. *The Journal of Biological Chemistry*, *282*(8), 5522–5528.

Narula, J., Haider, N., Virmani, R., DiSalvo, T. G., Kolodgie, F. D., Hajjar, R. J., et al. (1996). Apoptosis in myocytes in end-stage heart failure. *The New England Journal of Medicine*, *335*(16), 1182–1189.

Neely, G. G., Kuba, K., Cammarato, A., Isobe, K., Amann, S., Zhang, L., et al. (2010). A global in vivo Drosophila RNAi screen identifies NOT3 as a conserved regulator of heart function. *Cell*, *141*(1), 142–153.

Neri, M., Valli, E., Alfano, G., Bovolenta, M., Spitali, P., Rapezzi, C., et al. (2012). The absence of dystrophin brain isoform expression in healthy human heart ventricles explains the pathogenesis of 5' X-linked dilated cardiomyopathy. *BMC Medical Genetics*, *13*, 20.

Nichols, M., Townsend, N., Scarborough, P., & Rayner, M. (2013). European Cardiovascular Disease Statistics 4th edition 2012: EuroHeart II. *European Heart Journal*, *34*(39), 3007.

Nishikimia, T., Maeda, N., & Matsuokaa, H. (2006). The role of natriuretic peptides in cardioprotection. *Cardiovascular Research*, *69*(2), 318–328.

Noguchi, T., Takeda, K., Matsuzawa, A., Saegusa, K., Nakano, H., Gohda, J., et al. (2005). Recruitment of TRAF family proteins to the ASK1 signalosome is essential for oxidative stress-induced cell death. *The Journal of Biological Chemistry*, *280*(44), 37033–37040.

Norgett, E. E., Hatsell, S. J., Carvajal-Huerta, L., Cabezas, J. C., Common, J., Purkis, P. E., et al. (2000). Recessive mutation in desmoplakin disrupts desmoplakin-intermediate filament interactions and causes dilated cardiomyopathy, woolly hair and keratoderma. *Human Molecular Genetics*, *9*(18), 2761–2766.

Noseda, M., Peterkin, T., Simões, F. C., Patient, R., & Schneider, M. D. (2011). Cardiopoietic factors: Extracellular signals for cardiac lineage commitment. *Circulation Research*, *108*(1), 129–152.

Ocorr, K. A., Crawley, T., Gibson, G., & Bodmer, R. (2007). Genetic variation for cardiac dysfunction in Drosophila. *PLoS One*, *2*(7), e601.

Oerlemans, M. I., Koudstaal, S., Chamuleau, S. A., de Kleijn, D. P., Doevendans, P. A., & Sluijter, J. P. (2013). Targeting cell death in the reperfused heart: Pharmacological approaches for cardioprotection. *International Journal of Cardiology*, *165*(3), 410–422.

Oh, H., Wang, S. C., Prahash, A., Sano, M., Moravec, C. S., Taffet, G. E., et al. (2003). Telomere attrition and Chk2 activation in human heart failure. *Proceedings of the National Academy of Sciences of the United States of America*, *100*(9), 5378–5383.

Oka, T., Hikoso, S., Yamaguchi, O., Taneike, M., Takeda, T., Tamai, T., et al. (2012). Mitochondrial DNA that escapes from autophagy causes inflammation and heart failure. *Nature*, *485*(7397), 251–255.

Olivetti, G., Abbi, R., Quaini, F., Kajstura, J., Cheng, W., Nitahara, J. A., et al. (1997). Apoptosis in the failing human heart. *The New England Journal of Medicine*, *336*(16), 1131–1141.

Ortiz-Lopez, R., Li, H., Su, J., Goytia, V., & Towbin, J. A. (1997). Evidence for a dystrophin missense mutation as a cause for X-linked dilated cardiomyopathy. *Circulation*, *95*(10), 2434–2440.

Pass, J. M., Gao, J., Jones, W. K., Wead, W. B., Wu, X., Zhang, J., et al. (2001). Enhanced PKC beta II translocation and PKC beta II-RACK1 interactions in PKC epsilon-induced heart failure: A role for RACK1. *American Journal of Physiology. Heart and Circulatory Physiology*, *281*(6), H2500–H2510.

Pass, J. M., Zheng, Y., Wead, W. B., Zhang, J., Li, R. C., Bolli, R., et al. (2001). PKCepsilon activation induces dichotomous cardiac phenotypes and modulates PKCepsilon-RACK interactions and RACK expression. *American Journal of Physiology. Heart and Circulatory Physiology*, *280*(3), H946–H955.

Passier, R., Zeng, H., Frey, N., Naya, F. J., Nicol, R. L., McKinsey, T. A., et al. (2000). CaM kinase signalling induces cardiac hypertrophy and activates the MEF2 transcription factor in vivo. *The Journal of Clinical Investigation*, *105*(10), 1395–1406.

Pattingre, S., Tassa, A., Qu, X., Garuti, R., Liang, X. H., Mizushima, N., et al. (2005). Bcl-2 antiapoptotic proteins inhibit Beclin1-dependent autophagy. *Cell*, *122*(6), 927–939.

Pattison, J. S., Sanbe, A., Maloyan, A., Osinska, H., Klevitsky, R., & Robbins, J. (2008). Cardiomyocyte expression of a polyglutamine preamyloid oligomer causes heart failure. *Circulation*, *117*(21), 2743–2751.

Peter, M. E. (2004). The flip side of FLIP. *The Biochemical Journal*, *382*(2), e1–e3.

Petrich, B. G., Eloff, B. C., Lerner, D. L., Kovacs, A., Saffitz, J. E., Rosenbaum, D. S., et al. (2004). Targeted activation of c-Jun N-terminal kinase in vivo induces restrictive cardiomyopathy and conduction defects. *The Journal of Biological Chemistry*, *279*(15), 15330–15338.

Petrich, B. G., Molkentin, J. D., & Wang, Y. (2003). Temporal activation of c-Jun N-terminal kinase in adult transgenic heart via cre-loxP-mediated DNA recombination. *The FASEB Journal*, *17*(6), 749–751.

Piot, C., Croisille, P., Staat, P., Thibault, H., Rioufol, G., Mewton, N., et al. (2008). Effect of cyclosporine on reperfusion injury in acute myocardial infarction. *The New England Journal of Medicine*, *359*(5), 473–481.

Pomerantz, B. J., Reznikov, L. L., Harken, A. H., & Dinarello, C. A. (2001). Inhibition of caspase 1 reduces human myocardial ischemic dysfunction via inhibition of IL-18 and

IL-1beta. *Proceedings of the National Academy of Sciences of the United States of America*, *98*(5), 2871–2876.
Ponce de Leon, V., Merillat, A. M., Tesson, L., Anegon, I., & Hummler, E. (2014). Generation of TALEN-mediated GR(dim) knock-in rats by homologous recombination. *PLoS One*, *9*(2), e88146.
Potts, P. R., Singh, S., Knezek, M., Thompson, C. B., & Deshmukh, M. (2003). Critical function of endogenous XIAP in regulating caspase activation during sympathetic neuronal apoptosis. *The Journal of Cell Biology*, *163*(4), 789–799.
Prabhakar, R., Boivin, G. P., Grupp, I. L., Hoit, B., Arteaga, G., Solaro, R. J., et al. (2001). A familial hypertrophic cardiomyopathy α -tropomyosin mutation causes severe cardiac hypertrophy and death in mice. *Journal of Molecular and Cellular Cardiology*, *33*(10), 1815–1828.
Preuss, C., & Andelfinger, G. (2013). Genetics of heart failure in congenital heart disease. *The Canadian Journal of Cardiology*, *29*(7), 803–810.
Quaini, F., Cigola, E., Lagrasta, C., Saccani, G., Quaini, E., Rossi, C., et al. (1994). End-stage cardiac failure in humans is coupled with the induction of proliferating cell nuclear antigen and nuclear mitotic division in ventricular myocytes. *Circulation Research*, *75*(6), 1050–1063.
Ramirez, M. T., Sah, V. P., Zhao, X.-L., Hunter, J. J., Chien, K. R., & Brown, J. H. (1997). The MEKK-JNK pathway is stimulated by α1-adrenergic receptor and ras activation and is associated with in vitro and in vivo cardiac hypertrophy. *The Journal of Biological Chemistry*, *272*(22), 14057–14061.
Rampazzo, A., Nava, A., Malacrida, S., Beffagna, G., Bauce, B., Rossi, V., et al. (2002). Mutation in human desmoplakin domain binding to plakoglobin causes a dominant form of arrhythmogenic right ventricular cardiomyopathy. *The American Journal of Human Genetics*, *71*(5), 1200–1206.
Redfern, C. H., Degtyarev, M. Y., Kwa, A. T., Salomonis, N., Cotte, N., Nanevicz, T., et al. (2000). Conditional expression of a Gi-coupled receptor causes ventricular conduction delay and a lethal cardiomyopathy. *Proceedings of the National Academy of Sciences of the United States of America*, *97*(9), 4826–4831.
Rockman, H. A., Chien, K. R., Choi, D. J., Iaccarino, G., Hunter, J. J., Ross, J., Jr., et al. (1998). Expression of a beta-adrenergic receptor kinase 1 inhibitor prevents the development of myocardial failure in gene-targeted mice. *Proceedings of the National Academy of Sciences of the United States of America*, *95*(12), 7000–7005.
Rooney, J. E., Welser, J. V., Dechert, M. A., Flintoff-Dye, N. L., Kaufman, S. J., & Burkin, D. J. (2006). Severe muscular dystrophy in mice that lack dystrophin and alpha7 integrin. *Journal of Cell Science*, *119*(11), 2185–2195.
Rose, B. A., Force, T., & Wang, Y. (2010). Mitogen-activated protein kinase signalling in the heart: Angels versus demons in a heart-breaking tale. *Physiological Reviews*, *90*(4), 1507–1546.
Rothermel, B. A., & Hill, J. A. (2008). Autophagy in load-induced heart disease. *Circulation Research*, *103*(12), 1363–1369.
Sacco, A., Mourkioti, F., Tran, R., Choi, J., Llewellyn, M., Kraft, P., et al. (2010). Short telomeres and stem cell exhaustion model Duchenne muscular dystrophy in mdx/mTR mice. *Cell*, *143*(7), 1059–1071.
Sag, C. M., Wadsack, D. P., Khabbazzadeh, S., Abesser, M., Grefe, C., Neumann, K., et al. (2009). Calcium/calmodulin-dependent protein kinase II contributes to cardiac arrhythmogenesis in heart failure. *Circulation. Heart Failure*, *2*(6), 664–675.
Sartorius, T., Staiger, H., Ketterer, C., Heni, M., Machicao, F., Guilherme, A., et al. (2012). Association of common genetic variants in the MAP4K4 locus with prediabetic traits in humans. *PLoS One*, 7(10), e47647.

Sanbe, A. (2011). Molecular mechanisms of α-crystallinopathy and its therapeutic strategy. *Biological & Pharmaceutical Bulletin, 34*(11), 1653–1658.
Sanbe, A., Gulick, J., Hanks, M. C., Liang, Q., Osinska, H., & Robbins, J. (2003). Reengineering inducible cardiac-specific transgenesis with an attenuated myosin heavy chain promoter. *Circulation Research, 92*(6), 609–616.
Sanbe, A., Osinska, H., Saffitz, J. E., Glabe, C. G., Kayed, R., Maloyan, A., et al. (2004). Desmin-related cardiomyopathy in transgenic mice: A cardiac amyloidosis. *Proceedings of the National Academy of Sciences of the United States of America, 101*(27), 10132–10136.
Sanbe, A., Osinska, H., Villa, C., Gulick, J., Klevitsky, R., Glabe, C. G., et al. (2005). Reversal of amyloid-induced heart disease in desmin-related cardiomyopathy. *Proceedings of the National Academy of Sciences of the United States of America, 102*(38), 13592–13597.
Sano, M., Minamino, T., Toko, H., Miyauchi, H., Orimo, M., Qin, Y., et al. (2007). p53-induced inhibition of Hif-1 causes cardiac dysfunction during pressure overload. *Nature, 446*(7134), 444–448.
Sano, M., & Schneider, M. D. (2004). Cyclin-dependent kinase-9: An RNAPII kinase at the nexus of cardiac growth and death cascades. *Circulation Research, 95*(9), 867–876.
Sano, M., Wang, S. C., Shirai, M., Scaglia, F., Xie, M., Sakai, S., et al. (2004). Activation of cardiac Cdk9 represses PGC-1 and confers a predisposition to heart failure. *The EMBO Journal, 23*(17), 3559–3569.
Sato, Y., Kiriazis, H., Yatani, A., Schmidt, A. G., Hahn, H., Ferguson, D. G., et al. (2001). Rescue of contractile parameters and myocyte hypertrophy in calsequestrin overexpressing myocardium by phospholamban ablation. *The Journal of Biological Chemistry, 276*(12), 9392–9399.
Scarabelli, T. M., & Gottlieb, R. A. (2004). Functional and clinical repercussions of myocyte apoptosis in the multifaceted damage by ischemia/reperfusion injury: Old and new concepts after 10 years of contributions. *Cell Death and Differentiation, 11*(Suppl. 2), S144–S152.
Seo, K., Rainer, P. P., Lee, D. I., Hao, S., Bedja, D., Birnbaumer, L., et al. (2014). Hyperactive adverse mechanical-stress responses in dystrophic heart are coupled to TRPC6 and blocked by cGMP-PKG modulation. *Circulation Research, 114*(5), 823–832.
Seo, K., Rainer, P. P., Shalkey Hahn, V., Lee, D. I., Jo, S. H., Andersen, A., et al. (2014). Combined TRPC3 and TRPC6 blockade by selective small-molecule or genetic deletion inhibits pathological cardiac hypertrophy. *Proceedings of the National Academy of Sciences of the United States of America, 111*(4), 1551–1556.
Shareef, M. A., Anwer, L. A., & Poizat, C. (2013). Cardiac SERCA2A/B: Therapeutic targets for heart failure. *European Journal of Pharmacology, 724C*, 1–8.
Shimizu, S., Narita, M., & Tsujimoto, Y. (1999). Bcl-2 family proteins regulate the release of apoptogenic cytochrome c by the mitochondrial channel VDAC. *Nature, 399*(6735), 483–487.
Shirokova, N., & Niggli, E. (2013). Cardiac phenotype of Duchenne muscular dystrophy: Insights from cellular studies. *Journal of Molecular and Cellular Cardiology, 58*, 217–224.
Simon, M. M., Greenaway, S., White, J. K., Fuchs, H., Gailus-Durner, V., Wells, S., et al. (2013). A comparative phenotypic and genomic analysis of C57BL/6J and C57BL/6N mouse strains. *Genome Biology, 14*(7), R82.
Simoons, M. L., Serruys, P. W., van den Brand, M., Res, J., Verheugt, F. W., Krauss, X. H., et al. (1986). Early thrombolysis in acute myocardial infarction: Limitation of infarct size and improved survival. *Journal of the American College of Cardiology*, 7(4), 717–728.
Sinha, S., & Levine, B. (2008). The autophagy effector Beclin1: A novel BH3-only protein. *Oncogene, 27*(Suppl. 1), S137–S148.
Sivasubramanian, N., Coker, M. L., Kurrelmeyer, K. M., MacLellan, W. R., DeMayo, F. J., Spinale, F. G., et al. (2001). Left ventricular remodeling in transgenic mice with cardiac restricted overexpression of tumor necrosis factor. *Circulation, 104*(7), 826–831.

Sohal, D. S., Nghiem, M., Crackower, M. A., Witt, S. A., Kimball, T. R., Tymitz, K. M., et al. (2001). Temporally regulated and tissue-specific gene manipulations in the adult and embryonic heart using a tamoxifen-inducible Cre protein. *Circulation Research*, *89*(1), 20–25.

Song, W., Dyer, E., Stuckey, D. J., Copeland, O., Leung, M. C., Bayliss, C., et al. (2011). Molecular mechanism of the E99K mutation in cardiac actin (ACTC Gene) that causes apical hypertrophy in man and mouse. *The Journal of Biological Chemistry*, *286*(31), 27582–27593.

Sossalla, S., Fluschnik, N., Schotola, H., Ort, K. R., Neef, S., Schulte, T., et al. (2010). Inhibition of elevated Ca2+/calmodulin-dependent protein kinase II improves contractility in human failing myocardium. *Circulation Research*, *107*(9), 1150–1161.

Stuckey, D. J., Carr, C. A., Camelliti, P., Tyler, D. J., Davies, K. E., & Clarke, K. (2012). In vivo MRI characterization of progressive cardiac dysfunction in the mdx mouse model of muscular dystrophy. *PLoS One*, 7(1), e28569.

Sun, N., Yazawa, M., Liu, J., Han, L., Sanchez-Freire, V., Abilez, O. J., et al. (2012). Patient-specific induced pluripotent stem cells as a model for familial dilated cardiomyopathy. *Science Translational Medicine*, *4*(130), ra47.

Suzuki, M., Youle, R. J., & Tjandra, N. (2000). Structure of Bax: Coregulation of dimer formation and intracellular localization. *Cell*, *103*(4), 645–654.

Tang, X., Guilherme, A., Chakladar, A., Powelka, A. M., Konda, S., Virbasius, J. V., et al. (2006). An RNA interference-based screen identifies MAP4K4/NIK as a negative regulator of PPARgamma, adipogenesis, and insulin-responsive hexose transport. *Proceedings of the National Academy of Sciences of the United States of America*, *103*(7), 2087–2092.

Tannous, P., Zhu, H., Johnstone, J. L., Shelton, J. M., Rajasekaran, N. S., Benjamin, I. J., et al. (2008). Autophagy is an adaptive response in desmin-related cardiomyopathy. *Proceedings of the National Academy of Sciences of the United States of America*, *105*(28), 9745–9750.

Teekakirikul, P., Eminaga, S., Toka, O., Alcalai, R., Wang, L., Wakimoto, H., et al. (2010). Cardiac fibrosis in mice with hypertrophic cardiomyopathy is mediated by non-myocyte proliferation and requires Tgf-beta. *The Journal of Clinical Investigation*, *120*(10), 3520–3529.

Tesz, G. J., Guilherme, A., Guntur, K. V., Hubbard, A. C., Tang, X., Chawla, A., et al. (2007). Tumor necrosis factor alpha (TNFα) stimulates Map4k4 expression through TNFα receptor 1 signaling to c-Jun and activating transcription factor 2. *The Journal of Biological Chemistry*, *282*(27), 19302–19312.

Toko, H., Takahashi, H., Kayama, Y., Okada, S., Minamino, T., Terasaki, F., et al. (2010). ATF6 is important under both pathological and physiological states in the heart. *Journal of Molecular and Cellular Cardiology*, *49*(1), 113–120.

Towbin, J. A., Hejtmancik, J. F., Brink, P., Gelb, B., Zhu, X. M., Chamberlain, J. S., et al. (1993). X-linked dilated cardiomyopathy: Molecular genetic evidence of linkage to the Duchenne muscular dystrophy (dystrophin) gene at the Xp21 locus. *Circulation*, *87*(6), 1854–1865.

Traaseth, N. J., Ha, K. N., Verardi, R., Shi, L., Buffy, J. J., Masterson, L. R., et al. (2008). Structural and dynamic basis of phospholamban and sarcolipin inhibition of Ca(2+)-ATPase. *Biochemistry*, *47*(1), 3–13.

Tran, P., Vogelstein, B., Ho, S. M., Kinzler, K. W., Kim, B. G., Vuong, T. A., et al. (2012). TGF-beta activated kinase 1 (TAK1) and apoptosis signal-regulating kinase 1 (ASK1) interact with the promyogenic receptor cdo to promote myogenic differentiation via activation of p38MAPK pathway. *The Journal of Biological Chemistry*, *287*(15), 11602–11615.

Tsubata, S., Bowles, K. R., Vatta, M., Zintz, C., Titus, J., Muhonen, L., et al. (2000). Mutations in the human delta-sarcoglycan gene in familial and sporadic dilated cardiomyopathy. *The Journal of Clinical Investigation, 106*(5), 655–662.

Ursitti, J. A., Petrich, B. G., Lee, P. C., Resneck, W. G., Ye, X., Yang, J., et al. (2007). Role of an alternatively spliced form of alphaII-spectrin in localization of connexin 43 in cardiomyocytes and regulation by stress-activated protein kinase. *Journal of Molecular and Cellular Cardiology, 42*(3), 572–581.

van Berlo, J. H., Maillet, M., & Molkentin, J. D. (2013). Signalling effectors underlying pathologic growth and remodeling of the heart. *The Journal of Clinical Investigation, 123*(1), 37–45.

Varfolomeev, E. E., Schuchmann, M., Luria, V., Chiannilkulchai, N., Beckmann, J. S., Mett, I. L., et al. (1998). Targeted disruption of the mouse Caspase 8 gene ablates cell death induction by the TNF receptors, Fas/Apo1, and DR3 and is lethal prenatally. *Immunity, 9*(2), 267–276.

Vega, R. B., Harrison, B. C., Meadows, E., Roberts, C. R., Papst, P. J., Olson, E. N., et al. (2004). Protein kinases C and D mediate agonist-dependent cardiac hypertrophy through nuclear export of histone deacetylase 5. *Molecular and Cellular Biology, 24*(19), 8374–8385.

Vicart, P., Caron, A., Guicheney, P., Li, Z., Prevost, M. C., Faure, A., et al. (1998). A missense mutation in the αB-crystallin chaperone gene causes a desmin-related myopathy. *Nature Genetics, 20*(1), 92–95.

Vikstrom, K. L., Factor, S. M., & Leinwand, L. A. (1996). Mice expressing mutant myosin heavy chains are a model for familial hypertrophic cardiomyopathy. *Molecular Medicine, 2*(5), 556–567.

Vila-Petroff, M., Salas, M. A., Said, M., Valverde, C. A., Sapia, L., Portiansky, E., et al. (2007). CaMKII Inhibition protects against necrosis and apoptosis in irreversible ischemia–reperfusion injury. *Cardiovascular Research, 73*(4), 689–698.

Wakasaki, H., Koya, D., Schoen, F. J., Jirousek, M. R., Ways, D. K., Hoiy, B. D., et al. (1997). Targeted overexpression of protein kinase C β2 isoform in myocardium causes cardiomyopathy. *Proceedings of the National Academy of Sciences of the United States of America, 94*(17), 9320–9325.

Wan, Y. Y., Chi, H., Xie, M., Schneider, M. D., & Flavell, R. A. (2006). The kinase TAK1 integrates antigen and cytokine receptor signalling for T cell development, survival and function. *Nature Immunology, 7*(8), 851–858.

Wang, L., Du, Fenghe, & Wang, X. (2008). TNF-α induces two distinct caspase-8 activation pathways. *Cell, 133*(4), 693–703.

Wang, X., Klevitsky, R., Huang, W., Glasford, J., Li, F., & Robbins, J. (2003). αB-crystallin modulates protein aggregation of abnormal desmin. *Circulation Research, 93*(10), 998–1005.

Wang, G. S., Kuyumcu-Martinez, M. N., Sarma, S.; Mathur, N., Wehrens, X. H., & Cooper, T. A. (2009). PKC inhibition ameliorates the cardiac phenotype in a mouse model of myotonic dystrophy type 1. *The Journal of Clinical Investigation, 119*(12), 3797–3806.

Wang, X., Osinska, H., Dorn, G. W., 2nd, Nieman, M., Lorenz, J. N., Gerdes, A. M., et al. (2001a). Mouse model of desmin-related cardiomyopathy. *Circulation, 103*(19), 2402–2407.

Wang, X., Osinska, H., Klevitsky, R., Gerdes, A. M., Nieman, M., Lorenz, J., et al. (2001b). Expression of R120G-alphaB-crystallin causes aberrant desmin and alphaB-crystallin aggregation and cardiomyopathy in mice. *Circulation Research, 89*(1), 84–91.

Wang, Y. B., Su, B., Sah, V. P., Brown, J. H., Han, J. H., & Chien, K. R. (1998). Cardiac hypertrophy induced by mitogen-activated protein kinase kinase 7, a specific activator for c-Jun NH2-terminal kinase in ventricular muscle cells. *The Journal of Biological Chemistry, 273*(10), 5423–5426.

Wang, Y., Whelan, R. S., Kaplinskiy, V., & Kitsis, R. N. (2007). Mitogen-activated protein kinases in heart development and diseases. *Circulation, 116*(12), 1413–1423.

Wang, M., Amano, S. U., Flach, R. J., Chawla, A., Aouadi, M., & Czech, M. P. (2013). Identification of Map4k4 as a novel suppressor of skeletal muscle differentiation. *Molecular and Cellular Biology, 33*(4), 678–687.

Wei, M. C., Lindsten, T., Mootha, V. K., Weiler, S., Gross, A., Ashiya, M., et al. (2000). tBID, a membrane-targeted death ligand, oligomerizes BAK to release cytochrome c. *Genes & Development, 14*(16), 2060–2071.

Weisleder, N., Taffet, G. E., & Capetanaki, Y. (2004). Bcl-2 overexpression corrects mitochondrial defects and ameliorates inherited desmin null cardiomyopathy. *Proceedings of the National Academy of Sciences of the United States of America, 101*(3), 769–774.

Wencker, D., Chandra, M., Nguyen, K., Miao, W., Garantziotis, S., Factor, S. M., et al. (2003). A mechanistic role for cardiac myocyte apoptosis in heart failure. *The Journal of Clinical Investigation, 111*(10), 1497–1504.

Wessels, M. W., & Willems, P. J. (2010). Genetic factors in non-syndromic congenital heart malformations. *Clinical Genetics, 78*(2), 103–123.

Whelan, R. S., Kaplinskiy, V., & Kitsis, R. N. (2010). Cell death in the pathogenesis of heart disease: Mechanisms and significance. *Annual Review of Physiology, 72*, 19–44.

White, H. D., & Chew, D. P. (2008). Acute myocardial infarction. *Lancet, 372*(9638), 570–584.

Wong, L. S., Oeseburg, H., de Boer, R. A., van Gilst, W. H., van Veldhuisen, D. J., & van der Harst, P. (2009). Telomere biology in cardiovascular disease: The TERC −/− mouse as a model for heart failure and ageing. *Cardiovascular Research, 81*(2), 244–252.

Wu, G., Toyokawa, T., Hahn, H., & Dorn, G. W., 2nd. (2000). Epsilon protein kinase C in pathological myocardial hypertrophy. Analysis by combined transgenic expression of translocation modifiers and Galphaq. *The Journal of Biological Chemistry, 275*(39), 29927–29930.

Xie, M., Zhang, D., Dyck, J. R., Li, Y., Zhang, H., Morishima, M., et al. (2006). A pivotal role for endogenous TGF-beta-activated kinase-1 in the LKB1/AMP-activated protein kinase energy-sensor pathway. *Proceedings of the National Academy of Sciences of the United States of America, 103*(46), 17378–17383.

Xie, M., Zhang, D., Wang, S. C., Sato, M., Wang, X. Z., Pocius, J. S., et al. (2007). The protein kinase MAP4K4 is activated in failing human hearts and mediates cardiomyocyte apoptosis in experimental models, in vitro and in vivo. *Circulation, 116*, II-420.

Yamaguchi, O., Watanabe, T., Nishida, K., Kashiwase, K., Higuchi, Y., Takeda, T., et al. (2004). Cardiac-specific disruption of the c-raf-1 gene induces cardiac dysfunction and apoptosis. *The Journal of Clinical Investigation, 114*(7), 937–943.

Yamamoto, S., Yang, G., Zablocki, D., Liu, J., Hong, C., Kim, S. J., et al. (2003). Activation of Mst1 causes dilated cardiomyopathy by stimulating apoptosis without compensatory ventricular myocyte hypertrophy. *The Journal of Clinical Investigation, 111*(10), 1463–1474.

Yang, Y. M., Gupta, S. K., Kim, K. J., Powers, B. E., Cerqueira, A., Wainger, B. J., et al. (2013). A small molecule screen in stem-cell-derived motor neurons identifies a kinase inhibitor as a candidate therapeutic for ALS. *Cell Stem Cell, 12*(6), 713–726.

Yang, Y., Zhu, W.-Z., Joiner, M.-L., Zhang, R., Oddis, C. V., Hou, Y., et al. (2006). Calmodulin kinase II inhibition protects against myocardial cell apoptosis in vivo. *American Journal of Physiology. Heart and Circulatory Physiology, 291*(6), H3065–H3075.

Yao, Z., Zhou, G., Wang, X. S., Brown, A., Diener, K., Gan, H., et al. (1999). A novel human STE20-related protein kinase, HGK, that specifically activates the c- Jun N-terminal kinase signaling pathway. *The Journal of Biological Chemistry, 274*(4), 2118–2125.

Yeh, W. C., Itie, A., Elia, A. J., Ng, M., Shu, H. B., Wakeham, A., et al. (2000). Requirement for casper (c-FLIP) in regulation of death receptor-induced apoptosis and embryonic development. *Immunity, 12*(6), 633–642.

Yeh, W.-C., la Pompa, de J. L., McCurrach, M. E., Shu, H.-B., Elia, A. J., Shahinian, A., et al. (1998). FADD: Essential for embryo development and signalling from some, but not all, inducers of apoptosis. *Science, 279*(5358), 1954–1958.

Yu, X. C., Acehan, D., Menetret, J. F., Booth, C. R., Ludtke, S. J., Riedl, S. J., et al. (2005). A structure of the human apoptosome at 12.8 angstrom resolution provides insights into this cell death platform. *Structure, 13*(11), 1725–1735.

Yu, Y., Ge, N., Xie, M., Sun, W., Burlingame, S., Pass, A. K., et al. (2008). Phosphorylation of Thr-178 and Thr-184 in the TAK1 T-loop is required for interleukin (IL)-1-mediated optimal NFkappaB and AP-1 activation as well as IL-6 gene expression. *The Journal of Biological Chemistry, 283*(36), 24497–24505.

Yussman, M. G., Toyokawa, T., Odley, A., Lynch, R. A., Wu, G., Colbert, M. C., et al. (2002). Mitochondrial death protein Nix is induced in cardiac hypertrophy and triggers apoptotic cardiomyopathy. *Nature Medicine, 8*(7), 725–730.

Zhang, D., Gaussin, V., Taffet, G. E., Belaguli, N. S., Yamada, M., Schwartz, R. J., et al. (2000). TAK1 is activated in the myocardium after pressure overload and is sufficient to provoke heart failure in transgenic mice. *Nature Medicine, 6*(5), 556–563.

Zhang, T., Maier, L. S., Dalton, N. D., Miyamoto, S., Ross, J., Jr., Bers, D. M., et al. (2003). The deltaC isoform of CaMKII is activated in cardiac hypertrophy and induces dilated cardiomyopathy and heart failure. *Circulation Research, 92*(8), 912–919.

Zhang, L., Malik, S., Pang, J., Wang, H., Park, K. M., Yule, D. I., et al. (2013). Phospholipase Cε hydrolyzes perinuclear phosphatidylinositol 4-phosphate to regulate cardiac hypertrophy. *Cell, 153*(1), 216–227.

Zhu, H., Tannous, P., Johnstone, J. L., Kong, Y., Shelton, J. M., Richardson, J. A., et al. (2007). Cardiac autophagy is a maladaptive response to hemodynamic stress. *The Journal of Clinical Investigation, 117*(7), 1782–1793.

Zhu, X., Wheeler, M. T., Hadhazy, M., Lam, M. Y., & McNally, E. M. (2002). Cardiomyopathy is independent of skeletal muscle disease in muscular dystrophy. *The FASEB Journal, 16*(9), 1096–1098.

Zsebo, K. M., Yaroshinsky, A., Rudy, J. J., Wagner, K., Greenberg, B., Jessup, M., et al. (2013). Long term effects of AAV1/SERCA2a gene transfer in patients with severe heart failure: Analysis of recurrent cardiovascular events and mortality. *Circulation Research, 114*(1), 101–108.

CHAPTER FIVE

Mouse Models and Aging: Longevity and Progeria

Chen-Yu Liao, Brian K. Kennedy[1]
Buck Institute for Research on Aging, Novato, California, USA
[1]Corresponding author: e-mail address: bkennedy@buckinstitute.org

Contents

Abstract

Aging is a complex, multifactorial process that is likely influenced by the activities of a range of biological pathways. Genetic approaches to identify genes modulating longevity have been highly successful and recent efforts have extended these studies to mammalian aging. A variety of genetic models have been reported to have enhanced lifespan and, similarly, many genetic interventions lead to progeroid phenotypes. Here, we detail and evaluate both sets of models, focusing on the insights they provide about the molecular processes modulating aging and the extent to which mutations conferring progeroid pathologies really phenocopy accelerated aging.

1. INTRODUCTION

Aging is accelerated by a wide range of prevalent, chronic diseases, including cardiovascular disease, atherosclerosis, inflammation, and neurodegeneration (Gems & Partridge, 2013; Niccoli & Partridge, 2012).

Current Topics in Developmental Biology, Volume 109
ISSN 0070-2153
http://dx.doi.org/10.1016/B978-0-12-397920-9.00003-2

But is aging a causal factor? In other words, would slowing the aging process be a means to delay or prevent the onset or progression of aging pathologies? This is a critical question to address given that chronic diseases have emerged as the major causes of mortality and the global disease burden. Much of the evidence in favor of targeting aging has emerged from interventions that impact mouse aging.

Recent studies in mice have revealed that several aspects of aging are accelerated or delayed by single gene mutations (Liang et al., 2003; Selman & Withers, 2011; Yuan, Peters, & Paigen, 2011). These mutations include naturally occurring spontaneous mutations, genetically targeted mutations, and exogenous genes introduced into the mouse genome by using modern genetic and genomic techniques. Mouse models with increased longevity have become accepted approach to identify genes and molecular mechanisms that play a crucial role in modulation of aging.

In contrast, mouse models with a shortened lifespan can also result from intrinsic genetic deficits. A criterion for proposing that short-lived mouse models have accelerated aging is that multiple age-associated pathologies should be evident, not just that the mouse has a reduced lifespan. These models are termed progeroid and in some cases involve mutations in the same genes that have been linked to human progeria syndromes (Burtner & Kennedy, 2010; Cox & Faragher, 2007; Kudlow, Kennedy, & Monnat, 2007). It remains highly debatable, however, whether and to what extent the molecular events leading to progeria overlap with those driving normal aging (Hasty & Vijg, 2004; Kipling, Davis, Ostler, & Faragher, 2004; Miller, 2004). One goal of this review is to address this issue through analysis of the available mouse models of longevity and progeria.

Here, we summarize published reports using any intervention describing long-lived mouse models (Table 5.1, Suppl. Table 1 for more detailed information, http://dx.doi.org/10.1016/B978-0-12-397920-9.00003-2) and short-lived ones where a progeria phenotype was described (Table 5.2). We compare their phenotypes and discuss the potential common genes and pathways that modulate aging processes, as well as the extent to which they are shared between normal aging and progeria. While many aspects of aging remain to be uncovered, the aging field is beginning to reach consensus that the rate of aging can be modified and furthermore that interventions enhancing lifespan can delay the onset and/or progression of a range of chronic disease syndromes. These findings raise a unique possibility that derives from aging research—that by targeting aging pathways it will be possible to extend human healthspan.

Table 5.1 Long-lived mouse models

Signaling	Target gene or mouse model	References
Somatotropic (GH/IGF-1) axis	Ames Dwarf (*Prop1*$^{df/df}$)	Brown-Borg, Borg, Meliska, and Bartke (1996)
	Snell dwarf (*Pit1*$^{dwJ/dw}$, *Pit1*$^{dw/dw}$)	Flurkey, Papaconstantinou, Miller, and Harrison (2001) and Flurkey, Papaconstantinou, and Harrison (2002)
	Little mice (*ghrhr*$^{lit/lit}$)	Flurkey et al. (2001)
	GHR/BP$^{-/-}$	Coschigano, Clemmons, Bellush, and Kopchick (2000) and Coschigano et al. (2003)
	Irs1$^{-/-}$	Selman, Lingard, Choudhury, et al. (2008), Selman, Lingard, Gems, Partridge, and Withers (2008), and Selman, Partridge, and Withers (2011)
	Irs2$^{+/-}$, *bIrs2*$^{+/-}$, *bIrs2*$^{-/-}$	Taguchi, Wartschow, and White (2007)
	GHRH-KO	Sun et al. (2013)
	Igf1r$^{+/-}$	Holzenberger et al. (2003) Xu et al. (2013)
	bIGF1RKO$^{+/-}$	Kappeler et al. (2008)
	IGF-1 Tg	Li and Ren, (2007)
	FIRKO	Bluher, Kahn, and Kahn (2003)
	p110α$^{D933A/WT}$	Foukas et al. (2013)
	Pten^{tg}	Ortega-Molina et al. (2012)
	Akt$^{+/-}$	Nojima et al. (2013)
mTOR signaling	*S6K1*$^{-/-}$	Selman et al. (2009)
	Mtor$^{+/-}$ *Mlst8*$^{+/-}$	Lamming et al. (2012)
	mTOR$^{\Delta/\Delta}$	Wu et al. (2013)
	Atg5 Tg	Pyo et al. (2013)
Sirtuins	BRASTO	Satoh et al. (2013)
	Sirt6-tg	Kanfi et al. (2012)

Continued

Table 5.1 Long-lived mouse models—cont'd

Signaling	Target gene or mouse model	References
Oxidative stress	$p66^{sch-/-}$	Migliaccio et al. (1999)
	MCAT	Schriner et al. (2005)
	$Gpx4^{+/-}$	Ran et al. (2007)
	SOD-2 overexpression	Hu et al. (2007)
	mGsta4 null	Singh et al. (2010)
	hMTH1-Tg	De Luca et al. (2013)
	TRX-Tg	Mitsui et al. (2002)
	$Tg(TRX1)^{+/0}$	Perez et al. (2011)
	$mclk^{+/-}$	Liu, Jiang, et al. (2005), Liu, Wang, et al., 2005, and Lapointe, Stepanyan, Bigras, and Hekimi (2009)
Tumor suppressor	*Sp53/Sp16/Sarf/TgTert*	Tomas-Loba et al. (2008)
Metabolism	Hcrt-UCP2	Conti et al. (2006)
	UCP	Gates et al. (2007)
	$Dgat1^{-/-}$	Streeper et al. (2012)
	$RII\beta^{-/-}$	Enns et al. (2009)
	β/β	Chiu, Lin, Huang, and Lee (2004)
Inflammation	^{DN}IKK-α, $N/Ikbkb^{l/l}$	Zhang, Li, et al. (2013)
Hormones	Klotho overexpression	Kurosu et al. (2005)
	FGF-21-Tg	Zhang et al. (2012)
	PAPP-A KO	Conover and Bale (2007), Conover et al. (2010a), and Conover et al. (2010b)
Genomic integrity	BubR1 transgenic	Baker et al. (2013)
Others	MIF-KO	Harper et al. (2010)
	AC5-KO	Yan et al. (2007)

Table 5.1 Long-lived mouse models—cont'd

Signaling	Target gene or mouse model	References
	MT	Yang et al. (2006)
	Agtra$^{-/-}$	Benigni et al. (2009)
	αMUPA	Miskin and Masos (1997)
	Surf1^{loxP}$-/-$	Dell'agnello et al. (2007)
	ETAKO	Ceylan-Isik et al. (2013)
	AgRP$^{-/-}$	Redmann and Argyropoulos (2006)

Wild-type (WT; +/+), heterozygote (+/−), and null (−/−).

Table 5.2 Short-lived mouse models

Signaling	Target gene or mouse model	References
Lamin A maturation	*Zmpste24*$^{-/-}$	Bergo et al. (2002) Pendas et al. (2002)
	Lmna$^{H222P/H222P}$	Arimura et al. (2005)
	Lmna$^{G609G/G609G}$	Osorio et al. (2011)
	Lmna$^{L530P/L530P}$	Mounkes, Kozlov, Hernandez, Sullivan, and Stewart (2003)
	Lmna$^{-/-}$	Sullivan et al. (1999)
Somatotropic (GH/IGF-1) axis	bGH-Tg	Bartke (2003)
	hg/hg	Casellas and Medrano (2008)
DNA repair	*Xpd*$^{R722W/R722W}$	de Boer et al. (2002)
	ku86$^{-/-}$	Vogel, Lim, Karsenty, Finegold, and Hasty (1999)
	Ercc1$^{-/-}$	Niedernhofer et al. (2006)
	Csb$^{m/m}$/*Xpa*$^{-/-}$	van der Pluijm et al. (2007)
DNA replication	*top3β*$^{-/-}$	Kwan and Wang (2001)
	PolgA^{mut}/*PolgA*^{mut}	Trifunovic et al. (2004)
	Bub1b$^{H/H}$	Baker et al. (2004)
	Terc$^{-/-}$ *Atm*$^{-/-}$	Wong et al. (2003)

Continued

Table 5.2 Short-lived mouse models—cont'd

Signaling	Target gene or mouse model	References
Oxidative stress	*Prdx1*$^{-/-}$	Neumann et al. (2003)
	Sod1$^{-/-}$	Elchuri et al. (2005)
	Sod2$^{-/-}$	Li et al. (1995)
	MsrA$^{-/-}$	Moskovitz et al. (2001)
	p62$^{-/-}$	Kwon et al. (2012)
Sirtuins	*SIRT1*$^{-/-}$	Li, Xu, McBurney, and Longo (2008) Boily et al. (2008) Mercken et al. (2014)
	Sirt6$^{-/-}$	Mostoslavsky et al. (2006)
Hormone	*kl/kl*	Kuro-o et al. (1997)
Inflammation	^{CA}IKK-β	Zhang et al. (2014), Zhang, Kieckhaefer, and Cao (2013), and Zhang, Li, et al. (2013)
Tumor suppression	*p53*$^{+/m}$	Tyner et al. (2002)
	mTR$^{-/-}$	Rudolph et al. (1999)
	AIMP3 Tg	Oh et al. (2010)
	SnoN$^{m/m}$	Pan, Zhu, Conboy, Conboy, and Luo (2012)

2. MOUSE MODELS ANALYZING THE GH/IGF-1 AXIS

Loss-of-function mutations in the growth hormone/insulin-like growth factor-1 (GH/IGF-1) axis are the longest living mouse mutants discovered to date (Fig. 5.1). In mammals, spontaneous mutations that result in growth hormone (GH) deficiency, such as Ames (*Prop1*^{df}) and Snell (*Pit*^{dw}) dwarf mice (Brown-Borg et al., 1996; Flurkey et al., 2002), GH-releasing hormone (GHRH) defective little mice (*Ghrhr*$^{lit/lit}$) (Flurkey et al., 2001), GH receptor/binding protein null mice (*Ghr/bp*$^{-/-}$) (Coschigano et al., 2000, 2003), display dwarfism with reduced levels of IGF-1 and insulin serum concentrations, suppressed GH/IGF-1 signaling, and a significant increased lifespan compared to controls. Recently, mice with targeted

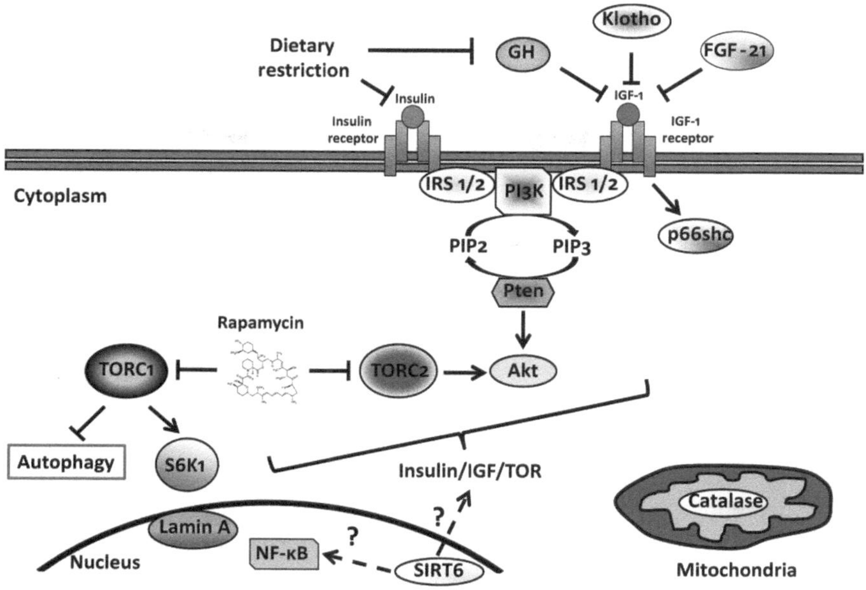

Figure 5.1 Major molecular pathways that regulate murine lifespan. The schematic represents an overview of the signaling pathways regulating lifespan. NF-κB, nuclear factor-κB; IRS-1/2, insulin receptor substrate-1/2; PI3K, phosphatidylinositol 3-kinase. (See the color plate.)

disruption of growth hormone-releasing hormone (GHRH-KO) were also reported as long lived (Sun et al., 2013).

While the actual survival period (lifespan) is perhaps the most direct measure of aging, this measure alone is generally not considered sufficient to conclude that aging itself is being delayed and increasing attention is being turned to healthspan, the functional and chronic disease-free period of life. Manipulations that reduce signaling through the GH/IGF-1 axis, at least to some extent meet these criteria. For example, Ames Dwarf mice have a lower incidence and delayed occurrence of lung adenocarcinoma (Ikeno, Bronson, Hubbard, Lee, & Bartke, 2003). GHR-BP knockout and GH-deficient mice have low-circulating IGF-1 levels, a lower incidence and delayed occurrence of tumors, increased insulin sensitivity, and a reduction in age-dependent cognitive impairment (Coschigano et al., 2000; Ikeno et al., 2009). A more complete description of these phenotypes can be found in a comprehensive review (Bartke, Sun, & Longo, 2013).

Conversely, overexpression of transgenic GH in mice (such as bGH-Tg) (Bartke, 2003; Wolf et al., 1993) and GH treatment in long-lived dwarf mice (Panici et al., 2010) reduces longevity. In another interesting mouse model, the high-growth (HG) phenotype mice (*hg/hg*) are characterized by a 30–50% increase in postweaning weight, without increased adiposity, and is due to a 500 kb deletion on chromosome 10 that includes the *suppressor of cytokine signaling* (*Socs2*) gene (Casellas & Medrano, 2008). Mice with the *hg* deletion, either on a C57BL/6J or a FVB background, showed a substantial increase in plasma IGF-1 levels and a shortened lifespan.

Homozygous deletion of the insulin-like growth factor type 1 receptor (IGF1R) leads to embryonic lethality (Liu, Baker, Perkins, Robertson, & Efstratiadis, 1993). However, heterozygous IGF1R mice ($Igf1r^{+/-}$) are longer-lived, specifically females and resistant to oxidative stress (Holzenberger et al., 2003). These mice display partial IGF-1 resistance, as indicated by higher serum IGF-1 levels compared to controls. However, a later contradictory study found that $Igf1r^{+/-}$ mice are not long-lived compared to control littermates (Bokov et al., 2011). This discrepancy may be due to differences in husbandry, given the mean lifespan of control females was improved from 568 days in Holzenberger et al. (2003) to 923 days in Bokov et al. (2011). The differences may also be due to the genetic background of $Igf1r^{+/-}$ mice because the mice used by Holzenberger et al. (2003) were on 129/J background, while those used by Bokov et al. (2011) were on C57BL/6 background. However, a recent report showed significant lifespan extension in female $Igf1r^{+/-}$ mice on a long-lived C57BL/6J mouse background (Xu et al., 2013), but the effect was smaller than previously published (Holzenberger et al., 2003). At present, the story still remains unresolved since another recent report conducted at three independent sites showed that a reduction in IGF-1 levels did not extend mean lifespan (Lorenzini et al., 2013). Nevertheless, circulating IGF-1 levels have been inversely correlated with median lifespan among 31 inbred mouse strains (Yuan et al., 2009), indicating that mice with lower IGF-1 levels tend to live longer than those with higher levels.

The effect of altering IGF-1 levels on lifespan may also be dependent on reduction of IGF signaling in specific tissues. For example, mice heterozygous for the IGF-1R in the central nervous system are hyposensitive to IGF-1 in the brain (bIGF1RKO$^{+/-}$) and have an increased lifespan (Kappeler et al., 2008). Similarly, mice bearing a fat-specific insulin receptor knockout (FIRKO mice) maintain glucose sensitivity with age (Bluher et al., 2002). Most interestingly, the lifespan of the FIRKO mice was significantly longer than that in control littermates (Bluher et al., 2003).

Global deletion of insulin receptor substrate 1 (*Irs1*), a key downstream mediator of insulin/IGF-1 signaling, increases lifespan in mice (*Irs1*$^{-/-}$ mice) (Selman, Lingard, Choudhury, et al., 2008; Selman et al., 2011). Similarly, mice heterozygous for a null mutation in the insulin receptor substrate 2 (Irs2, *Irs2*$^{+/-}$ mice) display a 17% increase in median lifespan (Taguchi et al., 2007). However, Selman, Lingard, Choudhury, et al. (2008) and Selman, Lingard, Gems, et al. (2008) using the same mouse model found no lifespan extension for reasons that are uncertain. Of note, male and female homozygous *Irs2*$^{-/-}$ mice are short-lived because of their diabetic phenotypes (Selman, Lingard, Choudhury, et al., 2008). Together, despite some contradictory findings, these studies suggest that reduced insulin receptor substrate levels are associated with enhanced longevity.

Consistent with aforementioned long-lived mouse models with decreased insulin/IGF-1 signaling, other genetic mutations in the insulin/IGF-1 pathway affect longevity. For instance, *Pten*-overexpressing mice (*Pten*^{tg}) (Ortega-Molina et al., 2012), hypomorphic PI3K mice (P110$\alpha^{D933A/WT}$) (Foukas et al., 2013), and hypomorphic Akt mice (Akt$^{+/-}$) (Nojima et al., 2013), all show extended lifespans. *Pten*^{tg} mice carrying an additional genomic copy of tumor suppressor *Pten* are protected from cancer and present a significant extension of lifespan that is independent of their lower cancer incidence. *Pten*^{tg} mice also exhibit a reduction of the insulin/IGF-1 signaling and increased energy expenditure that is associated with improved mitochondrial oxidative metabolism, as well as enhanced brown adipose activity. Male mice, heterozygous for P110$\alpha^{D933A/WT}$ mutation (homozygotes are embryonic lethals) (Foukas et al., 2006), with chronic p110α partial inactivation showed an extended lifespan through a mechanism likely associated with improved glucose homeostasis and reduced adiposity (Foukas et al., 2013). Haploinsufficiency for *Akt1* (Akt$^{+/-}$) also significantly promotes lifespan by a mechanism that reduces ribosomal biogenesis and mitochondrial DNA content, as well as oxidative stress (Nojima et al., 2013).

Altered levels of the Klotho protein are associated with both increased and decreased longevity. Klotho is a transmembrane protein expressed mainly in the distal convoluted tubules in the kidney and the choroid plexus in the brain. Mice bearing an insertional hypomorphic mutation in the *Klotho* gene (*kl/kl* mice) have accelerated human age-related traits, including arteriosclerosis, skin atrophy, and osteoporosis (Kuro-o et al., 1997). The *kl/kl* mice with accelerated age phenotypes die prematurely at ~8–9 weeks of age. Interestingly, all disorders observed in *kl/kl* mice were improved simply by exogenous *kl* gene expression. This rescue experiment supports the notion that *kl* is the gene responsible for accelerated aging and premature death.

Subsequently, it was determined that Klotho functions as a circulating hormone that binds to a cell-surface receptor and represses intracellular insulin/IGF-1 signals (Kurosu et al., 2005). The Klotho hormone inhibits IGF-1 signaling by binding to a cell-surface Klotho receptor that inhibits FOXO phosphorylation, promoting FOXO nuclear translocation. Interestingly, mice-overexpressing *Klotho* are insulin/IGF-1 resistant and resistant to oxidative stress but are normal in size (Kurosu et al., 2005). In parallel with mouse models that are deficient in insulin/IGF-1 signaling, such as $Igf1r^{+/-}$ and $Irs1^{-/-}$, mice-overexpressing *Klotho* have a longer lifespan than their littermates (Kurosu et al., 2005). These studies imply that the Klotho protein functions as an antiaging hormone.

3. MOUSE MODELS ANALYZING TARGET OF RAPAMYCIN SIGNALING

Target of rapamycin (TOR) signaling has emerged as a promising target for longevity because of its functions at the regulatory crossroad between nutrient sensing, energy metabolism, stress responses, and developmental growth (Johnson, Rabinovitch, & Kaeberlein, 2013). mTOR signaling is involved in most age-related diseases and rapamycin treatment shows efficacy in animal models with a wide range of chronic disease conditions (Johnson et al., 2013). Numerous reports suggest that decreasing mammalian TOR (mTOR) signaling, pharmacologically or genetically, extends lifespan in mice, as is the case for invertebrate models and yeast (see review, Johnson et al., 2013). Rapamycin, a FDA-approved inhibitor of the mammalian target of rapamycin (mTOR), is the only compound that has been repeatedly shown to extend lifespan in many different mouse strains. Treatment with rapamycin of genetically heterogeneous mice (UMHET3), beginning at 20 months of age, results in median and maximum lifespan extension by 10–15% (Harrison et al., 2009). Similar effects were seen when the same dose of rapamycin is provided starting at 9 months of age, suggesting that modulating mTOR activity may be more important later in life (Miller et al., 2011).

Even though the end-of-life pathology was not reported in the initial studies, more recent work strongly suggests that rapamycin delays multiple aspects of aging. For instance, rapamycin delays spontaneous cancer formation, both in normal (Anisimov et al., 2011) and cancer-prone mice

(Anisimov, Zabezhinski, et al., 2010). Other aspects, including alterations in heart, liver, adrenal gland, endometrium, and tendons, as well as an age-dependent decline in spontaneous activity, appear to be delayed (Flynn et al., 2013; Neff et al., 2013; Wilkinson et al., 2012). However, rapamycin does not delay all aspects of aging in mice and some conditions, such as cataract formation and testicular degeneration, appear to be accelerated (Neff et al., 2013; Wilkinson et al., 2012).

There are concerns with chronic rapamycin administration, as rapamycin is used clinically as an immunosuppressant and chronic treatment substantially impairs glucose tolerance and insulin action, an indicator for nascent diabetes (Lamming et al., 2012). At least in the metabolic context, a recent study has suggested that these side effects result from inhibition of the mTORC2 complex in the liver (Hughes & Kennedy, 2012; Lamming et al., 2012). Since mTORC1 inhibition is generally associated many of the efficacious aspects of rapamycin treatment, including lifespan extension, these findings suggest that drugs inhibiting mTORC1 and not mTORC2 would be of interest.

Several mouse genetic models involving altered mTOR signaling have been generated over the last decade (see Table 5.1). The S6 kinase is one of the major substrates of mTORC1 that in turn controls protein translation and other cellular processes (Laplante & Sabatini, 2012). Deletion of S6K1, one of two S6 kinase paralogs, extends female lifespan (Selman et al., 2009). There also was a trend toward increased lifespan in males as well, although it did not reach significance. Similar longevity effects have been observed in yeast, worms and flies lacking S6 kinase, indicating that this protein is a conserved modulator of longevity.

Eukaryotic initiation factor 4E binding protein (4E-BP), another substrate of mTORC1, is involved in regulating control of mRNA translation and protein synthesis (Laplante & Sabatini, 2012). mTORC1-mediated phosphorylation leads to inactivation of 4E-BP, which prevents it from inhibiting eIF4E activity. In *Drosophila*, 4E-BP is upregulated upon dietary restriction (DR) and is required for DR-mediated lifespan extension, as well as changes in mitochondrial activity (Zid et al., 2009). Mouse studies indicating whether increased 4E-BP activity is associated with enhanced longevity have yet to be reported.

As indicated above, mTOR is associated with two independent protein complexes, mTORC1 and mTORC2 (Alayev & Holz, 2013). Recently, mice with two hypomorphic alleles ($mTOR^{\Delta/\Delta}$ mice) that express mTOR at approximately 25% of wild-type levels were found to live longer than control littermates (Wu et al., 2013). As expected, these mice demonstrated reduced mTORC1 and mTORC2 activity and exhibited an approximately 20% increase in median lifespan. A more specific genetic targeting of the mTORC1, also lead to lifespan extension without interfering with glucose metabolism (Lamming et al., 2012). Female mice doubly heterozygous for *Mtor*$^{+/-}$ *Mlst8*$^{+/-}$ (a component of the mTORC1 complex) also had significant lifespan extension.

One of cellular functions mediated by mTORC1 signaling is autophagy (Laplante & Sabatini, 2012); active mTORC1 leads to suppression of autophagy through several mechanisms and inhibition of mTORC1 enhances autophagy (Fok et al., 2013; Zhang, Li, et al., 2013). Several reports have suggested that increased autophagy is necessary for lifespan extension in response to mTOR inhibition or DR. A recent study examined the longevity effects of enhancing autophagy through a more direct mechanism. Overexpression of Atg5, which encodes a protein essential for autophagosome formation, extends lifespan (Pyo et al., 2013). Conversely, mice lacking the p62 protein, which is important for removal of damaged protein aggregates and macromolecular complexes, exhibit phenotypes consistent with accelerated aging. Interestingly, tissues from these mice show elevated oxidative stress due to alterations in mitochondrial electron transport (Kwon et al., 2012).

4. MOUSE MODELS ANALYZING SIRTUIN FUNCTION

The family of Sirtuin (Sirt) genes encode multifunction enzymes that act as protein deacetylases (Imai, Armstrong, Kaeberlein, & Guarente, 2000), and ribosyltransferases (Tanny, Dowd, Huang, Hilz, & Moazed, 1999), that are involved in a variety of cellular functions ranging from genome maintenance to metabolism (Chang & Guarente, 2014). To date, seven human sirtuins (SIRT1–7) have been identified, with SIRT1 being the most extensively studied (Haigis & Sinclair, 2010). Sirt1, the homolog of Sir2 in mammals, localizes to the nucleus and involved in metabolism and stress response pathways (Guarente, 2013). These enzymes use nicotine adenine dinucleotide as a substrate for the deacetylase reaction. Moreover, resveratrol and other STACs (Sirtuin-activating compounds) have been extensively characterized. These molecular entities, which appear to be substrate-specific Sirt activators, have been reported as DR mimetics (Agarwal & Baur, 2011; Baur, 2010). Although *Sirt1*-overexpressing transgenic mice display phenotypes similar to DR mice (Bordone et al., 2007), they failed to show lifespan extension (Herranz et al., 2010). However, a more recent study demonstrated that brain-specific *Sirt1*-overexpressing (BRASTO) transgenic mice show significant lifespan extension in both males and females (Satoh et al., 2013). These phenotypes were attributed to increased SIRT1 hypothalamic activity in regulating tissue communication, potentially through altered metabolism and improved sleep quality. Conversely, mice lacking SIRT1 have been repeatedly shown to have a reduced lifespan under normal chow from different labs (Boily et al., 2008; Li et al., 2008; Mercken et al., 2014).

SIRT6 is a highly specific histone 3 deacetylase that targets H3K9 and H3K56 and plays an important role in DNA repair, telomerase function, genomic stability, and cellular senescence (Michishita et al., 2008; Mostoslavsky et al., 2006; Tennen & Chua, 2011). Recent studies link SIRT6 both to enhanced longevity and progeria (Liao & Kennedy, 2012). Mice lacking SIRT6 exhibit phenotypes resembling premature aging, including lordokyphosis, metabolic dysfunction, bone mineral density defects similar to osteoporosis, spinal curvature abnormalities, loss of subcutaneous fat, lymphocyte attrition, and severe metabolic defects with precipitous drops in serum glucose (Mostoslavsky et al., 2006). SIRT6 overexpression in male, leads to protection from diet-induced obesity and enhanced longevity, although the latter effect is seen only in males (Kanfi et al., 2012, 2010).

Possible connections exist between enhanced SIRT6 activity and the insulin/IGF/TOR pathways. For instance, long-lived SIRT6 transgenic males displayed lower levels of IGF-1 and IGF-1 signaling, higher levels of IGF-binding protein 1, and reduced mTORC1 and mTORC2 signaling. In contrast, SIRT6 transgenic females were not long-lived and did not show these phenotypes (Kanfi et al., 2012). SIRT6 deficiency causes metabolic imbalances, including postnatal growth retardation and premature death by 1 month (Mostoslavsky et al., 2006).

Notably, SIRT6 activity links to the transcription factor NF-κB, a key regulator of inflammation, discussed below. SIRT6 directly interacts with NF-κB and antagonizes NF-κB-induced gene expression, which has been proposed to accelerate aging. Suppression of NF-κB signaling partially restores lifespan and improves growth in the short-lived *Sirt6*$^{-/-}$ background (Kawahara et al., 2009). This study indicates that SIRT6 attenuates NF-κB signaling via H3K9 deacetylation and hyperactive NF-κB signaling may contribute to both premature and/or normal aging.

5. MOUSE MODELS OF INFLAMMATION

An emerging modulator of aging-related pathways in mammals is the NF-κB family of transcription factors, which control the activity of genes involved in apoptosis, cell senescence, inflammation, and immunity (Hayden & Ghosh, 2008). Molecularly, NF-κB activation is linked with many of the well-known longevity regulators and pathways, including insulin/IGF-1, FOXO, SIRT, mTOR, and DNA damage (Tilstra, Clauson, Niedernhofer, & Robbins, 2011). By modulating NF-κB activity in mice, a recent study illustrates the importance of NF-κB in mammalian aging (Zhang et al., 2014). Mice expressing a constitutively active IKK-β that stimulated NF-κB activity in the hypothalamus showed a reduction in longevity, whereas expression of dominant-negative IκB-α, which inhibits NF-κB, prolonged their lifespan, as did a hypothalamus-specific deletion of IKK-β (Zhang et al., 2014). This study suggests that NF-κB or the signaling mediated by NF-κB is one of the factors that both regulate short and long lifespans (see other sections for more about NF-κB in mammalian aging). Further studies of this pathway are clearly warranted.

6. MOUSE MODELS OF DR OR CALORIE RESTRICTION

Life expectancy is profoundly affected by caloric intake. DR (also known as calorie restriction, CR) is the most robust nongenetic approach to extend both mean and maximal lifespan in many organisms, including yeast, worms, flies, fish, rodents, and rhesus monkeys (Fontana, Partridge, & Longo, 2010; Masoro, 2005). In contrast, overnutrition (such as high-fat diet) is associated with a reduction in lifespan. In addition to extending lifespan, DR also slows the progression of most age-related changes simultaneously (Masoro, 2005; Weindruch & Walford, 1988); a wide variety of age-related pathological lesions are also significantly delayed and/or reduced in mouse models under DR. For instance, DR has been reported to prevent or delay the incidence of numerous age-related diseases, including type II diabetes, cancers, and cardiovascular disease. Variation in the DR response is observed in different mouse strains (Liao, Rikke, Johnson, Diaz, & Nelson, 2010), as well as different yeast genetic backgrounds (Schleit et al., 2013) and even in primates. One rhesus monkey study provided evidence that DR extends lifespan (Colman et al., 2009), while a more recent study reported that DR had no effect on animals' survival (Mattison et al., 2012). Even though longevity studies in primates have yielded conflicting results, DR did indeed reduce disease incidence in both cases (Colman et al., 2009; Mattison et al., 2012).

In 1935, Clive McCay and colleagues reported that reducing food intake extends lifespan in rats (McCay, Crowell, & Maynard, 1935). Since then, the efforts of DR studies have been focused on elucidating the mechanism by which DR enhances lifespan and developing genetic models or pharmaceutical DR mimetics. At the molecular level, the beneficial effects by DR appear to be mediated by multiple nutrient-sensing pathways including insulin/IGF-1, mTOR, and Sirt (Barzilai, Huffman, Muzumdar, & Bartke, 2012; Kenyon, 2010). Given the multiplicity of biological and physiological changes associated with DR, it is difficult to distinguish cause and effect. Applying DR in combination with genetic models or pharmacologic interventions can lead to identification of key metabolic regulators of lifespan (Curtis & de Cabo, 2013). For example, long-lived GHRKO mice subjected to DR feeding resemble that of long-lived GHRKO mice under *ad libitum* feeding (Bonkowski, Rocha, Masternak, Al Regaiey, & Bartke, 2006). Insulin sensitivity in GHRKO mice was greater than in normal mice and was not further increased by DR. This finding supported the notion that enhanced insulin sensitivity, one typical trait of DR mice, plays a prominent role in life extension. Interestingly, DR did increase the lifespan of long-lived GHRH-KO mice (Sun et al., 2013), suggesting that mechanisms underlying extended longevity in GHRH-KO mutants and DR mice may not overlap. The longevity effects of rapamycin and DR also do not completely overlap. In a study of short-term treatment with rapamycin and DR (40% DR), similar effects were seen such as enhanced autophagy, but many other responses differ including the levels of insulin sensitivity, and expression of a number of genes in pathways potentially linked to aging (Fok et al., 2013). Further elucidating the differential mechanisms influencing mouse longevity between genetic and DR interventions will likely lead to a better understanding of the DR response.

Metformin is also touted as a DR mimetic (Ingram et al., 2006). A recent study showed that metformin, a clinically approved drug used to treat both type II diabetes and cancer, modestly improves lifespan in mice (Martin-Montalvo et al., 2013). Metformin is most recognized for its ability to activate nutrient-responsive AMP kinase (Zhou et al., 2001), although other functions have been identified. The extent to which the longevity benefits of metformin come from tumor prevention remain to be fully determined, but this is possible given the known roles of the drug in cancer. For instance, metformin delayed tumor formation in cancer-prone mice (Anisimov et al., 2005; Anisimov, Egormin, et al., 2010).

So far, it is well known that inhibiting pathways promoting nutrient uptake or growth, or activating pathways helping the organism respond to stress or a shortage of food, can extend lifespan in animals. Hence, any proteins or hormones that can be activated by these kinds of stressors could, in theory, be used as a surrogate to elicit enhanced longevity. For example, fibroblast growth factor-21 (FGF-21) is an endocrine hormone secreted by the liver during fasting that elicits diverse aspects of the adaptive starvation response. One report indicated that mice, both males and females, with liver-specific overexpression of FGF-21 (FGF-21-Tg mice) are remarkably long lived (Zhang et al., 2012). FGF-21-Tg mice also share a number of phenotypes with long-lived mice including small size, enhanced insulin sensitivity, and a blunted GH/IGF-1 signaling axis (Inagaki et al., 2008; Kubicky, Wu, Kharitonenkov, & De Luca, 2012). Since FGF-21 activates known longevity pathways, it is possible that this hormone might also confer resistance against chronic diseases of aging as well. However, not all age-related changes are ameliorated, as 33- to 35-month-old FGF-21-Tg mice have lower bone density than wild-type littermates.

7. MOUSE MODELS OF PROGERIA

A number of mouse models have been created through genetic manipulation, which exhibit evidence, with varying degrees of severity, of accelerated or premature aging (termed progeroid). It remains highly debated whether genetic diseases that recapitulate aspects of aging at an accelerated pace can provide insights into the normal human aging process (Burtner & Kennedy, 2010). We list these genes (Table 5.2) whose mutations decrease longevity and appear to accelerate aging. Mouse models for Hutchinson–Gilford progeria syndrome (HGPS) (Schreiber & Kennedy, 2013; Stewart, Kozlov, Fong, & Young, 2007; Zhang et al., 2013), as well as other progeria models (Burtner & Kennedy, 2010), have been reviewed in depth. Here, we discuss some of those progeria mouse models, including their use to identify therapeutic approaches.

Among the mutations that are associated with progeria are dominant mutations in *LMNA* (encoding A-type nuclear lamins), which cause HGPS (De Sandre-Giovannoli et al., 2003; Eriksson et al., 2003; Mounkes et al., 2003). Children with HGPS show signs of accelerated aging, including chronic hair loss, joint pains, thinning of skin, and cardiovascular problems. Since 2003, a range of *Lmna* models have been generated to elucidate the molecular mechanisms underlying this rare disease and to find therapies. During protein maturation, prelamin A is processed to lamin A, a constituent of the nuclear lamina. Lamin A protein contains a unique C-terminus that ends in the residues CaaX (a = aliphatic), which result in farnesylation of the cysteine residue. Prelamin A is transiently farnesylated, after which the endopeptidase ZMPSTE24 removes the C-terminal 18 amino acids, including the CaaX motif, to generate mature lamin A. HGPS is most often caused by a point mutation in the *LMNA* gene that activates a cryptic splice donor site, yielding a truncated form of prelamin A called progerin (De Sandre-Giovannoli et al., 2003; Eriksson et al., 2003; Mounkes et al., 2003). Progerin, which lacks 50 amino acids as well as the internal ZMPSTE24 cleavage site, remains permanently farnesylated. Permanent farnesylation leads to progerin accumulation at the nuclear periphery and loss within internal regions of the nucleus. Although A-type lamins normally exist at the periphery, where they form a latticework-like intermediate filament structure along with B-type lamins, progerin accumulation introduces structure alterations that precipitate nuclear deformation as well as other malfunctions in a genetically dominant fashion (Dechat et al., 2008). Loss of lamin A in the interior may also contribute to defects in nuclear function(s). Inhibition of lamin A farnesylation with farnesyl-transferase inhibitors, modestly reduces the consequent pathology and extends survival of multiple mouse progeria models and is now undergoing clinical trials with HGPS patients (Gordon et al., 2012).

Mice with inactivating mutations in *ZMPSTE24* that result in permanent farnesylation of prelamin A give rise to a set of progeroid phenotypes due to the inability to process prelamin A. These overlap with mice bearing direct HGPS-associated mutations in *Lmna* and can be partly rescued by deletion of one copy of the *Lmna* gene (Navarro et al., 2004; Pendas et al., 2002).

Of note, complete loss of lamin A appears to lead to early postnatal lethality in mice and a multitude of severe, largely missense mutations result in a distinct set of phenotypes including dilated cardiomyopathy and muscular dystrophy. These phenotypes, which also model human laminopathies, will not be discussed in detail here (Stewart et al., 2007; Zhang et al., 2013).

Mice lacking Zmpste24 recapitulate many of the progeroid phenotypes in HGPS, including growth retardation, loss of adipose tissue, abnormality in nuclear architecture, and premature death, as well as phenotypes more associated with reduced A-type lamin function, including dilated cardiomyopathy and muscular dystrophy (Pendas et al., 2002). Most *Zmpste24*$^{-/-}$ mice die relatively early, having an average lifespan of 20 weeks. This mouse model has provided the link between progeria and GH/IGF-1 signaling, a key regulatory signaling in lifespan (see long-lived mouse models). *Zmpste24*$^{-/-}$ mice present profound transcriptional alternations in genes that regulate the somatotroph axis, together with extremely high circulating levels of GH and a reduction in IGF-1. IGF-1 treatment improved the survival of *Zmpste24*$^{-/-}$ mice, but only modestly (Marino et al., 2010). Whereas reduced IGF-1 signaling is often associated with lifespan extension, the opposite is appears to be the situation here; low IGF-1 signaling may promote progeroid phenotypes in this context. The reasons for this contradiction are unknown but may relate to a developmental requirement for IGF-1 signaling (particularly in a growth-retarded HGPS model) and/or are related to the levels of inhibition. For instance, modest reductions in IGF-1 signaling may be prolongevity, but severe reductions may be detrimental.

Recently, resveratrol, a potential DR mimetic, was shown to extend the lifespan in *Zmpste24*$^{-/-}$ mice (Liu et al., 2012). Resveratrol has been ascribed several activities, but the most studied involves its ability to activate the protein deacetylator, SIRT1, toward specific substrates (Baur, 2010; Baur & Sinclair, 2006; Hubbard et al., 2013). In these studies, A-type lamins were reported to mediate SIRT1 attachment to the nuclear matrix, promoting its activity. The presence of prelamin A or progerin was shown to interfere with SIRT1 binding and activity in a manner partly suppressed by resveratrol. In the mouse *Zmpste24*$^{-/-}$ model, resveratrol reversed the defects in some adult stem cell populations. Given that resveratrol is a natural product and that other more potent STACs exist, more studies of the links between A-type lamins and Sirt need to be performed.

Zmpste24 deficiency also elicits stress signaling, as demonstrated by a marked upregulation of p53 target genes (Varela et al., 2005). However, the aging phenotypes in *Zmpste24*$^{-/-}$ mice were partially reversed in Zmpste24 and p53 double knockout (*Zmpste24*$^{-/-}$ *p53*$^{-/-}$) mice (Varela et al., 2005).

Studies in *Zmpste24*$^{-/-}$ mice link progeria with genomic stability and DNA repair, since increased DNA damage and genomic instability are evident in this model (Liu, Jiang, et al., 2005). This genomic instability is associated with H4K16 hypoacetylation (Krishnan et al., 2011) and an impaired DNA damage response, which is linked to ATM-Kap-1-mediated chromatin remodeling (Liu, Wang, Ghosh, & Zhou, 2013). *Zmpste24*$^{-/-}$ mice treated with sodium butyrate and trichostatin A, both of which increase global histone acetylation, restored recruitment of the checkpoint protein 53BP1, and extended their lifespan (Krishnan et al., 2011). Furthermore, genetic inhibition of NF-κB signaling, which is activated by DNA damage, prevents age-associated features in *Zmpste24*$^{-/-}$ mice (Osorio et al., 2012), as is the case for other Lmna progeric models (see below).

The primary human lesion associated with HGPS is a heterozygous G608G mutation, which creates a splicing defect and leads to permanently farnesylated lamin A. While a number of mouse models have been created that express progerin directly or unprocessable versions of lamin A and these models have progeroid features, knock-in *Lmna*^{G609G} mice have recently been created that bear the analogous mutation to the human disease and are rapidly becoming the model of choice (Osorio et al., 2011). *Lmna*$^{G609G/G609G}$ mice displayed growth rate reduction after 3 weeks, progressive weight loss caused by bone deformation, infertility, abnormal posture, marked curvature of the spine, cardiovascular and metabolic anomalies that closely approximate to the human phenotype. These mice have a mean lifespan of around 100 days. In the same report, *Lmna*$^{G609G/G609G}$ mice that were treated with a combination of two antisense oligonucleotides that blocked aberrant splicing displayed reduced amounts of accumulated progerin, enhanced lifespan, and reduced or delayed pathology in many aspects of HGPS. Interestingly, genetic inhibition of NF-κB signaling also delays age-associated features in *Lmna*$^{G609G/G609G}$ mice (Osorio et al., 2012).

AIMP3/p18 is one of three subunits of a scaffold complex that with aminoacyl tRNA synthetases form the multisynthetase complex (Park, Choi, & Kim, 2010). It also acts as a tumor suppressor, is required for p53 activation in response to oncogenic stimuli (Park et al., 2006) and when overexpressed, results in lamin A destabilization and a higher ratio of prelamin A and progerin to mature lamin A (Oh et al., 2010). Consistent with this observation, a transgenic mouse-overexpressing AIMP3 exhibits progeria, including phenotypes such as alopecia, wrinkled skin with reduced fat, and reduced bone minerality and thickness (Oh et al., 2010).

One of the key areas of research has focused on understanding whether the mutated lamin A protein (Progerin) and normal aging are linked. Evidence has been presented that progerin may be produced with aging in some tissues, suggesting that progerin production may drive aspects of normal physiological aging (Cao, Blair, et al., 2011; Scaffidi & Misteli, 2006). It will be of interest to determine whether reducing progerin expression can extend lifespan and/or delay age-related pathologies in normal mice.

Finally, rapamycin has been reported to decrease progerin levels by the induction of autophagy, improving cellular defects associated with progerin expression (Cao, Graziotto, et al., 2011). This study demonstrated that rapamycin, which extends the murine lifespan in many genotypes, decreases the amount of the progerin by 50%, decreased the incidence of abnormal nuclear shape, and delayed cellular senescence. However, it has been previously noted that mTORC1 activity may already be reduced in progeria. This would be consistent with reduced IGF-1 signaling and is supported by the observation from *Zmpste24*$^{-/-}$ mice of reduced S6 kinase phosphorylation, a target of mTORC1 (Marino et al., 2008). It is also notable that mice with reduced A-type lamin function show increased mTORC1 signaling in pathologic tissues, with rapamycin reducing the pathology and extending longevity. Clearly, the mechanistic links between A-type lamins and mTOR signaling require a better understanding with regard to the different laminopathies (Choi et al., 2012; Ramos et al., 2012).

8. MOUSE MODELS ANALYZING DNA REPAIR AND GENOMIC STABILITY

The time-dependent accumulation of cellular and DNA damage is widely observed with age and has been proposed as a general cause of aging (Perez, Bokov, et al., 2009; Sohal & Weindruch, 1996). To counteract the effects of DNA damage, an intricate network of DNA repair pathways is employed. Numerous progeroid syndromes are linked to defects in the cellular response to DNA damage, including Cockayne syndrome (CS), Werner syndrome (WS), ataxia telangiectasia, and trichothiodystrophy (TTD) (Hasty, Campisi, Hoeijmakers, van Steeg, & Vijg, 2003), suggesting that defective genome maintenance may contribute to accelerated aging.

Nucleotide excision repair (NER) is a multistep mechanism that removes lesions affecting one strand of DNA. Two subpathways exist: transcription-coupled NER and global genome NER. Many mouse models with loss-of-function mutations present with a wide variety of age-related phenotypes, as have been reviewed in depth (Cox & Faragher, 2007; Freitas & de Magalhaes, 2011; Schumacher, Hoeijmakers, & Garinis, 2009). For instance, mice carrying an *XPD* point mutation [Arg722→Trp (R722W)] found in TTD patients display many symptoms of premature aging, including osteoporosis, osteosclerosis, graying, cachexia, infertility, and reduced lifespan (de Boer et al., 2002). *Ku86*$^{-/-}$ mice, which are defective for double-stranded DNA break repair, exhibit several characteristics of aging, including osteoporosis, growth plate closure, atrophic skin, liver pathology, sepsis, cancer, and shortened lifespan (Vogel et al., 1999).

Recent findings on progeroid mice deficient in NER led to the discovery of a novel link between inherent genome instability and an adaptive "survival" response similar to that seen in long-lived mutant dwarfs and wild-type mice. At least two mouse models have described the analysis of

DNA repair-deficient mice that identified a common mechanism by which exogenous and endogenous DNA damage triggers the development of progeroid changes (Niedernhofer et al., 2006; van der Pluijm et al., 2007). XPF-ERCC1 is an endonuclease required for NER. A series of mouse models were generated by mutation and deletion of *Ercc1*. ERCC1-null mice (*Ercc1*$^{-/-}$ *mice*) exhibit progeroid features, including neurodegeneration, sarcopenia, kyphosis, premature senescence, and short lifespan (Niedernhofer et al., 2006). Intriguingly, the *Ercc1*$^{-/-}$ mice showed evidence for increased cell death and induction of antioxidant defenses, as well as a shift toward anabolism and increased GH/IGF1 signaling. A second model was generated to mimic CS patients; $Csb^{m/m}/Xpa^{-/-}$ mice also show systemic suppression of the GH/IGF-1 somatotrophic axis and oxidative metabolism, increased antioxidant responses, and hypoglycemia (van der Pluijm et al., 2007). The working model is that the unrepaired cytotoxic DNA damage induces a highly conserved metabolic response mediated by the IGF1/insulin pathway, which reallocates resources from growth and proliferation to somatic preservation and protective maintenance, so minimizing further damage (Schumacher et al., 2009). Furthermore, *Ercc1*-deficient mice benefit from a reduction in NF-κB signaling. Genetic reduction in the level of the NF-κB subunit p65 (RelA) in *Ercc1*$^{-/-}$ and *Ercc1*$^{-/\Delta}$ mice delayed the onset of age-related pathology including muscle wasting, osteoporosis, and intervertebral disc degeneration (Tilstra et al., 2012). These results directly indicate that chronic NF-κB activation, in response to accumulated DNA damage, contributes to tissue degeneration and the accompanying pathology.

WS is a human inherited autosomal recessive condition that causes the premature onset of clinical signs of aging and aging-related diseases, including cardiovascular disease, cancer, and osteoporosis (Kipling & Faragher, 1997). It is caused by a mutation in *WRN*, a Rec-Q helicase and exonuclease, which hydrolyzes ATP to separate double-stranded DNA into single strands for replication, recombination, transcription, and repair (Monnat, 2010). Several mouse models of WS have been developed (Lebel & Leder, 1998; Lombard et al., 2000; Wang et al., 2000). In the first model, mice lacking the helicase unit of the *Wrn* gene (Lebel & Leder, 1998), exhibited many of the WS phenotypes including hypertriglyceridemia, insulin resistance, elevated levels of reactive oxygen species (ROS), oxidative DNA damage, an increased incidence of cancer, and a shortened lifespan (Lebel, Lavoie, Gaudreault, Bronsard, & Drouin, 2003; Massip et al., 2006). Some of these pathologies were ameliorated by long-term vitamin

C treatment, possibly through its antioxidant effects (Lebel, Massip, Garand, & Thorin, 2010). However, *WRN*-null mice (*Wrn*$^{-/-}$) present no obvious aging phenotypes (Lombard et al., 2000).

Knockout mice lacking the telomerase RNA component *Terc*$^{-/-}$ (also called *mTR*$^{-/-}$) display premature aging phenotypes after many generations, including skin lesions, alopecia, hair graying, and loss of subcutaneous fat (Rudolph et al., 1999). The generational delay is a consequence of the long telomeres present in commonly used mouse strains, creating a lag before they reach perilously short lengths. Interestingly, a *Terc*$^{-/-}$ *Wrn*$^{-/-}$ double mutant, that lacks the telomerase RNA template as well as Wrn, develops age-dependent pathologies that are very similar to those seen in WS patients (including grey hair, osteoporosis, type II diabetes, cataracts, an elevated frequency of nonepithelial malignancies, and premature death) at an accelerated rate compared to *Terc*$^{-/-}$ mice alone (Chang et al., 2004; Rudolph et al., 1999). This finding suggests a major role of Wrn is to maintain normal telomere structure, consistent with findings implicating Wrn in several aspects of telomere maintenance (Rossi, Ghosh, & Bohr, 2010).

9. MOUSE MODELS RESULTING IN TUMORIGENESIS

There are several progeric models associated with the p53 tumor suppressor. Mice with a truncated form of p53 (*p53*$^{+/m}$ mice) enhance wild-type p53 activity. This constitutive p53 activation resulted in a resistance to spontaneous tumors compared to wild-type littermates, but surprisingly exhibited early aging-related phenotypes, such as muscle and lymphoid atrophy, osteoporosis, a reduced subcutaneous adipose and dermal thickness, and shorter lifespan (Tyner et al., 2002). Mice that are homozygous for a mutation in SnoN (*SnoN*$^{m/m}$), a regulator of Tgf-β signaling, display decreased subcutaneous fat and muscle mass, osteoporosis, reduced regenerative capacity, reproductive capacity, and a shortened lifespan (Pan et al., 2012), similar to those found in the *p53*$^{+/m}$ mice. This model was consistent with the finding that SnoN both increases and activates p53 expression. However, mice carrying the p53$^{+/m}$ transgenic allele, in addition to the two endogenous alleles ("Super p53" mice), are not only protected from cancer but also age normally (Garcia-Cao et al., 2002). It would appear that mice with enhanced p53 activity under normal regulatory control experience lifespan extension, likely due to tumor resistance.

10. MOUSE MODELS ANALYZING OXIDATIVE STRESS

ROS, including superoxide and hydrogen peroxide, are constantly generated in aerobic organisms through multiple mechanisms, but most notably as a byproduct of cellular respiration. Excessive ROS causes oxidative damage to DNA, proteins, and lipids, which in turn impairs cellular function and almost certainly contributes to multiple diseases. For instance, oxidative stress is implicated in the pathogenesis of many cancers (Valko, Rhodes, Moncol, Izakovic, & Mazur, 2006) and neurodegenerative diseases (Andersen, 2004), as well as in the normal aging process (Sohal & Weindruch, 1996).

It has been suggested that mice either overexpressing antioxidant enzymes or that the administration of chemical ROS scavengers may mitigate oxidative stress and enhance longevity. However, the evidence to date is at best equivocal, but some examples exist that support this model. For instance, global overexpression of catalase in mitochondria, which catalyzes the decomposition of hydrogen peroxide to oxygen and water, extends lifespan in mice (MCAT mice) (Schriner et al., 2005). Mice-expressing human thioredoxin (TRX-Tg mice), which acts as a reductant, are resistant against oxidative stress, have enhanced telomerase activity and live longer (Mitsui et al., 2002; Perez et al., 2011). Finally, hMTH1 is a hydrolase that eliminates oxidized precursors from the dNTP pool, degrades 8-oxodGTP and 8-oxoGTP and prevents incorporation of 8-oxoguanine into either DNA or RNA. Mice-overexpressing hMTH-1 (hMTH1-Tg mice) have significantly lower levels of oxidized guanines and live significantly longer than wild-type littermates (De Luca et al., 2013).

However, overexpression of other antioxidant enzymes does not appear to increase the lifespan of mice. Several major antioxidant enzymes such as copper-zinc superoxide dismutase (CuZnSOD, SOD1), manganese superoxide dismutase (MnSOD, SOD2), and catalase fail to confer benefits (Jang et al., 2009; Perez, Van Remmen, et al., 2009). In contrast, mice lacking CuZnSOD, SOD1 (*Sod1*$^{-/-}$ mice) have a ~30% decrease in lifespan which is associated with a high incidence of hepatocellular carcinoma (Elchuri et al., 2005). Mice lacking MnSOD, SOD2 (*Sod2*$^{-/-}$ mice) display dilated cardiomyopathy and die within first 10 days (Li et al., 1995). Peroxiredoxins (Prdxs), a family of small antioxidant proteins/enzymes that contain essential catalytic cysteine residues and use thioredoxin as an electron donor, also scavenge peroxide and are thought to be involved in the cellular responses to ROS. Peroxiredoxin-deficient (*Prdx1*$^{-/-}$) mice have a shortened lifespan owing to the development at about 9 months of severe hemolytic anemia and several malignant cancers (Neumann et al., 2003). Whether these models can be classified as progeroid is debatable; however, they are worthy of mention.

Not all deletions of DNA repair enzymes lead to a short lifespan. A general antioxidant enzyme methionine sulfoxide reductase A (MsrA) repairs oxidized methionine residues in proteins. In the initial study, mice lacking methionine sulfoxide reductase A (*MsrA*$^{-/-}$ mice) were short-lived (Moskovitz et al., 2001). However, the same mice under optimal housing conditions did not show a diminished lifespan (Salmon et al., 2009). Intriguingly, mice heterozygous for the Sod2 gene (*Sod2*$^{+/-}$ mice) live as long as wild-type littermates, despite having increased DNA damage and a higher incidence of cancer (Van Remmen et al., 2003). Mice deficient in both Mn superoxide dismutase (*Sod2*) and glutathione peroxidase-1 (*Gpx1*) have increased oxidative damage and a higher incidence of various pathologies, but do not have a reduced lifespan (Zhang et al., 2009). Finally, a reduction in glutathione peroxidase 4 (*Gpx4*$^{+/-}$ mice), which is a key enzyme in the detoxification of lipid hydroperoxides, surprisingly increases the median lifespan, even though the increase is marginal (Ran et al., 2007). In summary, it is hard to draw a clear picture of the role of DNA damage or oxidative stress in aging from these animal models. Father studies are clearly necessary (Perez, Bokov, et al., 2009).

Reducing ROS production may be a better strategy to targeting aging and age-related diseases. Two examples, one with enhanced longevity and the other with progeria, support this assertion. The adaptor protein p66^{shc} acts as an oxidoreductase to generate ROS in mitochondria and activates the mitochondrial pathway of apoptosis by enhancing cytochrome C release. *p66*$^{shc-/-}$ mice display a 30% increase in longevity. In a second example, mice lacking *mclk1,* the murine ortholog of *clk-1*, a dual-specific kinase, which is involved in mitochondria respiration, also exhibited decreased ROS levels, and DNA damage, and had an increased lifespan (Liu, Wang, et al., 2005).

It has been hypothesized that cellular senescence, which halts the proliferation of damaged or dysfunctional cells, is implicated in age-related phenotypes and that removal of senescent cells may delay or prevent aging. Mice bearing a hypomorphic mutation of BubR1 (*BubR1*$^{H/H}$) exhibit a set of progeroid phenotypes and have been used to test this model. *Bub1b*

(Budding uninhibited by benzimidazoles 1 homolog, β) encodes the spindle assembly checkpoint protein BubR1, which is a mitotic regulator ensuring accurate segregation of chromosomes through the establishment of proper microtubule–kinetochore attachments. Natural aging of wild-type mice is marked by decreased expression of BubR1 in multiple tissues (Baker et al., 2004). Furthermore, the *BubR1*$^{H/H}$ mutation contributes to defects in chromosomal stability. These mutant mice develop progressive aneuploidy together with a variety of progeroid features, including short lifespan, dwarfism, kyphosis, cataracts, loss of subcutaneous fat, and impaired wound healing (Baker et al., 2004). Recently, this mouse model was used to test the hypothesis that clearance of senescence is beneficial for survival (Baker et al., 2011). Indeed, elimination of senescent cells expressing the tumor suppressor p16$^{\text{Ink4a}}$ in the *BubR1*$^{H/H}$ progeroid model delayed the onset of age-related pathologies, although it did not significantly affect lifespan. These studies have placed increased emphasis on the need to determine the extent to which senescent cells drive aspects of normal aging and it will be of interest to see whether the ablation of p16-expressing senescent cells can impact lifespan or age-related pathologies in mice with a normal lifespan.

11. SUMMARY

For years, scientists in the aging field have debated the molecular mechanisms that drive aging and affect longevity. A central point of debate has centered on whether progeroid mutations, either in the human population or in mouse models really accelerate aspects of normal aging. While the answers are not in, two things are evident. First, dramatic progress has been made using animal models to understand aging. The number of mutations that lead to enhanced lifespan or progeroid pathologies has increased dramatically and it is anticipated these numbers will continue to grow. Having an array of these mutations will lead to new theories about the causes of aging, in part by addressing what different mutations have in common. They also begin to provide a framework for the number and kinds of cellular pathways that impact normal aging and/or progeria.

Interestingly, many of the same pathways have now been linked to enhanced longevity and progeria (Fig. 5.2). While the drivers of normal aging and progeria are certainly not completely aligned, it is becoming harder to argue that there is no significant overlap. As this clarifies, it will provide increased opportunities to test longevity interventions in progeroid

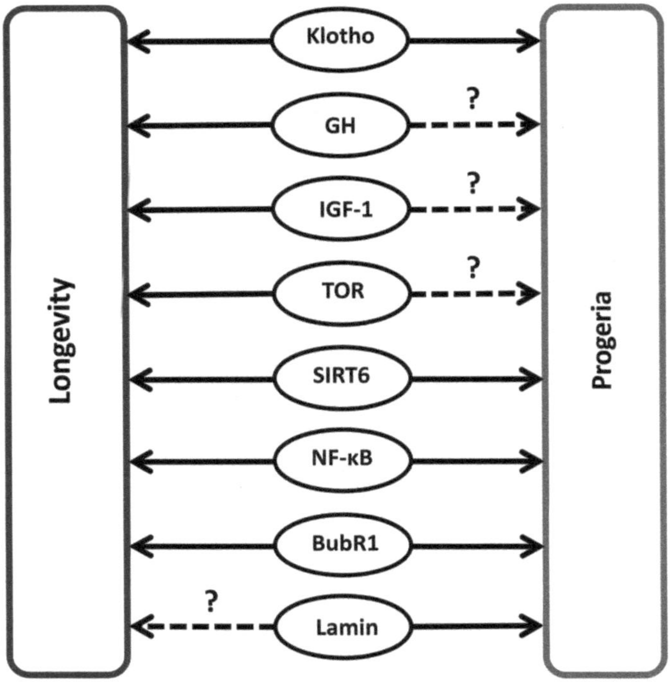

Figure 5.2 Same pathways that link to longevity and progeria. Pathways that mediate (or are proposed to mediate) both longevity and progeria. The dashed arrows indicate relationships that are yet to be firmly established. (See the color plate.)

models, leading hopefully to new therapeutic avenues for these rare but calamitous human diseases and providing a new lens for defining the pathways that control normal aging.

ACKNOWLEDGMENTS

The authors have tried to describe a collection of long-lived and progeroid models in as complete a fashion as possible and apologize if specific models have been unintentionally omitted. The authors would also like to thank Juniper Pennypacker for editorial comments and assistance with the manuscript. B. K. K. has a Senior Scholar Award in Aging from the Ellison Medical Foundation. Progeria research in the lab of B. K. K. is supported by the Progeria Research Foundation's Established Investigator Award and by NIH R01 AG033373.

REFERENCES

Agarwal, B., & Baur, J. A. (2011). Resveratrol and life extension. *Annals of the New York Academy of Sciences*, *1215*, 138–143.

Alayev, A., & Holz, M. K. (2013). mTOR signaling for biological control and cancer. *Journal of Cellular Physiology*, *228*, 1658–1664.

Andersen, J. K. (2004). Oxidative stress in neurodegeneration: Cause or consequence? *Nature Medicine, 10*(Suppl.), S18–S25.

Anisimov, V. N., Berstein, L. M., Egormin, P. A., Piskunova, T. S., Popovich, I. G., Zabezhinski, M. A., et al. (2005). Effect of metformin on life span and on the development of spontaneous mammary tumors in HER-2/neu transgenic mice. *Experimental Gerontology, 40*, 685–693.

Anisimov, V. N., Egormin, P. A., Piskunova, T. S., Popovich, I. G., Tyndyk, M. L., Yurova, M. N., et al. (2010). Metformin extends life span of HER-2/neu transgenic mice and in combination with melatonin inhibits growth of transplantable tumors in vivo. *Cell Cycle, 9*, 188–197.

Anisimov, V. N., Zabezhinski, M. A., Popovich, I. G., Piskunova, T. S., Semenchenko, A. V., Tyndyk, M. L., et al. (2010). Rapamycin extends maximal lifespan in cancer-prone mice. *The American Journal of Pathology, 176*, 2092–2097.

Anisimov, V. N., Zabezhinski, M. A., Popovich, I. G., Piskunova, T. S., Semenchenko, A. V., Tyndyk, M. L., et al. (2011). Rapamycin increases lifespan and inhibits spontaneous tumorigenesis in inbred female mice. *Cell Cycle, 10*, 4230–4236.

Arimura, T., Helbling-Leclerc, A., Massart, C., Varnous, S., Niel, F., Lacene, E., et al. (2005). Mouse model carrying H222P-Lmna mutation develops muscular dystrophy and dilated cardiomyopathy similar to human striated muscle laminopathies. *Human Molecular Genetics, 14*, 155–169.

Baker, D. J., Jeganathan, K. B., Cameron, J. D., Thompson, M., Juneja, S., Kopecka, A., et al. (2004). BubR1 insufficiency causes early onset of aging-associated phenotypes and infertility in mice. *Nature Genetics, 36*, 744–749.

Baker, D. J., Wijshake, T., Tchkonia, T., LeBrasseur, N. K., Childs, B. G., van de Sluis, B., et al. (2011). Clearance of p16Ink4a-positive senescent cells delays ageing-associated disorders. *Nature, 479*, 232–236.

Baker, D. J., Dawlaty, M. M., Wijshake, T., Jeganathan, K. B., Malureanu, L., van Ree, J. H., et al. (2013). Increased expression of BubR1 protects against aneuploidy and cancer and extends healthy lifespan. *Nature Cell Biology, 15*, 96–102.

Bartke, A. (2003). Can growth hormone (GH) accelerate aging? Evidence from GH-transgenic mice. *Neuroendocrinology, 78*, 210–216.

Bartke, A., Sun, L. Y., & Longo, V. (2013). Somatotropic signaling: Trade-offs between growth, reproductive development, and longevity. *Physiological Reviews, 93*, 571–598.

Barzilai, N., Huffman, D. M., Muzumdar, R. H., & Bartke, A. (2012). The critical role of metabolic pathways in aging. *Diabetes, 61*, 1315–1322.

Baur, J. A. (2010). Resveratrol, sirtuins, and the promise of a DR mimetic. *Mechanisms of Ageing and Development, 131*, 261–269.

Baur, J. A., & Sinclair, D. A. (2006). Therapeutic potential of resveratrol: The in vivo evidence. *Nature Reviews Drug Discovery, 5*, 493–506.

Benigni, A., Corna, D., Zoja, C., Sonzogni, A., Latini, R., Salio, M., et al. (2009). Disruption of the Ang II type 1 receptor promotes longevity in mice. *The Journal of Clinical Investigation, 119*, 524–530.

Bergo, M. O., Gavino, B., Ross, J., Schmidt, W. K., Hong, C., Kendall, L. V., et al. (2002). Zmpste24 deficiency in mice causes spontaneous bone fractures, muscle weakness, and a prelamin A processing defect. *Proceedings of the National Academy of Sciences of the United States of America, 99*, 13049–13054.

Bluher, M., Kahn, B. B., & Kahn, C. R. (2003). Extended longevity in mice lacking the insulin receptor in adipose tissue. *Science, 299*, 572–574.

Bluher, M., Michael, M. D., Peroni, O. D., Ueki, K., Carter, N., Kahn, B. B., et al. (2002). Adipose tissue selective insulin receptor knockout protects against obesity and obesity-related glucose intolerance. *Developmental Cell, 3*, 25–38.

Boily, G., Seifert, E. L., Bevilacqua, L., He, X. H., Sabourin, G., Estey, C., et al. (2008). SirT1 regulates energy metabolism and response to caloric restriction in mice. *PLoS One, 3*, e1759.

Bokov, A. F., Garg, N., Ikeno, Y., Thakur, S., Musi, N., DeFronzo, R. A., et al. (2011). Does reduced IGF-1R signaling in Igf1r+/− mice alter aging? *PLoS One, 6*, e26891.

Bonkowski, M. S., Rocha, J. S., Masternak, M. M., Al Regaiey, K. A., & Bartke, A. (2006). Targeted disruption of growth hormone receptor interferes with the beneficial actions of calorie restriction. *Proceedings of the National Academy of Sciences of the United States of America, 103*, 7901–7905.

Bordone, L., Cohen, D., Robinson, A., Motta, M. C., van Veen, E., Czopik, A., et al. (2007). SIRT1 transgenic mice show phenotypes resembling calorie restriction. *Aging Cell, 6*, 759–767.

Brown-Borg, H. M., Borg, K. E., Meliska, C. J., & Bartke, A. (1996). Dwarf mice and the ageing process. *Nature, 384*, 33.

Burtner, C. R., & Kennedy, B. K. (2010). Progeria syndromes and ageing: What is the connection? Nature reviews. *Molecular Cell. Biology, 11*, 567–578.

Cao, K., Blair, C. D., Faddah, D. A., Kieckhaefer, J. E., Olive, M., Erdos, M. R., et al. (2011). Progerin and telomere dysfunction collaborate to trigger cellular senescence in normal human fibroblasts. *The Journal of Clinical Investigation, 121*, 2833–2844.

Cao, K., Graziotto, J. J., Blair, C. D., Mazzulli, J. R., Erdos, M. R., Krainc, D., et al. (2011). Rapamycin reverses cellular phenotypes and enhances mutant protein clearance in Hutchinson-Gilford progeria syndrome cells. *Science Translational Medicine, 3*, 89ra58.

Casellas, J., & Medrano, J. F. (2008). Lack of Socs2 expression reduces lifespan in high-growth mice. *Age, 30*, 245–249.

Ceylan-Isik, A. F., Dong, M., Zhang, Y., Dong, F., Turdi, S., Nair, S., et al. (2013). Cardiomyocyte-specific deletion of endothelin receptor A rescues aging-associated cardiac hypertrophy and contractile dysfunction: Role of autophagy. *Basic Research in Cardiology, 108*, 335.

Chang, H. C., & Guarente, L. (2014). SIRT1 and other sirtuins in metabolism. *Trends in Endocrinology and Metabolism, 25*, 138–145.

Chang, S., Multani, A. S., Cabrera, N. G., Naylor, M. L., Laud, P., Lombard, D., et al. (2004). Essential role of limiting telomeres in the pathogenesis of Werner syndrome. *Nature Genetics, 36*, 877–882.

Chiu, C. H., Lin, W. D., Huang, S. Y., & Lee, Y. H. (2004). Effect of a C/EBP gene replacement on mitochondrial biogenesis in fat cells. *Genes & Development, 18*, 1970–1975.

Colman, R. J., Anderson, R. M., Johnson, S. C., Kastman, E. K., Kosmatka, K. J., Beasley, T. M., et al. (2009). Caloric restriction delays disease onset and mortality in rhesus monkeys. *Science, 325*, 201–204.

Conover, C. A., & Bale, L. K. (2007). Loss of pregnancy-associated plasma protein A extends lifespan in mice. *Aging Cell, 6*, 727–729.

Conover, C. A., Bale, L. K., Grell, J. A., Mader, J. R., & Mason, M. A. (2010a). Longevity is not influenced by prenatal programming of body size. *Aging Cell, 9*, 647–649.

Conover, C. A., Bale, L. K., Mader, J. R., Mason, M. A., Keenan, K. P., & Marler, R. J. (2010b). Longevity and age-related pathology of mice deficient in pregnancy-associated plasma protein-A. *The Journals of Gerontology. Series A, Biological Sciences and Medical Sciences, 65*, 590–599.

Conti, B., Sanchez-Alavez, M., Winsky-Sommerer, R., Morale, M. C., Lucero, J., Brownell, S., et al. (2006). Transgenic mice with a reduced core body temperature have an increased life span. *Science, 314*, 825–828.

Coschigano, K. T., Clemmons, D., Bellush, L. L., & Kopchick, J. J. (2000). Assessment of growth parameters and life span of GHR/BP gene-disrupted mice. *Endocrinology, 141*, 2608–2613.

Coschigano, K. T., Holland, A. N., Riders, M. E., List, E. O., Flyvbjerg, A., & Kopchick, J. J. (2003). Deletion, but not antagonism, of the mouse growth hormone receptor results in severely decreased body weights, insulin, and insulin-like growth factor I levels and increased life span. *Endocrinology, 144*, 3799–3810.

Cox, L. S., & Faragher, R. G. (2007). From old organisms to new molecules: Integrative biology and therapeutic targets in accelerated human ageing. *Cellular and Molecular Life Sciences, 64*, 2620–2641.

Curtis, J., & de Cabo, R. (2013). Utilizing calorie restriction to evaluate the role of sirtuins in healthspan and lifespan of mice. *Methods in Molecular Biology, 1077*, 303–311.

de Boer, J., Andressoo, J. O., de Wit, J., Huijmans, J., Beems, R. B., van Steeg, H., et al. (2002). Premature aging in mice deficient in DNA repair and transcription. *Science, 296*, 1276–1279.

De Luca, G., Ventura, I., Sanghez, V., Russo, M. T., Ajmone-Cat, M. A., Cacci, E., et al. (2013). Prolonged lifespan with enhanced exploratory behavior in mice overexpressing the oxidized nucleoside triphosphatase hMTH1. *Aging Cell, 12*, 695–705.

De Sandre-Giovannoli, A., Bernard, R., Cau, P., Navarro, C., Amiel, J., Boccaccio, I., et al. (2003). Lamin a truncation in Hutchinson-Gilford progeria. *Science, 300*, 2055.

Dechat, T., Pfleghaar, K., Sengupta, K., Shimi, T., Shumaker, D. K., Solimando, L., et al. (2008). Nuclear lamins: Major factors in the structural organization and function of the nucleus and chromatin. *Genes & Development, 22*, 832–853.

Dell'agnello, C., Leo, S., Agostino, A., Szabadkai, G., Tiveron, C., Zulian, A., et al. (2007). Increased longevity and refractoriness to Ca(2+)-dependent neurodegeneration in Surf1 knockout mice. *Human Molecular Genetics, 16*, 431–444.

Elchuri, S., Oberley, T. D., Qi, W., Eisenstein, R. S., Jackson Roberts, L., Van Remmen, H., et al. (2005). CuZnSOD deficiency leads to persistent and widespread oxidative damage and hepatocarcinogenesis later in life. *Oncogene, 24*, 367–380.

Enns, L. C., Morton, J. F., Treuting, P. R., Emond, M. J., Wolf, N. S., Dai, D. F., et al. (2009). Disruption of protein kinase A in mice enhances healthy aging. *PLoS One, 4*, e5963.

Eriksson, M., Brown, W. T., Gordon, L. B., Glynn, M. W., Singer, J., Scott, L., et al. (2003). Recurrent de novo point mutations in lamin A cause Hutchinson-Gilford progeria syndrome. *Nature, 423*, 293–298.

Flurkey, K., Papaconstantinou, J., & Harrison, D. E. (2002). The Snell dwarf mutation Pit1(dw) can increase life span in mice. *Mechanisms of Ageing and Development, 123*, 121–130.

Flurkey, K., Papaconstantinou, J., Miller, R. A., & Harrison, D. E. (2001). Lifespan extension and delayed immune and collagen aging in mutant mice with defects in growth hormone production. *Proceedings of the National Academy of Sciences of the United States of America, 98*, 6736–6741.

Flynn, J. M., O'Leary, M. N., Zambataro, C. A., Academia, E. C., Presley, M. P., Garrett, B. J., et al. (2013). Late-life rapamycin treatment reverses age-related heart dysfunction. *Aging Cell, 12*, 851–862.

Fok, W. C., Zhang, Y., Salmon, A. B., Bhattacharya, A., Gunda, R., Jones, D., et al. (2013). Short-term treatment with rapamycin and dietary restriction have overlapping and distinctive effects in young mice. *Journals of Gerontology Series A: Biological Sciences and Medical Sciences, 68*, 108–116.

Fontana, L., Partridge, L., & Longo, V. D. (2010). Extending healthy life span—From yeast to humans. *Science, 328*, 321–326.

Foukas, L. C., Bilanges, B., Bettedi, L., Pearce, W., Ali, K., Sancho, S., et al. (2013). Long-term p110alpha PI3K inactivation exerts a beneficial effect on metabolism. *EMBO Molecular Medicine, 5*, 563–571.

Foukas, L. C., Claret, M., Pearce, W., Okkenhaug, K., Meek, S., Peskett, E., et al. (2006). Critical role for the p110alpha phosphoinositide-3-OH kinase in growth and metabolic regulation. *Nature, 441*, 366–370.

Freitas, A. A., & de Magalhaes, J. P. (2011). A review and appraisal of the DNA damage theory of ageing. *Mutation Research, 728*, 12–22.
Gates, A. C., Bernal-Mizrachi, C., Chinault, S. L., Feng, C., Schneider, J. G., Coleman, T., et al. (2007). Respiratory uncoupling in skeletal muscle delays death and diminishes age-related disease. *Cell Metabolism, 6*, 497–505.
Garcia-Cao, I., Garcia-Cao, M., Martin-Caballero, J., Criado, L. M., Klatt, P., Flores, J. M., et al. (2002). "Super p53" mice exhibit enhanced DNA damage response, are tumor resistant and age normally. *The EMBO Journal, 21*, 6225–6235.
Gems, D., & Partridge, L. (2013). Genetics of longevity in model organisms: Debates and paradigm shifts. *Annual Review of Physiology, 75*, 621–644.
Gordon, L. B., Kleinman, M. E., Miller, D. T., Neuberg, D. S., Giobbie-Hurder, A., Gerhard-Herman, M., et al. (2012). Clinical trial of a farnesyltransferase inhibitor in children with Hutchinson-Gilford progeria syndrome. *Proceedings of the National Academy of Sciences of the United States of America, 109*, 16666–16671.
Guarente, L. (2013). Calorie restriction and sirtuins revisited. *Genes & Development, 27*, 2072–2085.
Haigis, M. C., & Sinclair, D. A. (2010). Mammalian sirtuins: Biological insights and disease relevance. *Annual Review of Pathology, 5*, 253–295.
Harper, J. M., Wilkinson, J. E., & Miller, R. A. (2010). Macrophage migration inhibitory factor-knockout mice are long lived and respond to caloric restriction. *The FASEB Journal, 24*, 2436–2442.
Harrison, D. E., Strong, R., Sharp, Z. D., Nelson, J. F., Astle, C. M., Flurkey, K., et al. (2009). Rapamycin fed late in life extends lifespan in genetically heterogeneous mice. *Nature, 460*, 392–395.
Hasty, P., Campisi, J., Hoeijmakers, J., van Steeg, H., & Vijg, J. (2003). Aging and genome maintenance: Lessons from the mouse? *Science, 299*, 1355–1359.
Hasty, P., & Vijg, J. (2004). Accelerating aging by mouse reverse genetics: A rational approach to understanding longevity. *Aging Cell, 3*, 55–65.
Hayden, M. S., & Ghosh, S. (2008). Shared principles in NF-kappaB signaling. *Cell, 132*, 344–362.
Herranz, D., Munoz-Martin, M., Canamero, M., Mulero, F., Martinez-Pastor, B., Fernandez-Capetillo, O., et al. (2010). Sirt1 improves healthy ageing and protects from metabolic syndrome-associated cancer. *Nature Communications, 1*, 3.
Holzenberger, M., Dupont, J., Ducos, B., Leneuve, P., Geloen, A., Even, P. C., et al. (2003). IGF-1 receptor regulates lifespan and resistance to oxidative stress in mice. *Nature, 421*, 182–187.
Hu, D., Cao, P., Thiels, E., Chu, C. T., Wu, G. Y., Oury, T. D., et al. (2007). Hippocampal long-term potentiation, memory, and longevity in mice that overexpress mitochondrial superoxide dismutase. *Neurobiology of Learning and Memory, 87*, 372–384.
Hubbard, B. P., Gomes, A. P., Dai, H., Li, J., Case, A. W., Considine, T., et al. (2013). Evidence for a common mechanism of SIRT1 regulation by allosteric activators. *Science, 339*, 1216–1219.
Hughes, K. J., & Kennedy, B. K. (2012). Cell biology. Rapamycin paradox resolved. *Science, 335*, 1578–1579.
Ikeno, Y., Bronson, R. T., Hubbard, G. B., Lee, S., & Bartke, A. (2003). Delayed occurrence of fatal neoplastic diseases in ames dwarf mice: Correlation to extended longevity. *Journals of Gerontology Series A: Biological Sciences and Medical Sciences, 58*, 291–296.
Ikeno, Y., Hubbard, G. B., Lee, S., Cortez, L. A., Lew, C. M., Webb, C. R., et al. (2009). Reduced incidence and delayed occurrence of fatal neoplastic diseases in growth hormone receptor/binding protein knockout mice. *Journals of Gerontology Series A: Biological Sciences and Medical Sciences, 64*, 522–529.
Imai, S., Armstrong, C. M., Kaeberlein, M., & Guarente, L. (2000). Transcriptional silencing and longevity protein Sir2 is an NAD-dependent histone deacetylase. *Nature, 403*, 795–800.

Inagaki, T., Lin, V. Y., Goetz, R., Mohammadi, M., Mangelsdorf, D. J., & Kliewer, S. A. (2008). Inhibition of growth hormone signaling by the fasting-induced hormone FGF21. *Cell Metabolism, 8*, 77–83.

Ingram, D. K., Zhu, M., Mamczarz, J., Zou, S., Lane, M. A., Roth, G. S., et al. (2006). Calorie restriction mimetics: An emerging research field. *Aging Cell, 5*, 97–108.

Jang, Y. C., Perez, V. I., Song, W., Lustgarten, M. S., Salmon, A. B., Mele, J., et al. (2009). Overexpression of Mn superoxide dismutase does not increase life span in mice. *Journals of Gerontology Series A: Biological Sciences and Medical Sciences, 64*, 1114–1125.

Johnson, S. C., Rabinovitch, P. S., & Kaeberlein, M. (2013). mTOR is a key modulator of ageing and age-related disease. *Nature, 493*, 338–345.

Kanfi, Y., Naiman, S., Amir, G., Peshti, V., Zinman, G., Nahum, L., et al. (2012). The sirtuin SIRT6 regulates lifespan in male mice. *Nature, 483*, 218–221.

Kanfi, Y., Peshti, V., Gil, R., Naiman, S., Nahum, L., Levin, E., et al. (2010). SIRT6 protects against pathological damage caused by diet-induced obesity. *Aging Cell, 9*, 162–173.

Kappeler, L., De Magalhaes Filho, C., Dupont, J., Leneuve, P., Cervera, P., Perin, L., et al. (2008). Brain IGF-1 receptors control mammalian growth and lifespan through a neuroendocrine mechanism. *PLoS Biology, 6*, e254.

Kawahara, T. L., Michishita, E., Adler, A. S., Damian, M., Berber, E., Lin, M., et al. (2009). SIRT6 links histone H3 lysine 9 deacetylation to NF-kappaB-dependent gene expression and organismal life span. *Cell, 136*, 62–74.

Kenyon, C. J. (2010). The genetics of ageing. *Nature, 464*, 504–512.

Kipling, D., Davis, T., Ostler, E. L., & Faragher, R. G. (2004). What can progeroid syndromes tell us about human aging? *Science, 305*, 1426–1431.

Kipling, D., & Faragher, R. G. (1997). Progeroid syndromes: Probing the molecular basis of aging? *Molecular Pathology, 50*, 234–241.

Krishnan, V., Chow, M. Z., Wang, Z., Zhang, L., Liu, B., Liu, X., et al. (2011). Histone H4 lysine 16 hypoacetylation is associated with defective DNA repair and premature senescence in Zmpste24-deficient mice. *Proceedings of the National Academy of Sciences of the United States of America, 108*, 12325–12330.

Kubicky, R. A., Wu, S., Kharitonenkov, A., & De Luca, F. (2012). Role of fibroblast growth factor 21 (FGF21) in undernutrition-related attenuation of growth in mice. *Endocrinology, 153*, 2287–2295.

Kudlow, B. A., Kennedy, B. K., & Monnat, R. J., Jr. (2007). Werner and Hutchinson-Gilford progeria syndromes: Mechanistic basis of human progeroid diseases. *Nature Reviews Molecular Cell Biology, 8*, 394–404.

Kuro-o, M., Matsumura, Y., Aizawa, H., Kawaguchi, H., Suga, T., Utsugi, T., et al. (1997). Mutation of the mouse klotho gene leads to a syndrome resembling ageing. *Nature, 390*, 45–51.

Kurosu, H., Yamamoto, M., Clark, J. D., Pastor, J. V., Nandi, A., Gurnani, P., et al. (2005). Suppression of aging in mice by the hormone Klotho. *Science, 309*, 1829–1833.

Kwan, K. Y., & Wang, J. C. (2001). Mice lacking DNA topoisomerase IIIbeta develop to maturity but show a reduced mean lifespan. *Proceedings of the National Academy of Sciences of the United States of America, 98*, 5717–5721.

Kwon, J., Han, E., Bui, C. B., Shin, W., Lee, J., Lee, S., et al. (2012). Assurance of mitochondrial integrity and mammalian longevity by the p62-Keap1-Nrf2-Nqo1 cascade. *EMBO Reports, 13*, 150–156.

Lamming, D. W., Ye, L., Katajisto, P., Goncalves, M. D., Saitoh, M., Stevens, D. M., et al. (2012). Rapamycin-induced insulin resistance is mediated by mTORC2 loss and uncoupled from longevity. *Science, 335*, 1638–1643.

Laplante, M., & Sabatini, D. M. (2012). mTOR signaling in growth control and disease. *Cell, 149*, 274–293.

Lapointe, J., Stepanyan, Z., Bigras, E., & Hekimi, S. (2009). Reversal of the mitochondrial phenotype and slow development of oxidative biomarkers of aging in long-lived Mclk1 +/− mice. *The Journal of Biological Chemistry*, *284*, 20364–20374.

Lebel, M., Lavoie, J., Gaudreault, I., Bronsard, M., & Drouin, R. (2003). Genetic cooperation between the Werner syndrome protein and poly(ADP-ribose) polymerase-1 in preventing chromatid breaks, complex chromosomal rearrangements, and cancer in mice. *The American Journal of Pathology*, *162*, 1559–1569.

Lebel, M., & Leder, P. (1998). A deletion within the murine Werner syndrome helicase induces sensitivity to inhibitors of topoisomerase and loss of cellular proliferative capacity. *Proceedings of the National Academy of Sciences of the United States of America*, *95*, 13097–13102.

Lebel, M., Massip, L., Garand, C., & Thorin, E. (2010). Ascorbate improves metabolic abnormalities in Wrn mutant mice but not the free radical scavenger catechin. *Annals of the New York Academy of Sciences*, *1197*, 40–44.

Li, Y., Huang, T. T., Carlson, E. J., Melov, S., Ursell, P. C., Olson, J. L., et al. (1995). Dilated cardiomyopathy and neonatal lethality in mutant mice lacking manganese superoxide dismutase. *Nature Genetics*, *11*, 376–381.

Li, Q., & Ren, J. (2007). Influence of cardiac-specific overexpression of insulin-like growth factor 1 on lifespan and aging-associated changes in cardiac intracellular Ca2+ homeostasis, protein damage and apoptotic protein expression. *Aging Cell*, *6*, 799–806.

Li, Y., Xu, W., McBurney, M. W., & Longo, V. D. (2008). SirT1 inhibition reduces IGF-I/IRS-2/Ras/ERK1/2 signaling and protects neurons. *Cell Metabolism*, *8*, 38–48.

Liang, H., Masoro, E. J., Nelson, J. F., Strong, R., McMahan, C. A., & Richardson, A. (2003). Genetic mouse models of extended lifespan. *Experimental Gerontology*, *38*, 1353–1364.

Liao, C. Y., & Kennedy, B. K. (2012). Will the real aging Sirtuin please stand up? *Cell Research*, *22*, 1215–1217.

Liao, C. Y., Rikke, B. A., Johnson, T. E., Diaz, V., & Nelson, J. F. (2010). Genetic variation in the murine lifespan response to dietary restriction: From life extension to life shortening. *Aging Cell*, *9*, 92–95.

Liu, J. P., Baker, J., Perkins, A. S., Robertson, E. J., & Efstratiadis, A. (1993). Mice carrying null mutations of the genes encoding insulin-like growth factor I (Igf-1) and type 1 IGF receptor (Igf1r). *Cell*, *75*, 59–72.

Liu, B., Ghosh, S., Yang, X., Zheng, H., Liu, X., Wang, Z., et al. (2012). Resveratrol rescues SIRT1-dependent adult stem cell decline and alleviates progeroid features in laminopathy-based progeria. *Cell Metabolism*, *16*, 738–750.

Liu, X., Jiang, N., Hughes, B., Bigras, E., Shoubridge, E., & Hekimi, S. (2005). Evolutionary conservation of the clk-1-dependent mechanism of longevity: Loss of mclk1 increases cellular fitness and lifespan in mice. *Genes & Development*, *19*, 2424–2434.

Liu, B., Wang, J., Chan, K. M., Tjia, W. M., Deng, W., Guan, X., et al. (2005). Genomic instability in laminopathy-based premature aging. *Nature Medicine*, *11*, 780–785.

Liu, B., Wang, Z., Ghosh, S., & Zhou, Z. (2013). Defective ATM-Kap-1-mediated chromatin remodeling impairs DNA repair and accelerates senescence in progeria mouse model. *Aging Cell*, *12*, 316–318.

Lombard, D. B., Beard, C., Johnson, B., Marciniak, R. A., Dausman, J., Bronson, R., et al. (2000). Mutations in the WRN gene in mice accelerate mortality in a p53-null background. *Molecular and Cellular Biology*, *20*, 3286–3291.

Lorenzini, A., Salmon, A. B., Lerner, C., Torres, C., Ikeno, Y., Motch, S., et al. (2013). Mice producing reduced levels of insulin-like growth factor type 1 display an increase in maximum, but not mean, life span. *Journals of Gerontology Series A: Biological Sciences and Medical Sciences*, *69*, 410–419.

Marino, G., Ugalde, A. P., Fernandez, A. F., Osorio, F. G., Fueyo, A., Freije, J. M., et al. (2010). Insulin-like growth factor 1 treatment extends longevity in a mouse model of

human premature aging by restoring somatotroph axis function. *Proceedings of the National Academy of Sciences of the United States of America, 107*, 16268–16273.
Marino, G., Ugalde, A. P., Salvador-Montoliu, N., Varela, I., Quiros, P. M., Cadinanos, J., et al. (2008). Premature aging in mice activates a systemic metabolic response involving autophagy induction. *Human Molecular Genetics, 17*, 2196–2211.
Martin-Montalvo, A., Mercken, E. M., Mitchell, S. J., Palacios, H. H., Mote, P. L., Scheibye-Knudsen, M., et al. (2013). Metformin improves healthspan and lifespan in mice. *Nature Communications, 4*, 2192.
Masoro, E. J. (2005). Overview of caloric restriction and ageing. *Mechanisms of Ageing and Development, 126*, 913–922.
Massip, L., Garand, C., Turaga, R. V., Deschenes, F., Thorin, E., & Lebel, M. (2006). Increased insulin, triglycerides, reactive oxygen species, and cardiac fibrosis in mice with a mutation in the helicase domain of the Werner syndrome gene homologue. *Experimental Gerontology, 41*, 157–168.
Mattison, J. A., Roth, G. S., Beasley, T. M., Tilmont, E. M., Handy, A. M., Herbert, R. L., et al. (2012). Impact of caloric restriction on health and survival in rhesus monkeys from the NIA study. *Nature, 489*, 318–321.
McCay, C. M., Crowell, M. F., & Maynard, L. A. (1935). The effect of retarded growth upon the length of life and upon the ultimate body size. *Journal of Nutrition, 10*, 63–79.
Mercken, E. M., Hu, J., Krzysik-Walker, S., Wei, M., Li, Y., McBurney, M. W., et al. (2014). SIRT1 but not its increased expression is essential for lifespan extension in caloric-restricted mice. *Aging Cell, 13*, 193–196.
Michishita, E., McCord, R. A., Berber, E., Kioi, M., Padilla-Nash, H., Damian, M., et al. (2008). SIRT6 is a histone H3 lysine 9 deacetylase that modulates telomeric chromatin. *Nature, 452*, 492–496.
Migliaccio, E., Giorgio, M., Mele, S., Pelicci, G., Reboldi, P., Pandolfi, P. P., et al. (1999). The p66shc adaptor protein controls oxidative stress response and life span in mammals. *Nature, 402*, 309–313.
Miller, R. A. (2004). 'Accelerated aging': A primrose path to insight? *Aging Cell, 3*, 47–51.
Miller, R. A., Harrison, D. E., Astle, C. M., Baur, J. A., Boyd, A. R., de Cabo, R., et al. (2011). Rapamycin, but not resveratrol or simvastatin, extends life span of genetically heterogeneous mice. *Journals of Gerontology Series A: Biological Sciences and Medical Sciences, 66*, 191–201.
Miskin, R., & Masos, T. (1997). Transgenic mice overexpressing urokinase-type plasminogen activator in the brain exhibit reduced food consumption, body weight and size, and increased longevity. *The Journals of Gerontology. Series A, Biological Sciences and Medical Sciences, 52*, B118–B124.
Mitsui, A., Hamuro, J., Nakamura, H., Kondo, N., Hirabayashi, Y., Ishizaki-Koizumi, S., et al. (2002). Overexpression of human thioredoxin in transgenic mice controls oxidative stress and life span. *Antioxidants & Redox Signaling, 4*, 693–696.
Monnat, R. J., Jr. (2010). Human RECQ helicases: Roles in DNA metabolism, mutagenesis and cancer biology. *Seminars in Cancer Biology, 20*, 329–339.
Moskovitz, J., Bar-Noy, S., Williams, W. M., Requena, J., Berlett, B. S., & Stadtman, E. R. (2001). Methionine sulfoxide reductase (MsrA) is a regulator of antioxidant defense and lifespan in mammals. *Proceedings of the National Academy of Sciences of the United States of America, 98*, 12920–12925.
Mostoslavsky, R., Chua, K. F., Lombard, D. B., Pang, W. W., Fischer, M. R., Gellon, L., et al. (2006). Genomic instability and aging-like phenotype in the absence of mammalian SIRT6. *Cell, 124*, 315–329.
Mounkes, L. C., Kozlov, S., Hernandez, L., Sullivan, T., & Stewart, C. L. (2003). A progeroid syndrome in mice is caused by defects in A-type lamins. *Nature, 423*, 298–301.
Navarro, C. L., De Sandre-Giovannoli, A., Bernard, R., Boccaccio, I., Boyer, A., Genevieve, D., et al. (2004). Lamin A and ZMPSTE24 (FACE-1) defects cause nuclear

disorganization and identify restrictive dermopathy as a lethal neonatal laminopathy. *Human Molecular Genetics*, *13*, 2493–2503.

Neff, F., Flores-Dominguez, D., Ryan, D. P., Horsch, M., Schroder, S., Adler, T., et al. (2013). Rapamycin extends murine lifespan but has limited effects on aging. *The Journal of Clinical Investigation*, *123*, 3272–3291.

Neumann, C. A., Krause, D. S., Carman, C. V., Das, S., Dubey, D. P., Abraham, J. L., et al. (2003). Essential role for the peroxiredoxin Prdx1 in erythrocyte antioxidant defence and tumour suppression. *Nature*, *424*, 561–565.

Niccoli, T., & Partridge, L. (2012). Ageing as a risk factor for disease. *Current Biology*, *22*, R741–R752.

Niedernhofer, L. J., Garinis, G. A., Raams, A., Lalai, A. S., Robinson, A. R., Appeldoorn, E., et al. (2006). A new progeroid syndrome reveals that genotoxic stress suppresses the somatotroph axis. *Nature*, *444*, 1038–1043.

Nojima, A., Yamashita, M., Yoshida, Y., Shimizu, I., Ichimiya, H., Kamimura, N., et al. (2013). Haploinsufficiency of akt1 prolongs the lifespan of mice. *PLoS One*, *8*, e69178.

Oh, Y. S., Kim, D. G., Kim, G., Choi, E. C., Kennedy, B. K., Suh, Y., et al. (2010). Down-regulation of lamin A by tumor suppressor AIMP3/p18 leads to a progeroid phenotype in mice. *Aging Cell*, *9*, 810–822.

Ortega-Molina, A., Efeyan, A., Lopez-Guadamillas, E., Munoz-Martin, M., Gomez-Lopez, G., Canamero, M., et al. (2012). Pten positively regulates brown adipose function, energy expenditure, and longevity. *Cell Metabolism*, *15*, 382–394.

Osorio, F. G., Barcena, C., Soria-Valles, C., Ramsay, A. J., de Carlos, F., Cobo, J., et al. (2012). Nuclear lamina defects cause ATM-dependent NF-kappaB activation and link accelerated aging to a systemic inflammatory response. *Genes & Development*, *26*, 2311–2324.

Osorio, F. G., Navarro, C. L., Cadinanos, J., Lopez-Mejia, I. C., Quiros, P. M., Bartoli, C., et al. (2011). Splicing-directed therapy in a new mouse model of human accelerated aging. *Science Translational Medicine*, *3*, 106ra107.

Pan, D., Zhu, Q., Conboy, M. J., Conboy, I. M., & Luo, K. (2012). SnoN activates p53 directly to regulate aging and tumorigenesis. *Aging Cell*, *11*, 902–911.

Panici, J. A., Harper, J. M., Miller, R. A., Bartke, A., Spong, A., & Masternak, M. M. (2010). Early life growth hormone treatment shortens longevity and decreases cellular stress resistance in long-lived mutant mice. *FASEB Journal*, *24*, 5073–5079.

Park, S. G., Choi, E. C., & Kim, S. (2010). Aminoacyl-tRNA synthetase-interacting multifunctional proteins (AIMPs): A triad for cellular homeostasis. *IUBMB Life*, *62*, 296–302.

Park, B. J., Oh, Y. S., Park, S. Y., Choi, S. J., Rudolph, C., Schlegelberger, B., et al. (2006). AIMP3 haploinsufficiency disrupts oncogene-induced p53 activation and genomic stability. *Cancer Research*, *66*, 6913–6918.

Pendas, A. M., Zhou, Z., Cadinanos, J., Freije, J. M., Wang, J., Hultenby, K., et al. (2002). Defective prelamin A processing and muscular and adipocyte alterations in Zmpste24 metalloproteinase-deficient mice. *Nature Genetics*, *31*, 94–99.

Perez, V. I., Bokov, A., Van Remmen, H., Mele, J., Ran, Q., Ikeno, Y., et al. (2009). Is the oxidative stress theory of aging dead? *Biochimica et Biophysica Acta*, *1790*, 1005–1014.

Perez, V. I., Cortez, L. A., Lew, C. M., Rodriguez, M., Webb, C. R., Van Remmen, H., et al. (2011). Thioredoxin 1 overexpression extends mainly the earlier part of life span in mice. *Journals of Gerontology Series A: Biological Sciences and Medical Sciences*, *66*, 1286–1299.

Perez, V. I., Van Remmen, H., Bokov, A., Epstein, C. J., Vijg, J., & Richardson, A. (2009). The overexpression of major antioxidant enzymes does not extend the lifespan of mice. *Aging Cell*, *8*, 73–75.

Pyo, J. O., Yoo, S. M., Ahn, H. H., Nah, J., Hong, S. H., Kam, T. I., et al. (2013). Overexpression of Atg5 in mice activates autophagy and extends lifespan. *Nature Communications, 4*, 2300.

Ramos, F. J., Chen, S. C., Garelick, M. G., Dai, D. F., Liao, C. Y., Schreiber, K. H., et al. (2012). Rapamycin reverses elevated mTORC1 signaling in lamin A/C-deficient mice, rescues cardiac and skeletal muscle function, and extends survival. *Science Translational Medicine, 4*, 144ra103.

Ran, Q., Liang, H., Ikeno, Y., Qi, W., Prolla, T. A., Roberts, L. J., 2nd., et al. (2007). Reduction in glutathione peroxidase 4 increases life span through increased sensitivity to apoptosis. *Journals of Gerontology Series A: Biological Sciences and Medical Sciences, 62*, 932–942.

Redmann, S. M., Jr., & Argyropoulos, G. (2006). AgRP-deficiency could lead to increased lifespan. *Biochemical and Biophysical Research Communications, 351*, 860–864.

Rossi, M. L., Ghosh, A. K., & Bohr, V. A. (2010). Roles of Werner syndrome protein in protection of genome integrity. *DNA Repair (Amst), 9*, 331–344.

Rudolph, K. L., Chang, S., Lee, H. W., Blasco, M., Gottlieb, G. J., Greider, C., et al. (1999). Longevity, stress response, and cancer in aging telomerase-deficient mice. *Cell, 96*, 701–712.

Salmon, A. B., Perez, V. I., Bokov, A., Jernigan, A., Kim, G., Zhao, H., et al. (2009). Lack of methionine sulfoxide reductase A in mice increases sensitivity to oxidative stress but does not diminish life span. *FASEB Journal, 23*, 3601–3608.

Satoh, A., Brace, C. S., Rensing, N., Cliften, P., Wozniak, D. F., Herzog, E. D., et al. (2013). Sirt1 extends life span and delays aging in mice through the regulation of Nk2 homeobox 1 in the DMH and LH. *Cell Metabolism, 18*, 416–430.

Scaffidi, P., & Misteli, T. (2006). Lamin A-dependent nuclear defects in human aging. *Science, 312*, 1059–1063.

Schleit, J., Johnson, S. C., Bennett, C. F., Simko, M., Trongtham, N., Castanza, A., et al. (2013). Molecular mechanisms underlying genotype-dependent responses to dietary restriction. *Aging Cell, 12*, 1050–1061.

Schreiber, K. H., & Kennedy, B. K. (2013). When lamins go bad: Nuclear structure and disease. *Cell, 152*, 1365–1375.

Schriner, S. E., Linford, N. J., Martin, G. M., Treuting, P., Ogburn, C. E., Emond, M., et al. (2005). Extension of murine life span by overexpression of catalase targeted to mitochondria. *Science, 308*, 1909–1911.

Schumacher, B., Hoeijmakers, J. H., & Garinis, G. A. (2009). Sealing the gap between nuclear DNA damage and longevity. *Molecular and Cellular Endocrinology, 299*, 112–117.

Selman, C., Lingard, S., Choudhury, A. I., Batterham, R. L., Claret, M., Clements, M., et al. (2008). Evidence for lifespan extension and delayed age-related biomarkers in insulin receptor substrate 1 null mice. *FASEB Journal, 22*, 807–818.

Selman, C., Lingard, S., Gems, D., Partridge, L., & Withers, D. J. (2008). Comment on "brain IRS2 signaling coordinates life span and nutrient homeostasis" *Science, 320*, 1012, author reply 1012.

Selman, C., Partridge, L., & Withers, D. J. (2011). Replication of extended lifespan phenotype in mice with deletion of insulin receptor substrate 1. *PLoS One, 6*, e16144.

Selman, C., Tullet, J. M., Wieser, D., Irvine, E., Lingard, S. J., Choudhury, A. I., et al. (2009). Ribosomal protein S6 kinase 1 signaling regulates mammalian life span. *Science, 326*, 140–144.

Selman, C., & Withers, D. J. (2011). Mammalian models of extended healthy lifespan. *Philosophical Transactions of the Royal Society of London. Series B: Biological Sciences, 366*, 99–107.

Singh, S. P., Niemczyk, M., Saini, D., Sadovov, V., Zimniak, L., & Zimniak, P. (2010). Disruption of the mGsta4 gene increases life span of C57BL mice. *The Journals of Gerontology. Series A, Biological Sciences and Medical Sciences*, *65*, 14–23.

Sohal, R. S., & Weindruch, R. (1996). Oxidative stress, caloric restriction, and aging. *Science*, *273*, 59–63.

Stewart, C. L., Kozlov, S., Fong, L. G., & Young, S. G. (2007). Mouse models of the laminopathies. *Experimental Cell Research*, *313*, 2144–2156.

Streeper, R. S., Grueter, C. A., Salomonis, N., Cases, S., Levin, M. C., Koliwad, S. K., et al. (2012). Deficiency of the lipid synthesis enzyme, DGAT1, extends longevity in mice. *Aging (Albany NY)*, *4*, 13–27.

Sullivan, T., Escalante-Alcalde, D., Bhatt, H., Anver, M., Bhat, N., Nagashima, K., et al. (1999). Loss of A-type lamin expression compromises nuclear envelope integrity leading to muscular dystrophy. *The Journal of Cell Biology*, *147*, 913–920.

Sun, L. Y., Spong, A., Swindell, W. R., Fang, Y., Hill, C., Huber, J. A., et al. (2013). Growth hormone-releasing hormone disruption extends lifespan and regulates response to caloric restriction in mice. *eLife*, *2*, e01098.

Taguchi, A., Wartschow, L. M., & White, M. F. (2007). Brain IRS2 signaling coordinates life span and nutrient homeostasis. *Science*, *317*, 369–372.

Tanny, J. C., Dowd, G. J., Huang, J., Hilz, H., & Moazed, D. (1999). An enzymatic activity in the yeast Sir2 protein that is essential for gene silencing. *Cell*, *99*, 735–745.

Tennen, R. I., & Chua, K. F. (2011). Chromatin regulation and genome maintenance by mammalian SIRT6. *Trends in Biochemical Sciences*, *36*, 39–46.

Tilstra, J. S., Clauson, C. L., Niedernhofer, L. J., & Robbins, P. D. (2011). NF-kappaB in aging and disease. *Aging and Disease*, *2*, 449–465.

Tilstra, J. S., Robinson, A. R., Wang, J., Gregg, S. Q., Clauson, C. L., Reay, D. P., et al. (2012). NF-kappaB inhibition delays DNA damage-induced senescence and aging in mice. *The Journal of Clinical Investigation*, *122*, 2601–2612.

Tomas-Loba, A., Flores, I., Fernandez-Marcos, P. J., Cayuela, M. L., Maraver, A., Tejera, A., et al. (2008). Telomerase reverse transcriptase delays aging in cancer-resistant mice. *Cell*, *135*, 609–622.

Trifunovic, A., Wredenberg, A., Falkenberg, M., Spelbrink, J. N., Rovio, A. T., Bruder, C. E., et al. (2004). Premature ageing in mice expressing defective mitochondrial DNA polymerase. *Nature*, *429*, 417–423.

Tyner, S. D., Venkatachalam, S., Choi, J., Jones, S., Ghebranious, N., Igelmann, H., et al. (2002). p53 mutant mice that display early ageing-associated phenotypes. *Nature*, *415*, 45–53.

Valko, M., Rhodes, C. J., Moncol, J., Izakovic, M., & Mazur, M. (2006). Free radicals, metals and antioxidants in oxidative stress-induced cancer. *Chemico-Biological Interactions*, *160*, 1–40.

van der Pluijm, I., Garinis, G. A., Brandt, R. M., Gorgels, T. G., Wijnhoven, S. W., Diderich, K. E., et al. (2007). Impaired genome maintenance suppresses the growth hormone–insulin-like growth factor 1 axis in mice with Cockayne syndrome. *PLoS Biology*, *5*, e2.

Van Remmen, H., Ikeno, Y., Hamilton, M., Pahlavani, M., Wolf, N., Thorpe, S. R., et al. (2003). Life-long reduction in MnSOD activity results in increased DNA damage and higher incidence of cancer but does not accelerate aging. *Physiological Genomics*, *16*, 29–37.

Varela, I., Cadinanos, J., Pendas, A. M., Gutierrez-Fernandez, A., Folgueras, A. R., Sanchez, L. M., et al. (2005). Accelerated ageing in mice deficient in Zmpste24 protease is linked to p53 signalling activation. *Nature*, *437*, 564–568.

Vogel, H., Lim, D. S., Karsenty, G., Finegold, M., & Hasty, P. (1999). Deletion of Ku86 causes early onset of senescence in mice. *Proceedings of the National Academy of Sciences of the United States of America*, *96*, 10770–10775.

Wang, L., Ogburn, C. E., Ware, C. B., Ladiges, W. C., Youssoufian, H., Martin, G. M., et al. (2000). Cellular Werner phenotypes in mice expressing a putative dominant-negative human WRN gene. *Genetics, 154*, 357–362.

Weindruch, R., & Walford, L. R. (1988). *The retardation of aging and disease by dietary restriction.* Springfield, III: Charles C Thomas Publisher.

Wilkinson, J. E., Burmeister, L., Brooks, S. V., Chan, C. C., Friedline, S., Harrison, D. E., et al. (2012). Rapamycin slows aging in mice. *Aging Cell, 11*, 675–682.

Wong, K. K., Maser, R. S., Bachoo, R. M., Menon, J., Carrasco, D. R., Gu, Y., et al. (2003). Telomere dysfunction and Atm deficiency compromises organ homeostasis and accelerates ageing. *Nature, 421*, 643–648.

Wolf, E., Kahnt, E., Ehrlein, J., Hermanns, W., Brem, G., & Wanke, R. (1993). Effects of long-term elevated serum levels of growth hormone on life expectancy of mice: Lessons from transgenic animal models. *Mechanisms of Ageing and Development, 68*, 71–87.

Wu, J. J., Liu, J., Chen, E. B., Wang, J. J., Cao, L., Narayan, N., et al. (2013). Increased mammalian lifespan and a segmental and tissue-specific slowing of aging after genetic reduction of mTOR expression. *Cell Reports, 4*, 913–920.

Xu, J., Gontier, G., Chaker, Z., Lacube, P., Dupont, J., & Holzenberger, M. (2013). Longevity effect of IGF-1R(+/−) mutation depends on genetic background-specific receptor activation. *Aging Cell, 13*, 19–28.

Yan, L., Vatner, D. E., O'Connor, J. P., Ivessa, A., Ge, H., Chen, W., et al. (2007). Type 5 adenylyl cyclase disruption increases longevity and protects against stress. *Cell, 130*, 247–258.

Yang, X., Doser, T. A., Fang, C. X., Nunn, J. M., Janardhanan, R., Zhu, M., et al. (2006). Metallothionein prolongs survival and antagonizes senescence-associated cardiomyocyte diastolic dysfunction: role of oxidative stress. *The FASEB Journal, 20*, 1024–1026.

Yuan, R., Peters, L. L., & Paigen, B. (2011). Mice as a mammalian model for research on the genetics of aging. *ILAR Journal/National Research Council, Institute of Laboratory Animal Resources, 52*, 4–15.

Yuan, R., Tsaih, S. W., Petkova, S. B., Marin de Evsikova, C., Xing, S., Marion, M. A., et al. (2009). Aging in inbred strains of mice: Study design and interim report on median lifespans and circulating IGF1 levels. *Aging Cell, 8*, 277–287.

Zhang, Y., Bokov, A., Gelfond, J., Soto, V., Ikeno, Y., Hubbard, G., et al. (2014). Rapamycin extends life and health in C57BL/6 mice. *The Journals of Gerontology. Series A, Biological Sciences and Medical Sciences, 69*, 119–130.

Zhang, Y., Ikeno, Y., Qi, W., Chaudhuri, A., Li, Y., Bokov, A., et al. (2009). Mice deficient in both Mn superoxide dismutase and glutathione peroxidase-1 have increased oxidative damage and a greater incidence of pathology but no reduction in longevity. *Journals of Gerontology Series A: Biological Sciences and Medical Sciences, 64*, 1212–1220.

Zhang, H., Kieckhaefer, J. E., & Cao, K. (2013). Mouse models of laminopathies. *Aging Cell, 12*, 2–10.

Zhang, G., Li, J., Purkayastha, S., Tang, Y., Zhang, H., Yin, Y., et al. (2013). Hypothalamic programming of systemic ageing involving IKK-beta, NF-kappaB and GnRH. *Nature, 497*, 211–216.

Zhang, Y., Xie, Y., Berglund, E. D., Coate, K. C., He, T. T., Katafuchi, T., et al. (2012). The starvation hormone, fibroblast growth factor-21, extends lifespan in mice. *eLife, 1*, e00065.

Zhou, G., Myers, R., Li, Y., Chen, Y., Shen, X., Fenyk-Melody, J., et al. (2001). Role of AMP-activated protein kinase in mechanism of metformin action. *The Journal of Clinical Investigation, 108*, 1167–1174.

Zid, B. M., Rogers, A. N., Katewa, S. D., Vargas, M. A., Kolipinski, M. C., Lu, T. A., et al. (2009). 4E-BP extends lifespan upon dietary restriction by enhancing mitochondrial activity in Drosophila. *Cell, 139*, 149–160.

CHAPTER SIX

LINC Complex Proteins in Development and Disease

Henning F. Horn[1]
Development and Regenerative Biology, Institute of Medical Biology, Singapore, Singapore
[1]Corresponding author: e-mail address: henning.horn@imb.a-star.edu.sg

Contents

Current Topics in Developmental Biology, Volume 109
ISSN 0070-2153
http://dx.doi.org/10.1016/B978-0-12-397920-9.00004-4

Abstract

The LINC complex spans the nuclear envelope and plays critical roles in coordinating nuclear and cytoplasmic activities and in organizing nuclear and cytoskeletal features. LINC complexes are assembled from KASH and SUN-domain proteins, which interact in the nuclear envelope and form a physical link between the cytoskeleton and the nuclear interior. A number of diseases have been associated with mutations in genes coding for LINC complex proteins. Mouse models of LINC complex protein have provided valuable insight into LINC complex functions and into how these proteins contribute to the various diseases with which they have been associated.

1. INTRODUCTION TO THE LINC COMPLEX

The nucleus holds the bulk of the genetic information of an organism and allows transcription and translation to be separated into different compartments, giving a cell greater control over both processes. The nuclear envelope (NE) forms the boundary between the cytoplasm and nucleoplasm and consists of an outer nuclear membrane (ONM) and an inner nuclear membrane (INM), which are separated by a 20–50 nm space known as the perinuclear space (PNS). Despite being separated by the PNS, the ONM and INM form a contiguous membrane system as they connect at sites of nuclear pore complex insertions. Nuclear pores allow for the regulated transport of macromolecules across the NE as well as diffusion of smaller molecules. Despite the connections at the nuclear pore complexes, the ONM and INM are biochemically distinct, largely due to the proteins that reside in each membrane. These proteins do not generally diffuse readily between the two membranes, and as a consequence, there are ONM proteins that are largely specific to the ONM and INM proteins that reside primarily in the INM. Underlying the INM is the nuclear lamina, a network of proteins that give the NE, and as a result the nucleus, some degree of rigidity and shape. Together, these three layers form the NE.

Positioning of the nucleus within the cell is important for several developmental processes and is regulated via interactions between the NE and the cytoskeleton (Burke & Roux, 2009; Gundersen & Worman, 2013; Luxton, Gomes, Folker, Worman, & Gundersen, 2011). A major protein structure responsible for correct nuclear placement is the LINC (linker of nucleoskeleton and cytoskeleton) complex, which forms a direct physical connection between the cytoskeleton and the nuclear interior (most recently reviewed in Isermann & Lammerding, 2013; Rothballer &

Kutay, 2013; Gundersen & Worman, 2013). The LINC complex consists of nesprins, or Klarsicht ANC-1 Syne Homology (KASH) domain proteins, which reside in the ONM and connect with SUN-domain proteins in the INM. Together, KASH and SUN-domain proteins form the core components of the LINC complex. KASH-domain proteins connect with various elements of the cytoskeleton, and SUN-domain proteins interact with various nucleoplasmic structures.

1.1. KASH- and SUN-domain proteins

Six different KASH-domain proteins have been identified in mammals (Rothballer & Kutay, 2013)(Fig. 6.1). KASH-domain proteins are single-pass transmembrane proteins that reside in the ONM. The bulk of these proteins protrude into the cytoplasm and different family members exhibit significant diversity in these cytoplasmic domains. However, the ~50 C-terminal amino acids that form the transmembrane and KASH

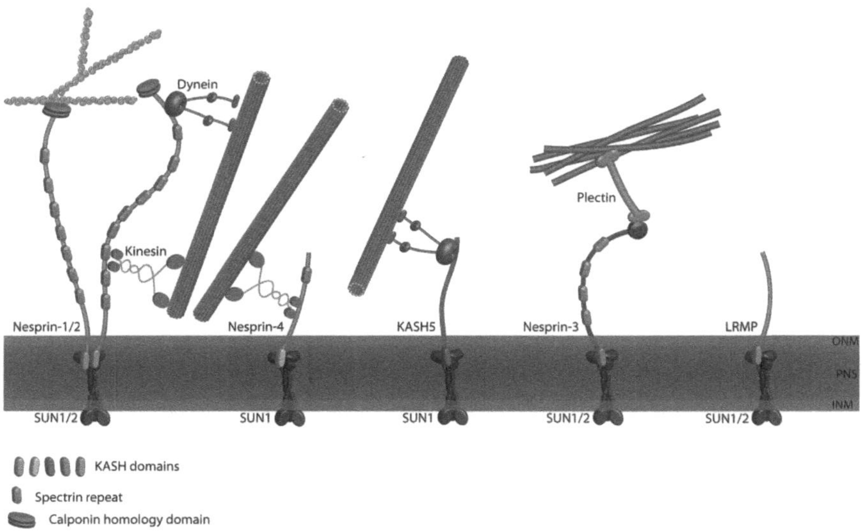

Figure 6.1 Mammalian LINC complex proteins. The six known KASH-domain proteins are shown with their respective SUN1/2 binding partners. Also shown are the cytoskeletal components with which the KASH-domain proteins have been shown to interact. Nesprin-1 and -2 interact with filamentous actin, and also with microtubules via the molecular motor proteins dynein and kinesin. Nesprin-4 interacts with the microtubule network via kinesin, and KASH5 interacts with microtubules via dynein. Nesprin-3 interacts with the intermediate filament network via the adapter protein plectin. No cytoskeletal interactions have been reported for LRMP. (See the color plate.)

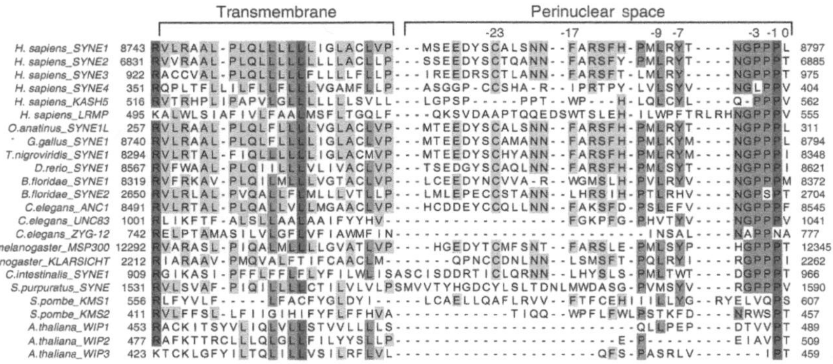

Current Opinion in Structural Biology

Figure 6.2 Alignment of the KASH and transmembrane domains from a large range of KASH-domain proteins. Organisms range from yeast and plants to humans. Residues that have been identified as important for SUN domain interactions are indicated with numbers. *Figure reproduced with permission from Sosa, Kutay, and Schwartz (2013).* (See the color plate.)

domains exhibit high homology, not only among KASH-domain protein family members of a single species but also across different species. Indeed some key residues are conserved across kingdoms (Fig. 6.2 from Sosa et al., 2013). The KASH-domain resides in the PNS where it interacts with SUN-domain proteins. The SUN (Sad-1 UNC-84) domain of these eponymous proteins is located at the C-terminus for most SUN proteins, while the N-terminus resides in the nucleoplasm and interacts with chromatin, other INM proteins, and components of the lamina. Five different SUN proteins have been identified in mammals, all of which are single-pass type II transmembrane proteins (Rothballer & Kutay, 2013).

1.2. The LINC complex

Structural biology recently provided significant insight not only into the structure of the SUN domain but also into the interaction of the SUN domain with KASH domain peptides (Sosa, Rothballer, Kutay, & Schwartz, 2012; Zhou, Du, et al., 2012). These studies demonstrated that SUN proteins assemble into trimers, which is essential for creating a KASH domain-binding pocket. The KASH domain-binding pocket runs across two adjacent SUN domains, forming a deep groove into which the KASH domain binds (Sosa et al., 2012). There appear to be three key points of contact between the SUN and KASH domains. The very C-terminus of the KASH domain forms several points of contact with the SUN protein and adding even a single additional residue to the KASH domain abolishes binding. Second, the hydrophobic residues located at −7 to −10 residues from the C-terminus bind to a structure called the KASH-lid on SUN domain one and at the same time to the core domain of SUN domain two. The KASH domain therefore fits snugly into the binding groove created by two SUN proteins. Third, most KASH domains have a conserved cysteine residue at position −23 from the C-terminus. This cysteine forms a disulfide bond with the SUN protein and significantly enhances the binding between KASH and SUN-domain proteins (Sosa et al., 2012). Since many of the KASH-domain proteins have dimerization motifs and are capable of homodimerization, these findings reveal an elegant mechanism for the formation of higher order SUN–KASH complexes that may be necessary to support the mechanical strain potentially exerted on these LINC complexes, such as in physical placement of the nucleus in various cell types (Sosa et al., 2012, 2013; Zhou, Du, et al., 2012).

While the structural studies on the SUN and KASH domain interactions have significantly advanced our understanding on the nature of the interaction, how this interaction is regulated remains poorly understood. Based on observations from ectopically expressed KASH and SUN-domain proteins in mammalian culture systems, it is clear that KASH-domain proteins localize to the ONM in a SUN-dependent manner. In addition, findings from several mouse models support the SUN-dependent localization of KASH-domain proteins to the NE (Horn, Brownstein, et al., 2013; Horn, Kim, et al., 2013; Lei et al., 2009). Removing SUN-domain proteins via RNA knockdown or genetic ablation prevents the accumulation of KASH-domain proteins in the NE (e.g., Horn, Brownstein, et al., 2013; Roux et al., 2009). Furthermore, ectopic expression of KASH-domain proteins in cultured cells often leads to an accumulation of these proteins in the ER, since the ER membrane is contiguous with the ONM. In these circumstances, increasing the expression of SUN-domain proteins pulls the KASH-domain proteins out of the ER and into the ONM. It is therefore clear that the INM SUN-domain proteins are essential for the localization of KASH-domain proteins to the ONM. While it seems unlikely that the formation of the LINC complexes is regulated by posttranslational modifications, something probably does regulate the interaction between SUN and KASH domains, since both proteins are made in the ER, but LINC complex formation has not been reported in this compartment. Sosa and colleagues speculate on a number of possibilities that prevent SUN–KASH interactions in the ER: the SUN domains may not trimerize until they reach the INM, chaperones prevent interactions in the ER, and/or prolines in the KASH domain may not all be in the correct *trans* conformation until they reach the ONM (Sosa et al., 2013).

Very little is also known about the regulation of LINC complex disassembly. The LINC complex needs to disassemble at least once during each cell cycle as the NE breaks down during mitosis, though this may not be the only time disassembly occurs. A recent study showed that TorsinA, a AAA+ ATPase, localizes to the NE and mediates displacement of some of the SUN and KASH-domain proteins, perhaps giving it a role in regulating the LINC complex disassembly (Sosa et al., 2013; Vander Heyden, Naismith, Snapp, Hodzic, & Hanson, 2009).

1.3. SUN protein localization

The formation of a functional LINC complex requires the proper localization of SUN-domain proteins to the INM. The shuttling of SUN protein from the ER synthesis sites to the NE requires a number of transport signals and pathways. Classical nuclear localization signals (cNLS), inner nuclear membrane sorting motifs (INM-SM), a SUN-nuclear envelope localization signal (SUN-NELS), the SUN domain, and Golgi retrieval signals all play a role in this process. In mammalian cells, SUN2 has a cNLS located in the amino terminal nucleoplasmic domain that helps target SUN2 to the INM by interacting with the nuclear importins (Turgay et al., 2010). Additional elements within the SUN domain itself also promote INM SUN2 localization, possibly through interactions with KASH-domain proteins in the ONM (Turgay et al., 2010). And finally, a stretch of four arginine residues interacts with the coatomer complex I protein and functions as a Golgi retrieval signal in retrieving SUN2 from the Golgi to the INM (Turgay et al., 2010). Findings from the *C. elegans* UNC-84 SUN-domain protein identified additional SUN protein INM-targeting sequences. While two cNLS were important, an INM localization of SUN-domain proteins sorting motif (INM-SM) and SUN-NELS both contributed to the NE localization (Tapley, Ly, & Starr, 2011). Mutations in any one of these motifs results in a delay, but not abrogation of UNC-84 NE localization. Only mutating all four sequences resulted in a complete failure for UNC-84 to localize to the NE, indicating that all four elements play a role in the correct targeting of SUN-domain proteins (Tapley et al., 2011). In addition, direct interactions with chromatin may also promote the INM localization of SUN proteins (Gardner et al., 2011).

2. THE LINC COMPLEX IN PLANTS

SUN-domain protein homologues have been discovered in several plant species, including rice, *Arabidopsis*, and maize (Graumann, Runions, & Evans, 2010; Moriguchi, 2005; Murphy, Simmons, & Bass, 2010). Indeed five SUN-domain proteins have been identified in maize (ZmSun1–5) of which two (ZmSun1 and ZmSun2) have good homology to mammalian SUN1 and SUN2 (Murphy et al., 2010). ZmSun3–5 have less homology to animal SUN-domain proteins, having three transmembrane domains (as opposed to one in animals), and the SUN domain in the middle of the proteins rather than at the C-terminus as for the canonical SUN-domain proteins (Murphy et al., 2010). Interestingly, ZmSun5 is preferentially expressed in pollen (Murphy et al., 2010), which is reminiscent of the testes-specific SUN isoforms in mammals. At a functional level, SUN-domain proteins in plants localize to the NE and contribute to nuclear shape in root hair cells (Moriguchi, 2005; Oda & Fukuda, 2011).

SUN-domain proteins in plants form a functional LINC complex with plant KASH-domain proteins. The KASH-domain proteins identified in plants (*Arabidopsis*) are WPP-interacting proteins which recruit a RanGTPase-activating protein to the NE (Xu, Meulia, & Meier, 2007). These proteins interact with the *Arabidopsis* SUN-domain proteins and together mediate nuclear shape in a variety of plant tissues, including leaf epidermal, trichome, and root hair cells (Zhou, Graumann, Evans, & Meier, 2012).

3. LINC COMPLEX IN YEAST

3.1. *Saccharomyces cerevisiae*

In the budding yeast, there are two distinct LINC complexes. Monopolar spindle 3 (Mps3)/Nep98p is the only identified SUN-domain protein, which interacts with either Mps2 or Csm4, the two ONM proteins with KASH-like functions. The Mps3/Mps2 LINC complex is important for spindle pole body (SPB—the yeast equivalent of the mammalian centrosome) duplication and insertion into the NE (Friederichs et al., 2011; Jaspersen, 2002; Nishikawa, 2002).

Msp3 can also interact with telomeres, anchoring them at the NE. This function not only protects telomere ends but is also necessary for rapid chromosome movement and the cohesion of sister chromatids during meiosis (Antoniacci, 2004; Bupp, Martin, Stensrud, & Jaspersen, 2007; Prasada

Rao, Shinohara, & Shinohara, 2011; Schober, Ferreira, Kalck, Gehlen, & Gasser, 2009). The Mps3/Csm4 LINC complex promotes chromosome movement, homologous recombination, and bouquet formation (Brown, Zanders, & Alani, 2011; Conrad et al., 2008; Conrad, Lee, Wilkerson, & Dresser, 2007; Kosaka, Shinohara, & Shinohara, 2008; Prasada Rao et al., 2011; Wanat et al., 2008).

3.2. *Schizossacharomayces pombe*

One SUN-domain protein (Sad1) and two KASH-domain proteins (Kms1 and Kms2) have been identified in fission yeast. Sad1 is found at the SPB and plays an important role in SPB assembly (Hagan & Yanagida, 1995). During meiosis, Sad1 localizes to telomeres and the Sad1/Kms1 LINC complex is essential for bouquet formation and progression through meiosis (Miki et al., 2004; Niwa, Shimanuki, & Miki, 2000; Shimanuki et al., 1997). The Sad1/Kms2 LINC complex links the nucleus to the microtubule cytoskeleton network and effects nuclear positioning (King, Drivas, & Blobel, 2008).

4. THE LINC COMPLEX IN *CAENORHABDITIS ELEGANS*

The nematode worm *C. elegans* expresses two SUN-domain proteins and four KASH-domain proteins. UNC-84 was the first SUN-domain protein identified and regulates nuclear migration during embryonic development (Malone, Fixsen, Horvitz, & Han, 1999). The first KASH-domain protein to be identified was ANC-1, which connects the nucleus to the actin cytoskeleton, and regulates nuclear positioning (Starr, 2002). ZYG-12, another KASH-domain protein that was identified shortly thereafter, was found to localize to centrosomes and the NE and is important for anchoring the centrosome to the NE (Malone et al., 2003). In nematodes expressing nonfunctional ZYG-12, centrosomes failed to attach to the NE, leading to a lethal phenotype that resulted from missegregation of DNA because of defective spindle formation (Malone et al., 2003). Matefin/SUN1 (SUN1) was identified as another SUN-domain protein in *C. elegans* (Fridkin et al., 2004). SUN1 is expressed in the embryo and in germ cells, where it is important for both embryonic survival and the survival of the germ cell lineage (Fridkin et al., 2004). SUN1/ZYG-12 forms a functional LINC complex (Minn, Rolls, Hanna-Rose, & Malone, 2009) that is important for homologous recombination in meiosis (Penkner et al., 2007; Sato et al., 2009), a process which requires that this LINC complex transmits forces generated by dynein and the microtubule network onto the chromatin (Sato et al., 2009). Phosphorylation-dependent modifications of SUN1 on CK2 consensus phosphorylation sites regulate the chromosome attachment sites on SUN1 during this process (Penkner et al., 2009).

An additional LINC complex is formed by UNC-84 and the KASH-domain protein UNC-83. UNC-83 localizes to the NE in an UNC-84-dependent manner and interacts with the microtubule molecular motor proteins kinesin-1 light chain KLC-2 (McGee, Rillo, Anderson, & Starr, 2006; Meyerzon, Fridolfsson, Ly, Mcnally, & Starr, 2009). Through the recruitment of KLC-2 to the NE, UNC-83 effectively turns the nucleus into a kinesin cargo, and the UNC-84/UNC-83 LINC complex is therefore important for nuclear migration (Meyerzon et al., 2009).

The final KASH-domain family member in *C. elegans* is KDP-1 (McGee, Stagljar, & Starr, 2009). KDP-1 can interact with both SUN-domain proteins, UNC-84 and SUN1, although it appears to form a functional LINC complex with SUN1 during embryonic and germline lineage development (McGee et al., 2009). KDP-1 appears to promote cell-cycle progression during both mitosis and meiosis from S phase to M, and interfering with KDP-1 expression leads to defects in larval development and gametogenesis (McGee et al., 2009).

5. THE LINC COMPLEX IN *DROSOPHILA MELANOGASTER*

The fruit fly *Drosophila* expresses two KASH-domain proteins and two SUN-domain proteins. MSP-300 is a KASH-domain protein that shows homology to the human Dystrophin protein and has a role in muscle development (Rosenberg-Hasson, Renert-Pasca, & Volk, 1996; Volk, 1992). MSP-300 also localizes to developing oocytes and nurse cells and promotes the anchoring of nuclei in an actin cytoskeleton-dependent manner (Yu et al., 2006). The second KASH-domain protein in *Drosophila* is Klarsicht (Klar). Klar links the centrosome to the nucleus and is important for the nuclear migration in cells of the developing eye (Kracklauer, Banks, Xie, Wu, & Fischer, 2007; Mosley-Bishop, Li, Patterson, & Fischer, 1999; Patterson et al., 2004). It also has a role in the transport of intracellular vesicles during embryo development (Welte, Gross, Postner, Block, & Wieschaus, 1998). Klar interacts with the microtubule network via its amino terminal sequences (Fischer, 2004), and it is likely through this interaction and with multiple motor proteins that Klar effects its transport roles (Welte et al., 1998). MSP-300 and Klar are required in muscle development, being important for the even distribution of nuclei in muscle fibers (Elhanany-Tamir et al., 2012). MSP-300 also contributes to the correct distribution of mitochondria and ER in muscle cells (Elhanany-Tamir et al., 2012).

The two SUN-domain proteins expressed in *Drosophila* are Klaroid and sperm-associated antigen 4 (SPAG4). Klaroid forms a functional LINC complex with Klar and is required for the eye development functions of Klar (Kracklauer et al., 2007). A *Klaroid* knockout has the same phenotype as *Klar*-null flies. In fact, even in a Klar-Klaroid double-knockout (DKO) fly, the only obvious phenotype is an eye defect (Kracklauer et al., 2007). SPAG4 is a testes-specific SUN-domain protein that interacts with a coiled-coil protein Yuri Gagarin and the dynein motor proteins and attaches the basal body to the nucleus during spermatogenesis (Kracklauer et al., 2010).

6. LINC COMPLEX PROTEINS IN MAMMALS

As organisms underwent evolutionarily development, the need to coordinate nuclear and cytoplasmic activities in a myriad of specialized cells was met by the development of additional LINC complexes. Six KASH-domain proteins and five SUN-domain proteins have been identified in mammals. The different LINC complexes that assemble out of these SUN and KASH components enable the nucleus to interact with all aspects of the cytoskeleton (i.e., microtubules, intermediate filaments, and actin microfilaments) and provide a variety of functional roles.

6.1. Nesprin-1/Nesprin-2

Nesprin-1 (Syne-1, Enaptin165, Myne1) and Nesprin-2 (Syne-2, NUANCE) were the first mammalian KASH-domain proteins, or nesprins (*N*uclear *e*nvelope *sp*ectrin *r*epeat) to be identified (Apel, 2000; Gough, Fan, Chu, Winnick, & Beck, 2003; Mislow et al., 2002; Padmakumar et al., 2004; Zhang, Ragnauth, Greener, Shanahan, & Roberts, 2002; Zhang et al., 2001; Zhen, Libotte, Munck, Noegel, & Korenbaum, 2002). Nesprin-1 and -2 are expressed in cardiac, skeletal, and smooth muscle cells, with elevated expression seen in nuclei of neuromuscular junctions (NMJs) (Apel, 2000; Zhang et al., 2001). The full-length isoforms of Nesprin-1 and -2 are ~1 mDa and ~800 kDa, respectively, though there are a large number of smaller splice variants that are functionally different from the full-length proteins. These isoforms localize to different cellular compartments and are expressed only in specific tissues (Rajgor, Mellad, Autore, Zhang, & Shanahan, 2012; Rajgor & Shanahan, 2013), which suggests that the smaller splice isoforms of Nesprin-1 and -2 may have important functions beyond their canonical roles at the NE (Rajgor et al., 2012). Though Nesprin-1 and -2 are categorically ONM proteins, there are reports that some isoforms function in the INM as well (recently reviewed in Morris & Randles, 2010).

Nesprin-1 and -2 both interact with F-actin through amino-terminal calponin homology domains (recently reviewed in Gundersen & Worman, 2013; Rajgor & Shanahan, 2013). Through this interaction with F-actin, Nesprin-2 regulates the nuclear movement in migrating fibroblasts (Luxton, Gomes, Folker, Vintinner, & Gundersen, 2010). Nesprin-1 and -2 have also been shown to interact with the microtubule network through interaction with molecular motor proteins kinesin-1 and dynein (Yu et al., 2011; Zhang et al., 2009). These interactions play an important role in interkinetic nuclear migration during retinal photoreceptor development, as well as in the migration of photoreceptors in the retina and neuronal migration in the cerebral cortex (Yu et al., 2011; Zhang et al., 2009).

6.2. Nesprin-3

Nesprin-3 was identified in 2005 as a novel ONM protein that interacts primarily with the intermediate filament network through the adapter protein plectin (Wilhelmsen, 2005). As is the case with Nesprin-1 and -2, Nesprin-3 is a type II transmembrane protein, though it is much smaller than either of the full-length Nesprin-1 or -2, with a size of around 110 kDa (Wilhelmsen, 2005). Nesprin-3 is expressed in a wide variety of tissues and cells, and expression of Nesprin-3 leads to the recruitment of plectin to the NE (Ketema, Kreft, Secades, Janssen, & Sonnenberg, 2013; Wilhelmsen, 2005). Two isoforms of Nesprin-3 have been identified, Nesprin-3α and Nesprin-3β. Nesprin-3α is the predominately expressed isoform and is also the one that binds plectin (Postel, Ketema, Kuikman, de Pereda, & Sonnenberg, 2011; Wilhelmsen, 2005). Nesprin-3 localizes to the ONM in a KASH domain-dependent manner and forms LINC complexes with both SUN1 and SUN2. Recently, a novel SUN1 isoform, SUN1η was found to interact with Nesprin-3. SUN1η is expressed during sperm head shaping in the forming acrosomal head, and Nesprin-3 forms a LINC complex with SUN1η at this location (Göb et al., 2010). Since it is not located at the NE, it will be interesting to determine whether this LINC complex forms via the same SUN–KASH domain interactions that have been elucidated from recent structural studies (Sosa et al., 2012; Zhou, Du, et al., 2012).

6.3. Nesprin-4

Nesprin-4 expression is largely restricted to secretory epithelia, (i.e., mammary tissue, exocrine pancreas and salivary glands) and to the hair cells of the cochlea (Horn, Brownstein, et al., 2013; Roux et al., 2009). Nesprin-4 is a ~42 kDa protein with a single spectrin repeat, and a well-conserved KASH domain at the C-terminus that promotes Nesprin-4 ONM localization in a SUN-dependent manner (Roux et al., 2009). Nesprin-4 interacts with the microtubule network through its interaction with the molecular motor kinesin, both the heavy chain Kif5B and the light chain (Roux et al., 2009). The microtubule-organizing center (MTOC) of the cell, or the centrosome, is normally juxtaposed to the NE. When Nesprin-4 is ectopically expressed in cultured cells, the MTOC and the Golgi are observed to migrate away from the NE, indicating that the Nesprin-4-mediated recruitment of kinesin to the NE effectively turns the nucleus into a kinesin cargo (Roux et al., 2009). This polarizing function of Nesprin-4 may contribute to the normal development of secretory epithelia.

6.4. Jaw1/LRMP/Futile Cycle

Jaw1/LRMP (lymphoid-restricted membrane protein) was first identified in B and T cells, where it is expressed on the ER membrane (Behrens et al., 1994). The localization of LRMP to the cytosolic side of the ER membrane is dependent on hydrophobic sequences at the C-terminus, which has sequence homology to the KASH domain of nesprin proteins (Behrens et al., 1994; Kearns, 1996; Lindeman & Pelegri, 2012). LRMP is expressed in taste receptor cells that require the type III inositol 1,4,5-triphosphate receptor (IP_3R3) for inducing the Ca^{2+} cascade in taste signaling. LRMP interacts with the IP_3R3, which indicates that it may also be important for IP_3R3-dependent Ca^{2+} signaling in taste receptors (Shindo et al., 2010), although this has not been directly demonstrated.

Recent studies in zebrafish implicate LRMP (Futile cycle) in pronuclear fusion (Lindeman & Pelegri, 2012). After fertilization of the egg, the male pronucleus contributes the MTOC/centrosome, which nucleates astral microtubules. These astral microtubules contact the female pronucleus, and in a dynein-dependent manner, the female and male pronuclei are pulled together to fuse. In the absence of LRMP, pronuclear fusion does not occur (Lindeman & Pelegri, 2012).

Interestingly, zebrafish *futile cycle* encodes for a protein that is nearly three times the size as its mammalian homolog (1437 amino acids vs. 535 (mouse) or 555 (human) amino acids) (Behrens et al., 1994; Lindeman & Pelegri, 2012). Protein domain analysis suggests that zebrafish Futile actually is the homolog of two different vertebrate KASH-domain proteins (Horn, Kim, et al., 2013). The C-terminal part of the zebrafish Futile corresponds to vertebrate LRMP, while the N-terminal domains appear to correspond to vertebrate KASH5 (Horn, Kim, et al., 2013; Lindeman & Pelegri, 2012). The corresponding SUN-domain protein for LRMP/futile cycle has not yet been identified, though LRMP colocalizes with both SUN1 and SUN2 in mammalian cells (Horn, Kim, et al., 2013).

6.5. KASH5

KASH5 is the most recently identified member of the mammalian KASH-domain family (Horn, Kim, et al., 2013; Morimoto et al., 2012). Using a BLASTP search with the mammalian LRMP identified a much larger zebrafish LRMP/Futile cycle. The N-terminal half was found to contain another putative KASH-domain sequence, which when used in a BLASTP search revealed a novel mammalian protein previously known as Ccdc155 (Horn, Kim, et al., 2013). This protein was also identified through a yeast-two-hybrid screen using Shugoshin 2, a cohesin protector during meiosis, as bait (Morimoto et al., 2012). While there is no evidence that KASH5 does in fact interact with Shugoshin 2, it was found to have a punctate localization pattern in meiosis (Morimoto et al., 2012). KASH5 interacts with the dynein and dynactin complex (Horn, Kim, et al., 2013; Morimoto et al., 2012), and when ectopically expressed in mammalian cells leads to the centrosome being drawn closer to the nucleus, a phenotype which is in contrast to that seen for Nesprin-4, which causes a separation of centrosome and nucleus (Horn, Kim, et al., 2013; Roux et al., 2009). KASH5 expression is most pronounced in testes, where its expression is found in developing spermatocytes (Horn, Kim, et al., 2013; Morimoto et al., 2012). During prophase I of meiosis, homologous chromosomes pair and bind in a process knows as synapsis (recently reviewed in Rog & Dernburg, 2013). This process allows for recombination and for the exchange of genetic information. Rapid chromosome movement during prophase I and the formation of a bouquet, which results from the clustering of telomeres near the MTOC, contribute to efficient synapsis (Scherthan, 2007). KASH5 forms a functional LINC complex with SUN1 in spermatocytes and connects the telomere ends of chromosomes to dynein motor protein to facilitate nuclear movement and bouquet formation (Horn, Kim, et al., 2013; Morimoto et al., 2012).

6.6. SUN1

SUN1 is expressed in many tissues and forms functional LINC complexes with at least five of the six known KASH-domain proteins (recently reviewed in Rothballer & Kutay, 2013). Mammalian SUN1 was first identified by proteomic analysis of NE proteins as a mammalian homologue of *C. elegans* UNC-84 (Dreger, Bengtsson, Schöneberg, Otto, & Hucho, 2001), and was subsequently confirmed to function as an INM LINC complex protein in mammalian cells, required for the proper NE localization of Nesprin-2 (Crisp, 2006; Padmakumar, 2005).

Several SUN1 isoforms have been identified, although the functions of most have not been elucidated. A recent study found at least seven different isoforms of SUN1, with most being expressed in testis, muscle, and brain (Göb et al., 2011). As mentioned earlier, one of these isoforms, SUN1η, is expressed in postmeiotic sperm and together with Nesprin-3 plays a role in shaping the sperm head (Göb et al., 2010).

6.7. SUN2

SUN2 is expressed in a variety of tissues and forms functional LINC complexes with Nesprin-1–3 (reviewed in Rothbauer et al., 2006). SUN2 appears to be the primary SUN protein responsible for anchoring transmembrane actin-associated nuclear (TAN) lines to the nucleus (Folker, Ostlund, Luxton, Worman, & Gundersen, 2011; Luxton et al., 2010). Fibroblasts that migrate into a scratch wound reposition the centrosome toward the leading edge by a rearward movement of the nucleus, away from the leading edge (Gomes, Jani, & Gundersen, 2005; Luxton et al., 2010). Nesprin-2G is the only actin-binding nesprin in migrating fibroblasts and together with SUN2 forms the TAN lines necessary for this nuclear movement (Folker et al., 2011; Luxton et al., 2010). SUN2 is essential for anchoring Nesprin-2G to the nucleus. In cells where SUN2 has been depleted, TAN lines can still assemble, but the nucleus is no longer coupled to the TAN lines and remains stationary (Folker et al., 2011).

SUN2 also interacts with Rab5, a small GTPase that has important roles in endocytosis and membrane dynamics (Epp, Rethmeier, Krämer, & Ungermann, 2011; Ewing et al., 2007; Liang, Chiu, Yip, & Chan, 2011). SUN2 interacts with Rab5 in a SUN domain-dependent manner, and a knockdown of SUN2 interferes with transferrin-mediated endocytosis. This finding suggests that SUN2 has cellular functions outside the INM and may be important for vesicle formation and endocytosis (Liang et al., 2011).

6.8. SUN3

SUN3 was first identified through database homology searches and is expressed primarily in testes (Crisp, 2006; Göb et al., 2010; Haque et al., 2006). SUN3 expression appears to be restricted to postmeiotic developing sperm, in a similar fashion to SUN1η. However, SUN3 has a different subcellular localization pattern from SUN1η. While SUN1η is strongly expressed at the anterior pole of the sperm head, SUN3 expression is located at the posterior side of the developing sperm head, at sites where microtubule bundles contact the NE (Göb et al., 2010). SUN3 appears to form a functional LINC complex with Nesprin-1 in these regions, as Nesprin-1 localizes to the same tubulin attachment sites as SUN3 (Göb et al., 2010). The functional significance of the Nesprin-1 SUN3 LINC complex needs to be determined. An interesting observation from these studies is that while two SUN and two KASH-domain proteins are expressed in the same cell, they do not appear to interact promiscuously, but form two distinct LINC complexes (Göb et al., 2010). Understanding how assembly of these different LINC complexes is regulated may give important insights into the regulation of LINC complex assembly in general.

6.9. SUN4/SPAG4

SUN4/SPAG4 was identified as a testes-specific protein in rat (Shao, Tarnasky, Lee, Oko, & van der Hoorn, 1999). It is specifically expressed in spermatids, where it interacts with Odf1, a protein found in the outer dense fibers (ODF) of the spermatozoon flagellum (Shao et al., 1999). Immuno EM studies localize SUN4 to the manchette microtubules and to axonemal microtubules. It is thought that SUN4 functions to recruit Odf1 to the axonemal microtubules and facilitates the assembly of the developing ODF (Shao et al., 1999). In humans, SUN4 is also found in testes, but its expression is not restricted to this tissue. Pancreas, stomach, lymph nodes, pituitary, and duodenum, all express SUN4, with the highest expression seen in the pancreas and the testes (Kennedy, Sebire, Kretser, & O'Bryan, 2003). It is worth noting that with the exception of the small intestine, which showed no SUN4 expression in the rat, none of the other tissues that showed SUN4 expression in humans were tested for SUN4 expression in rat (Shao et al., 1999). Thus, it is possible that SUN4 is expressed in other tissues in rodents. Nonetheless, what role SUN4 might play in the pancreas, or indeed any of the other tissues, has not been elucidated.

6.10. SUN5/SPAG4L

SUN5/SPAG4L/SRG4 (SPAG4L) was cloned from testicular tissue, with no expression detected in any of the other tissues that were analyzed in mouse (epididymis, ovary, kidney, muscle, heart, spleen, liver, lung, and brain) (Xing et al., 2004). SPAG4L is expressed in spermatocytes and round spermatids and expression decreases once spermatids elongate (Xing et al., 2004). An isoform of SPAG4L, SPAG4L-2, which differs from SPAG4L by the insertion of 25 amino acids near the N-terminus, was recently identified in testes (Frohnert, Schweizer, & Hoyer-Fender, 2011). Expression of both isoforms appears to be restricted to the round spermatid stage of sperm development, and both isoforms are expressed at roughly equal levels. When ectopically expressed in fibroblasts, SPAG4L and SPAG4L-2 are found in the ER. However, in the developing sperm, they are found in the NE, concentrating at the apical side of the nucleus, underneath the acrosome. This localization suggests a function that may be important for the correct positioning of the acrosome (Frohnert et al., 2011).

7. MOUSE MODELS OF LINC COMPLEX PROTEINS

7.1. Nesprin-1/Nesprin-2

Several different mouse models have been informative about the functions of Nesprin-1 and -2 *in vivo*. The first model published was a transgenic mouse that expressed a muscle-specific transgene of a C-terminal fragment of Nesprin-1 (Grady, Starr, Ackerman, Sanes, & Han, 2005). Specifically, this mouse expressed the C-terminal 344 residues of Nesprin-1, which includes the KASH domain, under the control of regulatory elements from the muscle creatine kinase gene (*Nesp1 DN KASH*). Transgenic expression of this KASH domain acts as a dominant negative and displaces endogenous Nesprin-1 from the NE of muscle cell nuclei, albeit not completely (Zhang, Xu, et al., 2007). Nevertheless, this decreased localization of Nesprin-1 in the NE of muscle cells leads to a decreased accumulation of nuclei under NMJs. This mouse was the first model to show that while Nesprin-1 is important for the positioning and maintenance of nuclei at NMJs, the function of NMJs is not affected, and that clustering of nuclei under NMJs is not essential for NMJ function (Grady et al., 2005) (Table 6.1).

To follow on from this study, Zhang and colleagues generated mice that lacked the KASH domains of Nesprin-1 and -2, (*Nesp1 ΔKASH* and *Nesp2 ΔKASH*, respectively) as well as a transgenic mouse expressing the 183

Table 6.1 Mouse models of LINC complex proteins

Mouse	LINC protein targeted	Method of targeting	Major phenotype(s) reported	References
Nesp1 DN KASH	Nesprin-1	C-terminal 344 aa of Nesprin-1 was expressed under the control of the mouse muscle creatine kinase promoter in transgenic animals	Failure to accumulate nuclei under NMJs	Grady et al. (2005)
Nesp1 ΔKASH	Nesprin-1	Targeted deletion of last exon, resulting in removal of KASH domain	Failure to accumulate nuclei under NMJs; improper organization of nuclei in muscle fibers	Zhang et al. (2007)
Nesp1 Δ/Δ	Nesprin-1	Floxed deletion of last exon resulting in a transcript that lacks the last two exons	50% neonatal lethality due to failure to breathe. Remaining 50% survive to adulthood but exhibit symptoms of muscle dystrophy, kyphoskoliosis, and heart conduction defects	Puckelwartz et al. (2008)
Nesprin-$1^{-/-}$	Nesprin-1	Floxed deletion of the last 16 exons to delete everything C-terminal to, and including the last spectrin repeat	60% neonatal lethality. Surviving KO animals show nuclear positioning defects in muscle fibers and a decreased capacity for exercise	Zhang et al. (2010)
Nesp2 DN KASH	Nesprin-2	C-terminal 183 aa of Nesprin-2 was expressed under the control of the mouse muscle creatine kinase promoter in transgenic animals	Displace Nesprin-2 as well as Nesprin-1 from nuclear envelope. Failure to accumulate nuclei under NMJs	Zhang, Xu, et al. (2007)
Nesp2 ΔKASH	Nesprin-2	Targeted deletion of last two exon resulting in removal of KASH domain	Poor retinal development due to defects in interkinetic nuclear migration. Defects in learning and memory	Yu et al. (2011), Zhang, Xu, et al. (2007), Zhang et al. (2009)

Continued

Table 6.1 Mouse models of LINC complex proteins—cont'd

Mouse	LINC protein targeted	Method of targeting	Major phenotype(s) reported	References
Nesp1/ *Nesp2* DKO	Nesprin-1 and Nesprin-2	Obtained by breeding *Nesp1* ΔKASH and *Nesp2* ΔKASH	Failure to breathe, neonatal lethality. Displaced nuclei at NMJs, but junctions are functional. Severe brain defects	Zhang, Xu, et al. (2007)
Nesprin-3 null	Nesprin-3	Removal of exon 2, which contains the start sequence for both isoforms of Nesprin-3	No overt phenotype	Ketema et al. (2013)
Nesprin-4 null	Nesprin-4	Targeted deletion of part of exon 1 and all of exons 2 through 6, removing most of the protein coding sequences	Hearing loss resulting from loss of auditory sensory cells	Horn, Brownstein, et al. (2013)
KASH5 null	KASH5	Floxed deletion of exons 5–8, resulting in premature stop codon	Sterility. Developing gametes arrested in meiotic prohase I	Horn, Kim, et al. (2013)
SUN1	SUN1	Floxed deletion of exons 10–11, coding for the transmembrane domain	Sterility. Developing gametes arrested in meiotic prohase I. Hearing loss resulting from loss of auditory sensory cells	Chi et al. (2009), Horn, Brownstein, et al. (2013), Horn, Kim, et al. (2013)
SUN1	SUN1	Targeted deletion of exons 10–13, coding for the transmembrane domain	Sterility. Developing gametes arrested in meiotic prohase I	Ding et al. (2007)
SUN2	SUN2	Targeted deletion of exons 11–16 coding for the SUN domain	No overt phenotype	Lei et al. (2009)
SUN1/ *SUN2* DKO	SUN1 and SUN2	Obtained by breeding *SUN1* (Ding et al., 2007) and *SUN2* mice	Failure to breathe, neonatal lethality. Displaced nuclei at NMJs, but junctions are functional. Severe brain defects. Defects in retinal development	Lei et al. (2009), Zhang et al. (2009)

The targeting strategy is listed for each mouse model along with any major phenotypes that have been reported.

C-terminal amino acids of Nesprin-2 under the control of the muscle creatine kinase promoter (*Nesp2 DN KASH*) (Zhang, Xu, et al., 2007). The *Nesp1 ΔKASH* mouse lines showed a near complete absence of nuclei under NMJs. The muscle fibers also exhibited an uneven distribution of nuclei with nuclear aggregates forming in the fibers, with a defect in nuclear anchoring (Zhang, Xu, et al., 2007). The NMJ phenotype of the *Nesp1 ΔKASH* mouse is similar, albeit more severe, than that seen in the *Nesprin-1 DN KASH* transgenic mouse. However, this is not surprising since the *Nesprin-1 KASH* transgene only caused partial displacement of Nesprin-1 from the NE (Grady et al., 2005; Zhang, Xu, et al., 2007). These results are in line with studies in *C. elegans*, which showed that ANC-1, the *C. elegans* homologue of Nesprin-1, is important for nuclear anchoring within cells (Hedgecock & Thomson, 1982; Starr, 2002).

The *Nesp2 ΔKASH* mouse did not display any defects in nuclear positioning, and nuclei at NMJs as well as nonsynaptic nuclei were properly positioned (Zhang, Xu, et al., 2007). However, the *Nesp2 ΔKASH* mouse suffers from retinal defects (Yu et al., 2011). In the process of retinal development, as photoreceptor cells divide and differentiate, the cells undergo interkinetic nuclear migration, where nuclei oscillate between basal and apical localization in each cell cycle. This nuclear movement is Nesprin-2 dependent, which links the molecular motor proteins kinesin and dynein to the NE and promotes nuclear movement (Yu et al., 2011). In *Nesp2 ΔKASH* mice, interkinetic nuclear migration and photoreceptor development is impaired and mice have structural and functional defects in the retina (Yu et al., 2011).

While mice lacking the KASH domain of Nesprin-2 do not appear to be affected by the aberrantly placed nuclei in NMJs, the *Nesp2 DN KASH* transgenic mouse displayed a phenotype similar to the one seen in the *Nesp1 DN KASH* transgenic mouse (Grady et al., 2005; Zhang, Xu, et al., 2007). The homology between KASH domains of Nesprin-1 and -2 is high and it is therefore not surprising that expressing the KASH domain of Nesprin-2 affects not only the nuclear localization of Nesprin-2 but also of Nesprin-1 (Zhang, Xu, et al., 2007). Thus, it is likely that the phenotype of the *Nesp2 DN KASH* is largely due to perturbing Nesprin-1 function.

Nesp1 ΔKASH and *Nesp2 ΔKASH* are viable and fertile. However, mice lacking the KASH domains of both Nesprin-1 and -2 (*Nesp1/Nesp2* DKO) die within minutes of birth due to respiratory failure (Zhang, Xu, et al., 2007). *Nesp1/Nesp2* DKO pups are cyanotic, show no signs of breathing, and examination of the lungs reveals that the aveoli are not inflated. As is the case in *Nesp1 ΔKASH* mice, the NMJs are affected, and NMJs have nearly no nuclei associated with them. Despite this fact, the NMJs are functional and the only muscle-associated phenotype in these animals is the displaced nuclei, both at NMJ and within the muscle fibers. The phrenic nerves, which innervate the diaphragm and are important for breathing, display longer branching in the *Nesp1/Nesp2* DKO mice, but appear to function normally (Zhang, Xu, et al., 2007).

Further analysis of the *Nesp1/Nesp2* DKO mice revealed a severe central nervous system defect, with significantly smaller brains in the *Nesp1/Nesp2* DKO mice compared to wild-type littermates (Zhang et al., 2009). The brains of the DKO mice showed inversions of some cell layers and complete absence of other cell layers. Furthermore they exhibited enlarged lateral ventricles as well as laminary defects in many regions of the brain (Zhang et al., 2009). At a cellular level, neurons from *Nesp1/Nesp2* DKO mice exhibited a detached centrosome and defects in nuclear movement. Primarily Nesprin-2 was found to play an important role in interkinetic nuclear migration of developing neurons and fewer neural progenitor cells were found in *Nesp1/Nesp2* DKO as well as in $Nesp1^{+/\Delta KASH}$ *Nesp2 ΔKASH* mice (Zhang et al., 2009). In addition, Nesprin-2 appears to colocalize with the molecular motor proteins dynein and kinesin and play an important role in linking the nucleus to the microtubule network via these molecular motors (Zhang et al., 2009). Thus, Nesprin-1 and -2 are required for normal brain development, with Nesprin-2 having a more dominant role in this process. Indeed, consistent with the role for Nesprin-2 in brain development, *Nesp2 ΔKASH* exhibited significant defects in learning and memory (Zhang et al., 2009).

Puckelwartz and colleagues generated another *Nesprin-1* knockout mouse model (*Nesp1 Δ/Δ*) using a similar approach to that used by Zhang and colleagues. However, instead of making a conventional knockout mouse, the last exon of *Nesprin-1* was flanked by *loxP* sites and the mouse was crossed with a *Protamine-Cre* to delete the last exon, effectively deleting the KASH domain (Puckelwartz et al., 2008). In the *Nesp1 Δ/Δ* mouse model, 50% of the null offspring died perinatally due to the inability to breathe. The mice that died showed a similar phenotype to that reported by Zhang and colleagues for the *Nesp1/Nesp2* DKO mice (Zhang, Xu, et al., 2007), with pups being cyanotic and aveoli not inflated. The other 50% of the *Nesp1 Δ/Δ* pups survived but were smaller and exhibited muscle wasting (Puckelwartz et al., 2008). The phenotypic discrepancy between the two *Nesprin-1*-null models may be due to a slightly different targeting strategy. Even though only the last exon was floxed in the *Nesp1 Δ/Δ* model, the resulting mRNA after deletion of the floxed sequence was missing not only the last, but also the penultimate exon (Puckelwartz et al., 2008). Thus, it is possible that the *Nesp1 ΔKASH* retains Nesprin-1 isoforms that are important for survival, which are eliminated by removing the penultimate exon (Zhang et al., 2010). Indeed, a Nesprin-1 isoform that would be expressed in the *Nesp1 ΔKASH* but not in the *Nesp1 Δ/Δ* was found in the brain (Zhang et al., 2010).

The muscle-wasting phenotype of the *Nesp1 Δ/Δ* mice, as evidenced by weaker hind limbs, was initially only significant in females. While some of the male *Nesp1 Δ/Δ* also showed muscle wasting, it was not significantly different to control mice (Puckelwartz et al., 2008). However, with age, both male and female mice showed significant muscle deterioration, which was linked with a failure to groom, increased incidents of eye infections, and kyphoscoliosis (Puckelwartz et al., 2008). The muscles of *Nesp1 Δ/Δ* mice had thinner myofibers and improperly organized synaptic and nonsynaptic nuclei. In addition, these mice developed cardiac conduction defects (Puckelwartz et al., 2008).

A third *Nesprin-1* knockout mouse model was generated lacking all Nesprin-1 isoforms containing the C-terminal spectrin repeat region and the KASH domain (Zhang et al., 2010). These mice ($Nesp1^{-/-}$) exhibited around 60% mortality in the first 2 weeks of life. Surviving mice had lower body weight at an early age, but by 3 months, there was no longer a significant weight difference between wild-type and $Nesp1^{-/-}$ mice. $Nesp1^{-/-}$ mice show abnormally positioned nuclei in muscle fibers, with nuclear anchoring defects. $Nesp1^{-/-}$ nuclei also failed to deform substantially when muscle fibers were stretched, indicating a disconnection between the nucleus and the cytoskeleton. The $Nesp1^{-/-}$ mice showed a decreased capacity for exercise compared to wild-type mice. This decreased capacity was not due to any cardiac defects, as heart function in $Nesp1^{-/-}$ mice was normal (Zhang et al., 2010).

7.2. Nesprin-3

The *Nesprin-3*-null mouse is viable and fertile with no overt detectable phenotype (Ketema et al., 2013). Unpublished observations from our own *Nesprin-3*-null mouse model confirm these findings (H. Horn and C. Stewart, unpublished data). Detailed analysis of the testes from *Nesprin-3*-null males indicate that Nesprin-3 is expressed in Sertoli cells, where it functions to recruit both plectin and vimentin to the nuclear periphery. However, this recruitment appears to be dispensable for normal testicular function (Ketema et al., 2013). In humans, Nesprin-3 may function in regulating cell morphology and flow-induced polarization of human aortic endothelial cells (Morgan et al., 2011). Since both zebrafish (Postel et al., 2011) and mouse models lacking Nesprin-3 do not show any overt phenotypes, it is likely that any critical Nesprin-3 function will only be discovered under situations of organ or organismal stress.

7.3. Nesprin-4

Nesprin-4-null mice are viable and fertile. Despite the fact that Nesprin-4 is expressed in secretory epithelia (Roux et al., 2009), there are no overt defects in mammary tissue, exocrine pancreas, or salivary glands (Horn, Brownstein, et al., 2013). The major phenotypic abnormality found in *Nesprin-4*-null mice is a loss of sensory cells in the cochlea and a concomitant loss of hearing (Horn, Brownstein, et al., 2013). The sensory cells in the ear are divided into outer hair cells (OHCs) and inner hair cells with the former acting as cochlear amplifiers and the latter as primary sound sensors (Richardson, de Monvel, & Petit, 2011). In normal OHCs, the nucleus is positioned basally, the localization of which requires Nesprin-4 (Horn, Brownstein, et al., 2013). In *Nesprin-4*-null mice, the nuclei in OHCs are found near the apex of the cell, and OHCs are lost after the onset of hearing around 2 weeks of age (Horn, Brownstein, et al., 2013). Auditory brainstem recordings show that the hearing loss in *Nesprin-4*-null mice is progressive, showing significant hearing loss at 2 weeks old, and becoming more pronounced as mice get older (Horn, Brownstein, et al., 2013).

7.4. KASH5

The *KASH5* knockout mouse is viable and phenotypically normal with the exception of having smaller testes or very poorly formed ovaries, and both males and females being sterile (Horn, Kim, et al., 2013). The function of KASH5 in gametogenesis is to link the chromosome ends inside the nucleus with dynein at the outer nuclear surface to facilitate the rapid chromosome movement associated with synapsis (Horn, Kim, et al., 2013). In the absence of KASH5, the chromosome ends do not recruit dynein and synapsis does not occur. Thus, the developing sperm in the testes from a *KASH5*-null mouse do not progress beyond the spermatocyte stage, where they are arrested in prophase I (leptotene/zygotene stage) of meiosis (Horn, Kim, et al., 2013).

7.5. SUN1

Because SUN1 forms the primary binding partner for KASH-domain proteins in a number of different tissues, it has a critical role in these tissues in establishing a functional LINC complex. Therefore, it is not surprising that *SUN1*-null mice share some of the same phenotypes as has been reported for the corresponding KASH domain LINC complex partner. For example, SUN1 forms a functional LINC complex with KASH5 in the process of spermatogenesis. Knockout of *KASH5* results in the arrest of spermatogenesis and infertility in knockout mice (Horn, Kim, et al., 2013). Indeed, knockout of *SUN1* results in a virtually identical phenotype, with arrest in the same stage of spermatogenesis (Chi et al., 2009; Ding et al., 2007; Horn, Kim, et al., 2013). A subtle difference between *SUN1*- and *KASH5*-null mice is that in the absence of KASH5, there are no pachytene spermatocytes, while in the absence of SUN1, the occasional pachytene spermatocyte was identified (Ding et al., 2007; Horn, Kim, et al., 2013). This likely reflects the fact that when chromosome ends are immobilized at the NE via SUN1, but are not moved around by KASH5/dynein, they are less likely to pair than in a scenario where chromosome ends are not attached to the NE (as is the case in SUN1-null spermatocytes) and are free to move around, even if not actively so.

SUN1 also forms a functional LINC complex with Nesprin-4 in the sensory cells of the ear. In these cells, loss of either *Nesprin-4* or *SUN1* results in hearing loss with an identical cellular phenotype (Horn, Brownstein, et al., 2013). Nesprin-4 localization to the ONM of the OHCs is dependent on SUN1, and in the absence of SUN1, Nesprin-4 does not localize to the NE of the OHCs (Horn, Brownstein, et al., 2013). Lastly, SUN1 forms a LINC complex with Nesprin-1 in muscle cells. Loss of *SUN1* in muscle cells results in a slight modification of the NMJ, with fewer nuclei found under the NMJ in *SUN1*-null mice than in wild-type control animals, reminiscent of the phenotype observed in *Nesp1 ΔKASH* mice (Lei et al., 2009). Indeed, Nesprin-1 localization to the NE in muscle cells is partially dependent on SUN1. SUN2 also contributes to the Nesprin-1 localization, but since loss of *SUN2* in muscle cells does not affect nuclear positioning or anchorage, it would seem that the SUN1/Nesprin-1 LINC complex is the functionally predominant LINC complex in nuclear positioning in muscle cells (Lei et al., 2009).

Recent studies have found a surprising relationship between SUN1 and the nuclear lamina. Mice lacking lamin A/C ($Lmna^{-/-}$) are a model for

AD-EDMD (autosomal dominant Emery–Dreifuss muscular dystrophy) (Sullivan et al., 1999), and *LmnaΔ9* mice exhibit progeroid features (Mounkes, Kozlov, Hernandez, Sullivan, & Stewart, 2003). Both mouse lines die within a few weeks of birth with defective growth, and in the *Lmna*$^{-/-}$ mice, cardiac and muscle defects. Surprisingly, removing SUN1 in either of these backgrounds significantly ameliorates the pathologies of these mice. They live significantly longer and gain significantly more weight than either *Lmna*$^{-/-}$ or *LmnaΔ9* do on their own (Chen et al., 2012). At a cellular level, SUN1 accumulates at the NE and Golgi in the absence of LaminA/C. This accumulation may be responsible for some of the LaminA/C-null phenotypes, as targeting of SUN1 to the Golgi in a wild-type background causes nuclear herniation, much like what is seen in LaminA/C-null mice. Interestingly, knocking down SUN1 in HGPS patient fibroblast decreases the incidence of senescence and increases cellular proliferation compared to controls (Chen et al., 2012).

One possible explanation for why SUN1 may have such an affect was suggested by a study showing that SUN1 and SUN2 play a role in the DNA damage response pathway (Lei et al., 2012). *SUN1/SUN2* DKO MEFs proliferate poorly and show a higher incidence of apoptosis compared to WT MEFs (Lei et al., 2012). While *SUN1/SUN2* DKO MEFs show a higher DNA damage level compared to WT MEFs, they do not activate the DNA damage response pathway, with decreased activation of ATM and γH2AX. SUN1 and SUN2 interact with the DNAPK, a holoenzyme well established to function in nonhomologous end joining DNA repair mechanisms (Lei et al., 2012 and references therein). Lei et al. (2012) propose that the reason too much SUN1 in *LaminA*-null mice results in a progeric phenotype may be due to an overactive DNA damage response pathway.

7.6. SUN2

Unlike the *SUN1*-knockout mouse, no phenotype has been discovered for the *SUN2*-knockout mouse (Lei et al., 2009), suggesting that its function is more redundant than that of SUN1. *SUN2*-null mice are overtly normal and do not display any defect in NMJs, hearing loss or fertility (Ding et al., 2007; Horn, Brownstein, et al., 2013; Lei et al., 2009).

7.7. SUN1/SUN2 DKO

While the *SUN1*-knockout mouse shows slight defects in NMJs, and a *SUN2*-knockout mouse has no detectable phenotype in muscle fibers and at NMJs, removing both *SUN1* and *SUN2* results in a severe defect in NMJs

(Chi et al., 2009; Ding et al., 2007; Lei et al., 2009), indicating that SUN1 and SUN2 have partially overlapping roles. In many tissues, this appears to be the case, given that single knockout animals either have no phenotype (*SUN2*) or phenotypes that are compatible with life (*SUN1*). However, SUN-domain proteins are essential for viability, as animals lacking both *SUN1* and *SUN2* die perinatally because of a failure to breathe (Lei et al., 2009; Zhang et al., 2009). DKO animals for *SUN1* and *SUN2* have a similar phenotype to the *Nesp1/Nesp2* DKO mice. In brief, they have severe defects in brain development, with brain size from DKO animals significantly smaller than wild-type controls or animals with even one remaining intact allele of either *SUN1* or *SUN2* (Lei et al., 2009; Zhang et al., 2009). SUN1 and SUN2 are essential for anchoring the centrosome at the nucleus. They are also necessary, through the anchoring of Nesprin1/2 proteins in the ONM, for recruiting cytoplasmic dynein/dynactin to the nucleus, which is necessary for nuclear movement and neuronal migration during brain development (Zhang et al., 2009). *SUN1/SUN2* DKO mice also display defects in retinal development very similar to what is seen in the *Nesp2 ΔKASH* mice (Yu et al., 2011).

8. DISEASES OF THE LINC COMPLEX

A number of human diseases are associated with mutations in genes coding for NE proteins, including some specifically coding for LINC complex members. One condition that has mutations associated with several NE proteins is EDMD. EDMD is a late-onset disease with initial manifestations of tightening of the elbows and Achilles tendon within the first decade of life, followed by progressive muscle wasting, including the hamstring, quadriceps, biceps, and triceps (Emery & Dreifuss, 1966). These patients also develop cardiac conduction defects later in life, which is usually the cause of death (Burke & Stewart, 2006; Emery & Dreifuss, 1966). EDMD is distinguished from other forms of muscular dystrophy in the fact that muscle contractures develop before there are any significant signs of muscle weakness (Bione et al., 1994). Because patients usually die from cardiac arrest, the survival can be greatly improved with the fitting of a pacemaker. There are several forms of EDMD, including X-linked and AD. The X-linked EDMD has been associated with mutations in the NE protein Emerin (Bione et al., 1994), whereas the AD form has been linked to mutations in the Lamin A/C protein (Bonne et al., 1999). Both diseases have similar phenotypes. However, only around 40% of the EDMD cases can be attributed to mutations in *Lamin A/C* and *Emerin* (Bonne et al., 2003; Zhang, Bethmann, et al., 2007).

Mutations in *Nesprin-1* and *-2* have recently been shown to have a role in EDMD. These mutations cause defects in nuclear morphology and mislocalization of other NE proteins, such as SUN2 and Emerin (Zhang, Bethmann, et al., 2007).

Other diseases linked to Nesprin-1 or -2 include autosomal recessive arthrogryposis, Meckel–Gruber syndrome, autosomal recessive cerebellar ataxia, and depression and bipolar disorder (Attali et al., 2009; Dawe et al., 2009; Dupré et al., 2007; Green et al., 2012; Gros-Louis et al., 2007). Autosomal recessive arthrogryposis is a group of conditions that all exhibit joint contracture due to reduced fetal movement. The underlying causes can be varied and include myopathic conditions, neurogenic processes involving both peripheral and central nervous system, and connective tissue disorders (Pakkasjärvi et al., 2006). A splice acceptor site mutation in *Nesprin-1* was recently associated with arthrogryposis (Attali et al., 2009). This mutation causes the retention of intron 136, introduction of a premature stop codon, and expression of a truncated Nesprin-1 protein lacking the KASH domain (Attali et al., 2009). Nesprin-1 mouse models exhibit both muscle and neurological defects, and this human *Nesprin-1* mutation may thus contribute to the arthrogryposis phenotype through both myopathic and neurogenic avenues (Attali et al., 2009).

Autosomal recessive cerebellar ataxia type 1 (ARCA1) is a late-onset cerebellar ataxia that results in an impaired ability to walk due to uncoordinated limb movement. Several *Nesprin-1* mutations were associated with ARCA1, and ARCA1 patients presented with cerebellar atrophy and decreased nuclei concentration under NMJs (Dupré et al., 2007; Gros-Louis et al., 2007). These phenotypes are very similar to the NMJ defects and brain disorders observed in the Nesprin-1 mouse models. Interestingly, a *Nesprin-1* single nucleotide polymorphism was recently linked to bipolar disorder and depression (Green et al., 2012).

Meckel–Gruber syndrome is an inherited disorder associated with defects in the primary cilium and basal body, resulting in renal dysplasia and defects in the central nervous system (Dawe et al., 2009). Nesprin-1 and -2 interact with components of the basal body, promoting basal body migration and ciliogenesis. In the absence of either Nesprin-1 or -2, basal bodies are not positioned correctly and ciliogenesis is affected (Dawe et al., 2009). While no Meckel–Gruber syndrome-associated mutations have been reported in *Nesprin-1* or *-2*, mutations in either will likely contribute to the phenotypes associated with Meckel–Gruber syndrome (Dawe et al., 2009).

No disease-associated mutations have been discovered in *Nesprin-3*. However, Nesprin-3 (under the name NET 53) was identified in a proteomic screen identifying NE proteins that mapped to chromosomal regions associated with a variety of dystrophies (Schirmer, 2003).

Nesprin-4 mutations in humans are associated with hearing loss (Horn, Brownstein, et al., 2013). Our collaborators identified two families with early-onset nonsyndromic progressive high-frequency hearing loss. The inheritance pattern was consistent with an autosomal recessive disease and a mutation was mapped to the *Nesprin-4* gene. A two base-pair deletion in the second exon leads to the expression of a truncated protein in *in vitro* experiments. This truncated protein does not localize properly, as it is missing the KASH domain, and is therefore predicted to represent a functional *Nesprin-4*-null phenotype (Horn, Brownstein, et al., 2013).

Mutations in *SUN1* have been associated with muscular dystrophies, both Duchenne muscular dystrophy as well as EDMD (Li, Meinke, Huong, Wehnert, & Noegel, 2013; Taranum et al., 2012). In all cases, the *SUN1* mutations were found in addition to other mutations that were likely the primary disease-causing mutation. However, it appears that the *SUN1* mutations modulate the disease phenotype. For example, the *SUN1 A203V* and *A614V* mutations found in EDMD patients were mutated on a background of mutations in *Emerin (A203V)* and *LAP2alpha (A614V)*. Both mutations affect the interaction of SUN1 with Emerin and caused a slightly more severe phenotype in patient fibroblasts cultured *in vitro* (Li et al., 2013).

SUN1 has recently been associated with DYT1 dystonia, a neurological condition that arises from a single glutamine deletion (ΔE) of the AAA+ protein torsinA. Torsin A normally localizes to the NE via interactions with the LaminA-associated-polypeptide 1. However, the ΔE mutant appears to form a novel interaction with SUN1, which is required to localize ΔE torsinA to the NE. This torsinA ΔE-SUN1 interaction suggests that the LINC complex may be involved in dystonia (Jungwirth, Kumar, Jeong, & Goodchild, 2011).

While SUN1 has potential associations with a number of human diseases, only a single *SUN2* mutation has been reported. A patient exhibiting clinical signs of EDMD and Charcot-Marie-Tooth syndrome was found to have a mutation in *SUN2* that led to an amino acid exchange in the N-terminus of SUN2 (T33A). This mutation was found to exist in conjunction with a *Nesprin-1* mutation (Taranum et al., 2012).

SUN4/SPAG4 expression levels have been linked to a variety of human cancer tissues (Kennedy et al., 2003). No direct mutations have been identified in *SUN4*, and while the role of SUN4 in cancer is unknown, it may serve to be a useful clinical marker (Kennedy et al., 2003).

9. CONCLUSIONS

Mouse models provide a valuable tool for studying human diseases. This has certainly been true for mouse models of LINC complex proteins and their associated diseases. The various Nesprin-1 mouse models have augmented our understanding of the underlying biology of muscular dystrophy, autosomal recessive arthrogryposis, and autosomal recessive cerebellar ataxia. The mouse models for the Nesprin-4-associated hearing loss were critical in elucidating the cell biology and provided key insights into this entirely novel class of human hearing loss. And mouse models of SUN1 and KASH5 have allowed us a greater appreciation for the importance of chromosomal movement in the development of gametes. Indeed, our mouse model-generated understanding of LINC complex functions may even prove to be predictive for human diseases. For example, a novel human disorder was recently described that has features of mandibular acral dysplasia but also includes deafness and male hypogonadism as prominent associated features (Shastry et al., 2010). Several candidate genes were examined, but no mutations were found to cause this genetic condition. However, given our knowledge of the roles of SUN1, and the phenotypes of the SUN1 mouse models (hearing loss and hypogonadism), it would be interesting to check the function of SUN1 in these patients.

While a number of mouse models now exist for LINC complex proteins, the field is still relatively young. Indeed, we are still discovering novel LINC complex functions and novel variants of LINC complexes. We therefore expect that future mouse models will continue to augment our understanding of the LINC complex in normal as well as pathophysiological roles.

ACKNOWLEDGMENTS

I would like to thank Drs. Colin Stewart and Brian Burke for helpful discussions and comments in preparing this chapter, and the Institute of Medical Biology and the Singapore Agency for Science, Technology, and Research (A*STAR) for funding.

REFERENCES

Antoniacci, L. M. (2004). The spindle pole body assembly component Mps3p/Nep98p functions in sister chromatid cohesion. *Journal of Biological Chemistry, 279*, 49542–49550.

Apel, E. D. (2000). Syne-1, A dystrophin- and klarsicht-related protein associated with synaptic nuclei at the neuromuscular junction. *Journal of Biological Chemistry, 275*, 31986–31995.

Attali, R., Warwar, N., Israel, A., Gurt, I., McNally, E., Puckelwartz, M., et al. (2009). Mutation of SYNE-1, encoding an essential component of the nuclear lamina, is responsible for autosomal recessive arthrogryposis. *Human Molecular Genetics, 18*, 3462–3469.

Behrens, T. W., Jagadeesh, J., Scherle, P., Kearns, G., Yewdell, J., & Staudt, L. M. (1994). Jaw1, A lymphoid-restricted membrane protein localized to the endoplasmic reticulum. *Journal of Immunology, 153*, 682–690.

Bione, S., Maestrini, E., Rivella, S., Mancini, M., Regis, S., Romeo, G., et al. (1994). Identification of a novel X-linked gene responsible for Emery-Dreifuss muscular dystrophy. *Nature Genetics, 8*, 323–327.

Bonne, G., Di Barletta, M. R., Varnous, S., Bécane, H. M., Hammouda, E. H., Merlini, L., et al. (1999). Mutations in the gene encoding lamin A/C cause autosomal dominant Emery-Dreifuss muscular dystrophy. *Nature Genetics, 21*, 285–288.

Bonne, G., Yaou, R. B., Béroud, C., Boriani, G., Brown, S., de Visser, M., et al. (2003). 108th ENMC International Workshop, 3rd Workshop of the MYO-CLUSTER project: EUROMEN, 7th International Emery-Dreifuss Muscular Dystrophy (EDMD) Workshop, 13–15 September 2002, Naarden, The Netherlands. *Neuromuscular Disorders, 13*, 508–515.

Brown, M. S., Zanders, S., & Alani, E. (2011). Sustained and rapid chromosome movements are critical for chromosome pairing and meiotic progression in budding yeast. *Genetics, 188*, 21–32.

Bupp, J. M., Martin, A. E., Stensrud, E. S., & Jaspersen, S. L. (2007). Telomere anchoring at the nuclear periphery requires the budding yeast Sad1-UNC-84 domain protein Mps3. *The Journal of Cell Biology, 179*, 845–854.

Burke, B., & Roux, K. J. (2009). Nuclei take a position: Managing nuclear location. *Developmental Cell, 17*, 587–597.

Burke, B., & Stewart, C. L. (2006). The laminopathies: The functional architecture of the nucleus and its contribution to disease. *Annual Review of Genomics and Human Genetics, 7*, 369–405.

Chen, C.-Y., Chi, Y.-H., Mutalif, R. A., Starost, M. F., Myers, T. G., Anderson, S. A., et al. (2012). Accumulation of the inner nuclear envelope protein Sun1 is pathogenic in progeric and dystrophic laminopathies. *Cell, 149*, 565–577.

Chi, Y. H., Cheng, L. I., Myers, T., Ward, J. M., Williams, E., Su, Q., et al. (2009). Requirement for Sun1 in the expression of meiotic reproductive genes and piRNA. *Development, 136*, 965–973.

Conrad, M. N., Lee, C.-Y., Chao, G., Shinohara, M., Kosaka, H., Shinohara, A., et al. (2008). Rapid telomere movement in meiotic prophase is promoted by NDJ1, MPS3, and CSM4 and is modulated by recombination. *Cell, 133*, 1175–1187.

Conrad, M. N., Lee, C.-Y., Wilkerson, J. L., & Dresser, M. E. (2007). MPS3 mediates meiotic bouquet formation in Saccharomyces cerevisiae. *Proceedings of the National Academy of Sciences of the United States of America, 104*, 8863–8868.

Crisp, M. (2006). Coupling of the nucleus and cytoplasm: role of the LINC complex. *The Journal of Cell Biology, 172*, 41–53.

Dawe, H. R., Adams, M., Wheway, G., Szymanska, K., Logan, C. V., Noegel, A. A., et al. (2009). Nesprin-2 interacts with meckelin and mediates ciliogenesis via remodelling of the actin cytoskeleton. *Journal of Cell Science, 122*, 2716–2726.

Ding, X., Xu, R., Yu, J., Xu, T., Zhuang, Y., & Han, M. (2007). SUN1 is required for telomere attachment to nuclear envelope and gametogenesis in mice. *Developmental Cell, 12*, 863–872.

Dreger, M., Bengtsson, L., Schöneberg, T., Otto, H., & Hucho, F. (2001). Nuclear envelope proteomics: novel integral membrane proteins of the inner nuclear membrane. *Proceedings of the National Academy of Sciences of the United States of America, 98*, 11943–11948.

Dupré, N., Gros-Louis, F., Chrestian, N., Verreault, S., Brunet, D., de Verteuil, D., et al. (2007). Clinical and genetic study of autosomal recessive cerebellar ataxia type 1. *Annals of Neurology, 62*, 93–98.

Elhanany-Tamir, H., Yu, Y. V., Shnayder, M., Jain, A., Welte, M., & Volk, T. (2012). Organelle positioning in muscles requires cooperation between two KASH proteins and microtubules. *The Journal of Cell Biology, 198*, 833–846.

Emery, A. E., & Dreifuss, F. E. (1966). Unusual type of benign x-linked muscular dystrophy. *Journal of Neurology, Neurosurgery and Psychiatry, 29*, 338–342.

Epp, N., Rethmeier, R., Krämer, L., & Ungermann, C. (2011). Membrane dynamics and fusion at late endosomes and vacuoles—Rab regulation, multisubunit tethering complexes and SNAREs. *European Journal of Cell Biology, 90*, 779–785.

Ewing, R. M., Chu, P., Elisma, F., Li, H., Taylor, P., Climie, S., et al. (2007). Large-scale mapping of human protein–protein interactions by mass spectrometry. *Molecular Systems Biology, 3*, 1–17.

Fischer, J. A. (2004). Drosophila klarsicht has distinct subcellular localization domains for nuclear envelope and microtubule localization in the eye. *Genetics, 168*, 1385–1393.

Folker, E. S., Ostlund, C., Luxton, G. W. G., Worman, H. J., & Gundersen, G. G. (2011). Lamin A variants that cause striated muscle disease are defective in anchoring transmembrane actin-associated nuclear lines for nuclear movement. *PNAS, 108*, 131–136.

Fridkin, A., Mills, E., Margalit, A., Neufeld, E., Lee, K. K., Feinstein, N., et al. (2004). Matefin, a Caenorhabditis elegans germ line-specific SUN-domain nuclear membrane protein, is essential for early embryonic and germ cell development. *Proceedings of the National Academy of Sciences of the United States of America, 101*, 6987–6992.

Friederichs, J. M., Ghosh, S., Smoyer, C. J., McCroskey, S., Miller, B. D., Weaver, K. J., et al. (2011). The SUN protein Mps3 is required for spindle pole body insertion into the nuclear membrane and nuclear envelope homeostasis. *PLoS Genetics, 7*, e1002365.

Frohnert, C., Schweizer, S., & Hoyer-Fender, S. (2011). SPAG4L/SPAG4L-2 are testis-specific SUN domain proteins restricted to the apical nuclear envelope of round spermatids facing the acrosome. *Molecular Human Reproduction, 17*, 207–218.

Gardner, J. M., Smoyer, C. J., Stensrud, E. S., Alexander, R., Gogol, M., Wiegraebe, W., et al. (2011). Targeting of the SUN protein Mps3 to the inner nuclear membrane by the histone variant H2A.Z. *The Journal of Cell Biology, 193*, 489–507.

Göb, E., Meyer-Natus, E., Benavente, R., & Alsheimer, M. (2011). Expression of individual mammalian Sun1 isoforms depends on the cell type. *Communicative & Integrative Biology, 4*, 440–442.

Göb, E., Schmitt, J., Benavente, R., & Alsheimer, M. (2010). Mammalian sperm head formation involves different polarization of two novel LINC complexes. *PLoS One, 5*, e12072.

Gomes, E. R., Jani, S., & Gundersen, G. G. (2005). Nuclear movement regulated by Cdc42, MRCK, myosin, and actin flow establishes MTOC polarization in migrating cells. *Cell, 121*, 451–463.

Gough, L. L., Fan, J., Chu, S., Winnick, S., & Beck, K. A. (2003). Golgi localization of Syne-1. *Molecular Biology of the Cell, 14*, 2410–2424.

Grady, R. M., Starr, D. A., Ackerman, G. L., Sanes, J. R., & Han, M. (2005). Syne proteins anchor muscle nuclei at the neuromuscular junction. *Proceedings of the National Academy of Sciences of the United States of America, 102*, 4359–4364.

Graumann, K., Runions, J., & Evans, D. E. (2010). Characterization of SUN-domain proteins at the higher plant nuclear envelope. *The Plant Journal, 61*, 134–144.

Green, E. K., Grozeva, D., Forty, L., Gordon-Smith, K., Russell, E., Farmer, A., et al. (2012). mp201248a. *Molecular Psychiatry, 18*, 614–617.

Gros-Louis, F., Dupré, N., Dion, P., Fox, M. A., Laurent, S., Verreault, S., et al. (2007). Mutations in SYNE1 lead to a newly discovered form of autosomal recessive cerebellar ataxia. *Nature Genetics, 39*, 80–85.

Gundersen, G. G., & Worman, H. J. (2013). Nuclear positioning. *Cell, 152*, 1376–1389.

Hagan, I., & Yanagida, M. (1995). The product of the spindle formation gene sad1+ associates with the fission yeast spindle pole body and is essential for viability. *Journal of Cell Biology, 129*, 1033–1047.

Haque, F., Lloyd, D. J., Smallwood, D. T., Dent, C. L., Shanahan, C. M., Fry, A. M., et al. (2006). SUN1 interacts with nuclear lamin A and cytoplasmic nesprins to provide a physical connection between the nuclear lamina and the cytoskeleton. *Molecular and Cellular Biology, 26*, 3738–3751.

Hedgecock, E. M., & Thomson, J. N. (1982). A gene required for nuclear and mitochondrial attachment in the nematode Caenorhabditis elegans. *Cell, 30*, 321–330.

Horn, H. F., Brownstein, Z., Lenz, D. R., Shivatzki, S., Dror, A. A., Dagan-Rosenfeld, O., et al. (2013). The LINC complex is essential for hearing. *123*, 740–750.

Horn, H. F., Kim, D. I., Wright, G. D., Wong, E. S. M., Stewart, C. L., Burke, B., et al. (2013). A mammalian KASH domain protein coupling meiotic chromosomes to the cytoskeleton. *The Journal of Cell Biology, 202*, 1023–1039.

Isermann, P., & Lammerding, J. (2013). Nuclear mechanics and mechanotransduction in health and disease. *Current Biology, 23*, R1113–R1121.

Jaspersen, S. L. (2002). Mps3p is a novel component of the yeast spindle pole body that interacts with the yeast centrin homologue Cdc31p. *The Journal of Cell Biology, 159*, 945–956.

Jungwirth, M. T., Kumar, D., Jeong, D. Y., & Goodchild, R. E. (2011). 1471-2121-12-24. *BMC Cell Biology, 12*, 24.

Kearns, G. M. (1996). Carboxyl-terminal targeting and novel post-translational processing of JAW1, a lymphoid protein of the endoplasmic reticulum. *Journal of Biological Chemistry, 271*, 23528–23534.

Kennedy, C., Sebire, K., Kretser, D. M., & O'Bryan, M. K. (2003). Human sperm associated antigen 4 (SPAG4) is a potential cancer marker. *Cell and Tissue Research, 315*, 279–283.

Ketema, M., Kreft, M., Secades, P., Janssen, H., & Sonnenberg, A. (2013). Nesprin-3 connects plectin and vimentin to the nuclear envelope of Sertoli cells but is not required for Sertoli cell function in spermatogenesis. *Molecular Biology of the Cell, 24*, 2454–2466.

King, M., Drivas, T., & Blobel, G. (2008). A network of nuclear envelope membrane proteins linking centromeres to microtubules.pdf. *Cell, 134*, 427–438.

Kosaka, H., Shinohara, M., & Shinohara, A. (2008). Csm4-dependent telomere movement on nuclear envelope promotes meiotic recombination. *PLoS Genetics, 4*, e1000196.

Kracklauer, M. P., Banks, S. M. L., Xie, X., Wu, Y., & Fischer, J. A. (2007). Drosophila klaroid encodes a SUN domain protein required for Klarsicht localization to the nuclear envelope and nuclear migration in the eye. *Fly (Austin), 1*, 75–85.

Kracklauer, M. P., Wiora, H. M., Deery, W. J., Chen, X., Bolival, B., Romanowicz, D., et al. (2010). The Drosophila SUN protein Spag4 cooperates with the coiled-coil protein Yuri Gagarin to maintain association of the basal body and spermatid nucleus. *Journal of Cell Science, 123*, 2763–2772.

Lei, K., Zhang, X., Ding, X., Guo, X., Chen, M., Zhu, B., et al. (2009). SUN1 and SUN2 play critical but partially redundant roles in anchoring nuclei in skeletal muscle cells in mice. *PNAS, 106*, 10207–10212.
Lei, K., Zhu, X., Xu, R., Shao, C., Xu, T., Zhuang, Y., et al. (2012). Inner nuclear envelope proteins SUN1 and SUN2 play a prominent role in the DNA damage response. *Current Biology, 22*, 1609–1615.
Li, P., Meinke, P., Huong, L. T. T., Wehnert, M., & Noegel, A. A. (2013). Contribution of SUN1 Mutations to the Pathomechanism in Muscular Dystrophies. *Human Mutation, 35*, 452–461.
Liang, Y., Chiu, P. H., Yip, K. Y., & Chan, S. Y. (2011). Subcellular localization of SUN2 is regulated by lamin A and Rab5. *PLoS One, 6*, e20507.
Lindeman, R. E., & Pelegri, F. (2012). Localized products of futile cycle/lrmp promote centrosome-nucleus attachment in the zebrafish zygote. *Current Biology, 22*, 843–851.
Luxton, G. W. G., Gomes, E. R., Folker, E. S., Vintinner, E., & Gundersen, G. G. (2010). Linear arrays of nuclear envelope proteins harness retrograde actin flow for nuclear movement. *Science, 329*, 956–959.
Luxton, G. W. G., Gomes, E. R., Folker, E. S., Worman, H., & Gundersen, G. G. (2011). TAN lines: A novel nuclear envelope structure involved in nuclear positioning. *Nucleus, 2*, 173–181.
Malone, C. J., Fixsen, W. D., Horvitz, H. R., & Han, M. (1999). UNC-84 localizes to the nuclear envelope and is required for nuclear migration and anchoring during C. elegans development. *Development, 126*, 3171–3181.
Malone, C. J., Misner, L., Le Bot, N., Tsai, M.-C., Campbell, J. M., Ahringer, J., et al. (2003). The C. elegans hook protein, ZYG-12, mediates the essential attachment between the centrosome and nucleus. *Cell, 115*, 825–836.
McGee, M. D., Rillo, R., Anderson, A. S., & Starr, D. A. (2006). UNC-83 IS a KASH protein required for nuclear migration and is recruited to the outer nuclear membrane by a physical interaction with the SUN protein UNC-84. *Molecular Biology of the Cell, 17*, 1790–1801.
McGee, M. D., Stagljar, I., & Starr, D. A. (2009). KDP-1 is a nuclear envelope KASH protein required for cell-cycle progression. *Journal of Cell Science, 122*, 2895–2905.
Meyerzon, M., Fridolfsson, H. N., Ly, N., Mcnally, F. J., & Starr, D. A. (2009). UNC-83 is a nuclear-specific cargo adaptor for kinesin-1-mediated nuclear migration. *Development, 136*, 2725–2733.
Miki, F., Kurabayashi, A., Tange, Y., Okazaki, K., Shimanuki, M., & Niwa, O. (2004). Two-hybrid search for proteins that interact with Sad1 and Kms1, two membrane-bound components of the spindle pole body in fission yeast. *Molecular Genetics and Genomics, 270*, 449–461.
Minn, I. L., Rolls, M. M., Hanna-Rose, W., & Malone, C. J. (2009). SUN-1 and ZYG-12, mediators of centrosome-nucleus attachment, are a functional SUN/KASH pair in Caenorhabditis elegans. *Molecular Biology of the Cell, 20*, 4586–4595.
Mislow, J. M. K., Holaska, J. M., Kim, M. S., Lee, K. K., Segura-Totten, M., Wilson, K. L., et al. (2002). Nesprin-1alpha self-associates and binds directly to emerin and lamin A in vitro. *FEBS Letters, 525*, 135–140.
Morgan, J. T., Pfeiffer, E. R., Thirkill, T. L., Kumar, P., Peng, G., Fridolfsson, H. N., et al. (2011). Nesprin-3 regulates endothelial cell morphology, perinuclear cytoskeletal architecture, and flow-induced polarization. *Molecular Biology of the Cell, 22*, 4324–4334.
Moriguchi, K. (2005). Functional isolation of novel nuclear proteins showing a variety of subnuclear localizations. *The Plant Cell Online, 17*, 389–403.
Morimoto, A., Shibuya, H., Zhu, X., Kim, J., Ishiguro, K. I., Han, M., et al. (2012). A conserved KASH domain protein associates with telomeres, SUN1, and dynactin during mammalian meiosis. *The Journal of Cell Biology, 198*, 165–172.

Morris, G. E., & Randles, K. N. (2010). Nesprin isoforms: Are they inside or outside the nucleus? *Biochemical Society Transactions*, *38*, 278.
Mosley-Bishop, K. L., Li, Q., Patterson, L., & Fischer, J. A. (1999). Molecular analysis of the klarsicht gene and its role in nuclear migration within differentiating cells of the Drosophila eye. *Current Biology*, *9*, 1211–1220.
Mounkes, L. C., Kozlov, S., Hernandez, L., Sullivan, T., & Stewart, C. L. (2003). A progeroid syndrome in mice is caused by defects in A-type lamins. *Nature*, *423*, 298–301.
Murphy, S. P., Simmons, C. R., & Bass, H. W. (2010). 1471-2229-10-269. *BMC Plant Biology*, *10*, 269.
Nishikawa, S. I. (2002). Nep98p is a component of the yeast spindle pole body and essential for nuclear division and fusion. *Journal of Biological Chemistry*, *278*, 9938–9943.
Niwa, O., Shimanuki, M., & Miki, F. (2000). Telomere-led bouquet formation facilitates homologous chromosome pairing and restricts ectopic interaction in fission yeast meiosis. *EMBO Journal*, *19*, 3831–3840.
Oda, Y., & Fukuda, H. (2011). Dynamics of Arabidopsis SUN proteins during mitosis and their involvement in nuclear shaping. *The Plant Journal*, *66*, 629–641.
Padmakumar, V. C. (2005). The inner nuclear membrane protein Sun1 mediates the anchorage of Nesprin-2 to the nuclear envelope. *Journal of Cell Science*, *118*, 3419–3430.
Padmakumar, V. C., Abraham, S., Braune, S., Noegel, A. A., Tunggal, B., Karakesisoglou, I., et al. (2004). Enaptin, a giant actin-binding protein, is an element of the nuclear membrane and the actin cytoskeleton. *Experimental Cell Research*, *295*, 330–339.
Pakkasjärvi, N., Ritvanen, A., Herva, R., Peltonen, L., Kestilä, M., & Ignatius, J. (2006). Lethal congenital contracture syndrome (LCCS) and other lethal arthrogryposes in Finland—An epidemiological study. *American Journal of Medical Genetics*, *140A*, 1834–1839.
Patterson, K., Molofsky, A. B., Robinson, C., Acosta, S., Cater, C., & Fischer, J. A. (2004). The functions of Klarsicht and nuclear lamin in developmentally regulated nuclear migrations of photoreceptor cells in the Drosophila eye. *Molecular Biology of the Cell*, *15*, 600–610.
Penkner, A. M., Fridkin, A., Gloggnitzer, J., Baudrimont, A., Machacek, T., Woglar, A., et al. (2009). Meiotic chromosome homology search involves modifications of the nuclear envelope protein Matefin/SUN-1. *Cell*, *1–14*.
Penkner, A., Tang, L., Novatchkova, M., Ladurner, M., Fridkin, A., Gruenbaum, Y., et al. (2007). The nuclear envelope protein Matefin/SUN-1 is required for homologous pairing in C. elegans meiosis. *Developmental Cell*, *12*, 873–885.
Postel, R., Ketema, M., Kuikman, I., de Pereda, J. M., & Sonnenberg, A. (2011). Nesprin-3 augments peripheral nuclear localization of intermediate filaments in zebrafish. *Journal of Cell Science*, *124*, 755–764.
Prasada Rao, H. B. D., Shinohara, M., & Shinohara, A. (2011). Mps3 SUN domain is important for chromosome motion and juxtaposition of homologous chromosomes during meiosis. *Genes to Cells*, *16*, 1081–1096.
Puckelwartz, M. J., Kessler, E., Zhang, Y., Hodzic, D., Randles, K. N., Morris, G., et al. (2008). Disruption of nesprin-1 produces an Emery Dreifuss muscular dystrophy-like phenotype in mice. *Human Molecular Genetics*, *18*, 607–620.
Rajgor, D., Mellad, J. A., Autore, F., Zhang, Q., & Shanahan, C. M. (2012). Multiple novel Nesprin-1 and Nesprin-2 variants act as versatile tissue-specific intracellular scaffolds. *PLoS One*, 7, e40098.
Rajgor, D., & Shanahan, C. M. (2013). Nesprins: from the nuclear envelope and beyond. *Expert Reviews in Molecular Medicine*, *15*, e5.
Richardson, G. P., de Monvel, J. B., & Petit, C. (2011). How the genetics of deafness illuminates auditory physiology. *Annual Review of Physiology*, *73*, 311–334.

Rog, O., & Dernburg, A. F. (2013). Chromosome pairing and synapsis during Caenorhabditis elegans meiosis. *Current Opinion in Cell Biology*, *25*, 349–356.
Rosenberg-Hasson, Y., Renert-Pasca, M., & Volk, T. (1996). A Drosophila dystrophin-related protein, MSP-300, is required for embryonic muscle morphogenesis. *Mechanisms of Development*, *60*, 83–94.
Rothballer, A., & Kutay, U. (2013). The diverse functional LINCs of the nuclear envelope to the cytoskeleton and chromatin. *Chromosoma*, *122*, 415–429.
Rothbauer, U., Zolghadr, K., Tillib, S., Nowak, D., Schermelleh, L., Gahl, A., et al. (2006). Targeting and tracing antigens in live cells with fluorescent nanobodies. *Nature Methods*, *3*, 887–889.
Roux, K. J., Crisp, M. L., Liu, Q., Kim, D., Kozlov, S., Stewart, C. L., et al. (2009). Nesprin 4 is an outer nuclear membrane protein that can induce kinesin-mediated cell polarization. *PNAS*, *106*, 2194–2199.
Sato, A., Isaac, B., Phillips, C. M., Rillo, R., Carlton, P. M., Wynne, D. J., et al. (2009). Cytoskeletal forces span the nuclear envelope to coordinate meiotic chromosome pairing and synapsis. *Cell*, *139*, 907–919.
Scherthan, H. (2007). Telomeres and meiosis in health and disease. *Cellular and Molecular Life Sciences*, *64*, 117–124.
Schirmer, E. C. (2003). Nuclear membrane proteins with potential disease links found by subtractive proteomics. *Science*, *301*, 1380–1382.
Schober, H., Ferreira, H., Kalck, V., Gehlen, L. R., & Gasser, S. M. (2009). Yeast telomerase and the SUN domain protein Mps3 anchor telomeres and repress subtelomeric recombination. *Genes & Development*, *23*, 928–938.
Shao, X., Tarnasky, H. A., Lee, J. P., Oko, R., & van der Hoorn, F. A. (1999). Spag4, a novel sperm protein, binds outer dense-fiber protein Odf1 and localizes to microtubules of manchette and axoneme. *Developmental Biology*, *211*, 109–123.
Shastry, S., Simha, V., Godbole, K., Sbraccia, P., Melancon, S., Yajnik, C. S., et al. (2010). A novel syndrome of mandibular hypoplasia, deafness, and progeroid features associated with lipodystrophy, undescended testes, and male hypogonadism. *Journal of Clinical Endocrinology & Metabolism*, *95*, E192–E197.
Shimanuki, M., Miki, F., Ding, D. Q., Chikashige, Y., Hiraoka, Y., Horio, T., et al. (1997). A novel fission yeast gene, kms1+, is required for the formation of meiotic prophase-specific nuclear architecture. *Molecular and General Genetics*, *254*, 238–249.
Shindo, Y., Kim, M. R., Miura, H., Yuuki, T., Kanda, T., Hino, A., et al. (2010). Lrmp/Jaw1 is expressed in sweet, bitter, and umami receptor-expressing cells. *Chemical Senses*, *35*, 171–177.
Sosa, B. A., Kutay, U., & Schwartz, T. U. (2013). Structural insights into LINC complexes. *Current Opinion in Structural Biology*, *23*, 285–291.
Sosa, B. A., Rothballer, A., Kutay, U., & Schwartz, T. U. (2012). LINC complexes form by binding of three KASH peptides to domain interfaces of trimeric SUN proteins. *Cell*, *149*, 1035–1047.
Starr, D. A. (2002). Role of ANC-1 in tethering nuclei to the actin cytoskeleton. *Science*, *298*, 406–409.
Sullivan, T., Escalante-Alcalde, D., Bhatt, H., Anver, M., Bhat, N., Nagashima, K., et al. (1999). Loss of A-type lamin expression compromises nuclear envelope integrity leading to muscular dystrophy. *Journal of Cell Biology*, *147*, 913–920.
Tapley, E. C., Ly, N., & Starr, D. A. (2011). Multiple mechanisms actively target the SUN protein UNC-84 to the inner nuclear membrane. *Molecular Biology of the Cell*, *22*, 1739–1752.
Taranum, S., Vaylann, E., Meinke, P., Abraham, S., Yang, L., Neumann, S., et al. (2012). LINC complex alterations in DMD and EDMD/CMT fibroblasts. *European Journal of Cell Biology*, *91*, 614–628.
Turgay, Y., Ungricht, R., Rothballer, A., Kiss, A., Csucs, G., Horvath, P., et al. (2010). emboj2010119a. *EMBO Journal*, *29*, 2262–2275.

Vander Heyden, A. B., Naismith, T. V., Snapp, E. L., Hodzic, D., & Hanson, P. I. (2009). LULL1 retargets TorsinA to the nuclear envelope revealing an activity that is impaired by the DYT1 dystonia mutation. *Molecular Biology of the Cell, 20*, 2661–2672.

Volk, T. (1992). A new member of the spectrin superfamily may participate in the formation of embryonic muscle attachments in Drosophila. *Development, 116*, 721–730.

Wanat, J. J., Kim, K. P., Koszul, R., Zanders, S., Weiner, B., Kleckner, N., et al. (2008). Csm4, in collaboration with Ndj1, mediates telomere-led chromosome dynamics and recombination during yeast meiosis. *PLoS Genetics, 4*, e1000188.

Welte, M. A., Gross, S. P., Postner, M., Block, S. M., & Wieschaus, E. F. (1998). Developmental regulation of vesicle transport in Drosophila embryos: Forces and kinetics. *Cell, 92*, 547–557.

Wilhelmsen, K. (2005). Nesprin-3, a novel outer nuclear membrane protein, associates with the cytoskeletal linker protein plectin. *The Journal of Cell Biology, 171*, 799–810.

Xing, X.-W., Li, L.-Y., Liu, G., Fu, J.-J., Tan, X.-J., & Lu, G.-X. (2004). Identification of a novel gene SRG4 expressed at specific stages of mouse spermatogenesis. *Acta Biochimica et Biophysica Sinica (Shanghai), 36*, 351–359.

Xu, X. M., Meulia, T., & Meier, I. (2007). Anchorage of plant RanGAP to the nuclear envelope involves novel nuclear-pore-associated proteins. *Current Biology, 17*, 1157–1163.

Yu, J., Lei, K., Zhou, M., Craft, C. M., Xu, G., Xu, T., et al. (2011). KASH protein Syne-2/Nesprin-2 and SUN proteins SUN1/2 mediate nuclear migration during mammalian retinal development. *Human Molecular Genetics, 20*, 1061–1073.

Yu, J., Starr, D. A., Wu, X., Parkhurst, S. M., Zhuang, Y., Xu, T., et al. (2006). The KASH domain protein MSP-300 plays an essential role in nuclear anchoring during Drosophila oogenesis. *Developmental Biology, 289*, 336–345.

Zhang, Q., Bethmann, C., Worth, N. F., Davies, J. D., Wasner, C., Feuer, A., et al. (2007). Nesprin-1 and -2 are involved in the pathogenesis of Emery Dreifuss muscular dystrophy and are critical for nuclear envelope integrity. *Human Molecular Genetics, 16*, 2816–2833.

Zhang, J., Felder, A., Liu, Y., Guo, L. T., Lange, S., Dalton, N. D., et al. (2010). Nesprin 1 is critical for nuclear positioning and anchorage. *Human Molecular Genetics, 19*, 329–341.

Zhang, X., Lei, K., Yuan, X., Wu, X., Zhuang, Y., Xu, T., et al. (2009). SUN1/2 and Syne/Nesprin-1/2 complexes connect centrosome to the nucleus during neurogenesis and neuronal migration in mice. *Neuron, 64*, 173–187.

Zhang, Q., Ragnauth, C., Greener, M. J., Shanahan, C. M., & Roberts, R. G. (2002). The Nesprins are giant actin-binding proteins, orthologous to Drosophila melanogaster muscle protein MSP-300. *Genomics, 80*, 473–481.

Zhang, Q., Skepper, J. N., Yang, F., Davies, J. D., Hegyi, L., Roberts, R. G., et al. (2001). Nesprins: a novel family of spectrin-repeat-containing proteins that localize to the nuclear membrane in multiple tissues. *Journal of Cell Science, 114*, 4485–4498.

Zhang, X., Xu, R., Zhu, B., Yang, X., Ding, X., Duan, S., et al. (2007). Syne-1 and Syne-2 play crucial roles in myonuclear anchorage and motor neuron innervation. *Development, 134*, 901–908.

Zhen, Y.-Y., Libotte, T., Munck, M., Noegel, A. A., & Korenbaum, E. (2002). NUANCE, a giant protein connecting the nucleus and actin cytoskeleton. *Journal of Cell Science, 115*, 3207–3222.

Zhou, Z., Du, X., Cai, Z., Song, X., Zhang, H., Mizuno, T., et al. (2012). Structure of Sad1-UNC84 homology (SUN) domain defines features of molecular bridge in nuclear envelope. *Journal of Biological Chemistry, 287*, 5317–5326.

Zhou, X., Graumann, K., Evans, D. E., & Meier, I. (2012). Novel plant SUN-KASH bridges are involved in RanGAP anchoring and nuclear shape determination. *The Journal of Cell Biology, 196*, 203–211.

INDEX

Note: Page numbers followed by "*f*" indicate figures and "*t*" indicate tables.

D

E

F

G

T

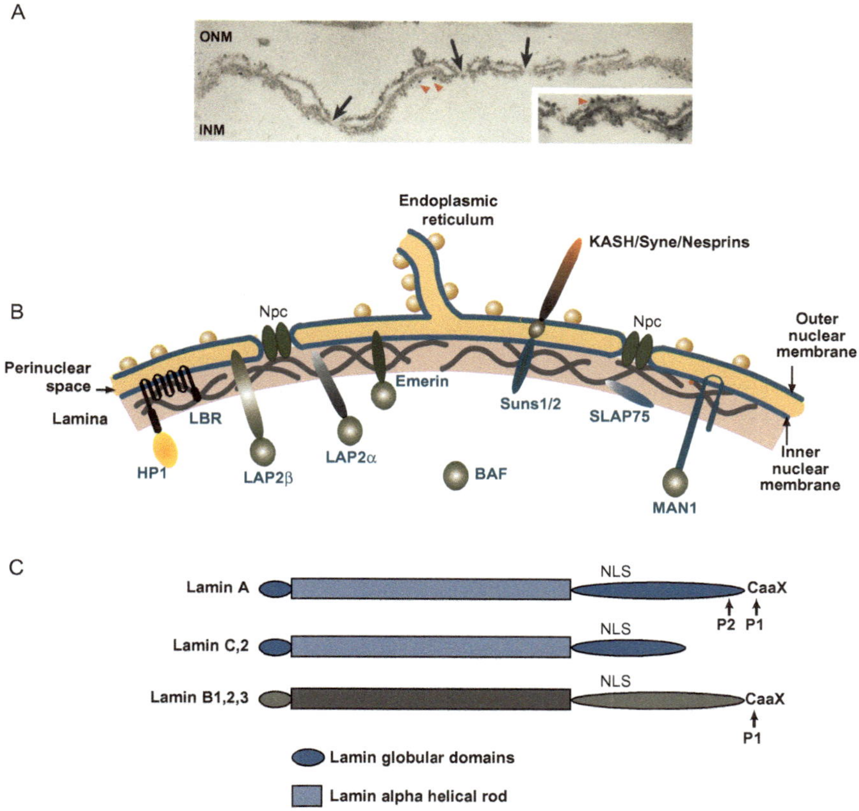

Brian Burke and Colin L. Stewart, Figure 1.1 The nuclear envelope (NE), lamina, and some of the more extensively characterized proteins associated with the lamina and NE. Panel A shows an electron micrograph of isolated rat liver NEs. The NE consists of the contiguous inner and outer nuclear membranes (INM and ONM, respectively), which are connected where the nuclear pore complexes traverse the two membranes. The arrows show the location of nuclear pore complexes. The red (dark gray in print) arrowheads indicate immunogold-labeled antibodies to Lamin A decorating the underside of the INM. Panel B is a diagram of the NE/lamina with the localization of some on the proteins associated with the NE. Underlying the inner nuclear membrane is the 20–50-nm thick lamina, comprised of the lamins. Associated with the ONM, INM, and lamina are various proteins, with many connecting the INM to chromatin and DNA, for example, BAF and HP1. Panel C illustrates the structures of the mammalian lamins consisting of a central rod domain flanked at either end by globular domains. Only Lamin A undergoes a second proteolytic cleavage step (P2) that removes the farnesylated cysteine. P, protease cleavage site; NLS, nuclear localization sequence.

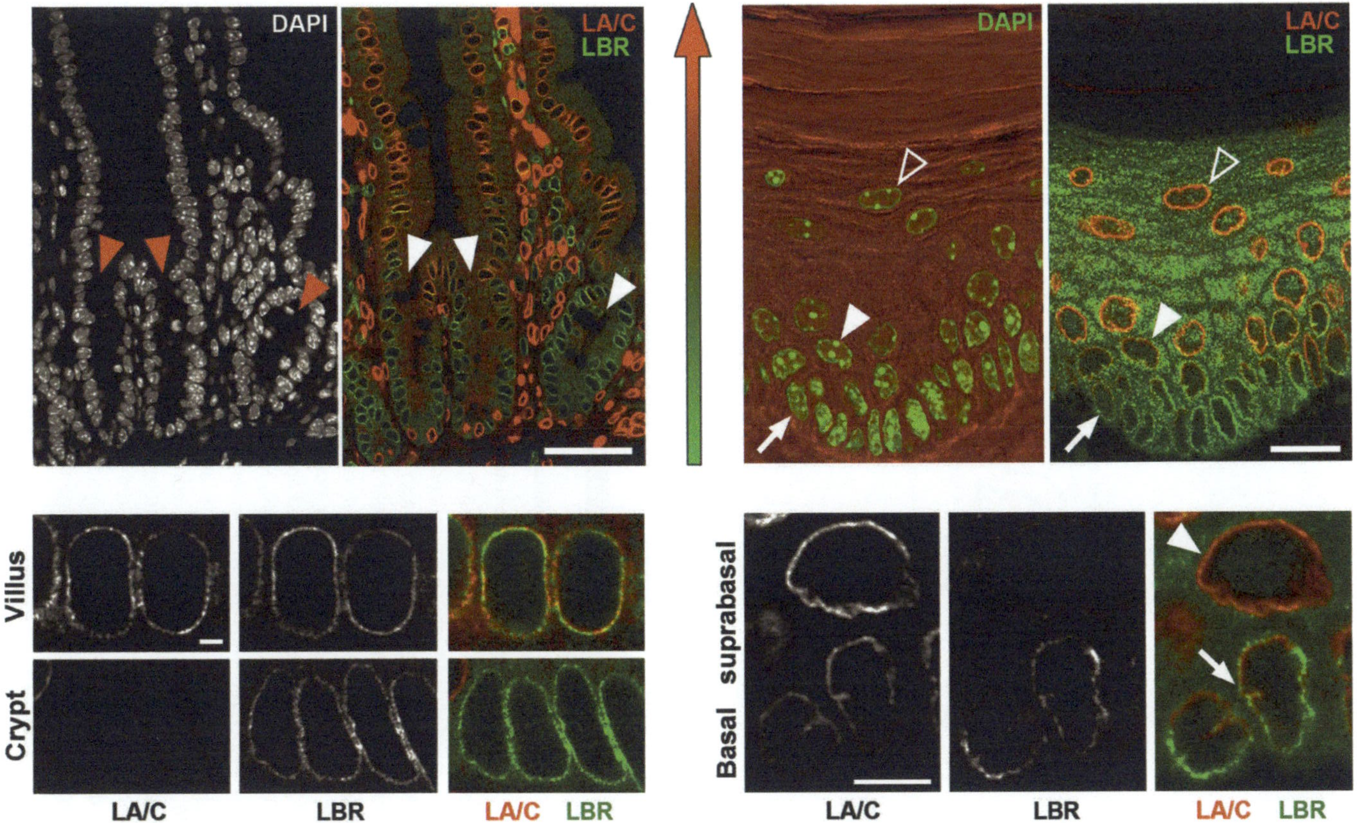

Brian Burke and Colin L. Stewart, Figure 1.3 Sequential changes in expression of the LBR and Lamin A proteins in the nuclei of the gut villous epithelium epidermal layer of the skin. In the left panels, the nuclei in the intestinal crypt epithelial cells express the Lamin B receptor (LBR). As the cells move up the villus the epithelial nuclei coexpress LBR and Lamin A and then further up the villus LBR expression is lost and with lamin A being exclusively expressed. Similarly, in the skin epidermis (right panels), the nuclei in the basal layer of the epidermis only express the LBR. In the suprabasal layer, the nuclei express both LBR and Lamin A and the most distal nuclei express only lamin A (Solovei et al., 2013).

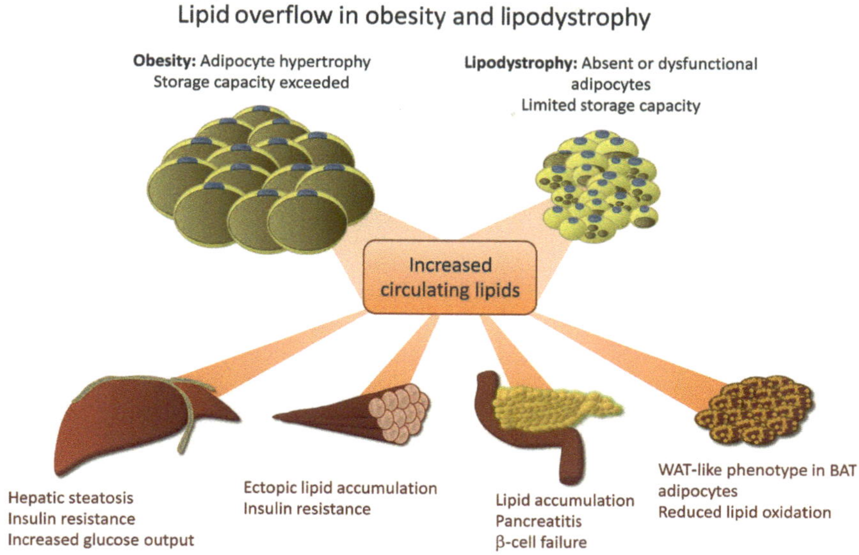

Justin J. Rochford, Figure 2.1 Lipid overflow and lipotoxicity may be a common driver of metabolic disease in both obesity and lipodystrophy. Excessive nutrient storage in the adipocytes of obese individuals leads the adipose tissue to reach a limit of expansion. Further nutrients, particularly lipids, can then no longer be stored safely in adipocytes. Hence, while fat tissue is abundant, there is a state of relative adipose tissue insufficiency. This is similar to the case in lipodystrophy where there is a primary defect of adipose tissue development and/or function. In both cases, the lipids spill over into the circulation and subsequently accumulate ectopically in other tissues where their effects are harmful. Accumulation in the liver and skeletal muscle may cause insulin resistance and so impaired glucose metabolism. Lipid accumulation in the pancreas can cause β-cell dysfunction, particularly when allied to increased demand for insulin secretion from this organ as the body attempts to overcome peripheral insulin resistance. In rodents, brown adipose tissue (BAT) can also become less able to oxidize lipids and accumulate more and larger lipid droplets, adopting an appearance more like white adipose tissue. The relevance of this in humans is not yet clear although significant BAT depots can be detected in healthy individuals. Overall, the lipotoxic effects deriving from an inability to store lipids safely in adipose tissue may explain the similarities in metabolic disease observed in obese and lipodystrophic individuals despite the disparity in adiposity.

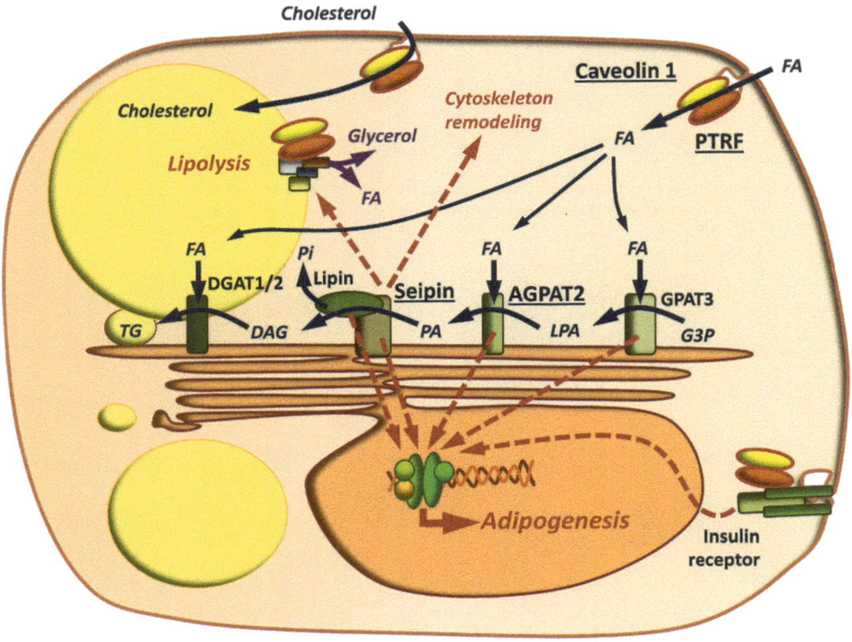

Justin J. Rochford, Figure 2.2 Molecular mechanisms via which disruption of CGL genes may cause generalized lipodystrophy. The proteins encoded by genes whose disruption causes CGL (whose names are underlined) have multiple roles in adipocyte development and function. AGPAT2 converts lysophosphatidic acid (LPA) to phosphatidic acid (PA) by the addition of a fatty acid (FA) moiety. This is an important intermediate in the generation of triacylglycerol (TG). Initially, glycerol 3-phosphate (G3P) is converted to LPA by glycerol 3-phosphate acyltransferase 3 (GPAT3), then LPA to PA by AGPAT2. PA is dephosphorylated by lipin enzymes, which are PA phosphatases, to generate diacylglycerol (DAG). DAG is then converted to TG by diacylglycerol acyltransferases (DGATs). Seipin, the product of the *BSCL2* gene, can recruit lipin to the ER membrane, potentiating its function and thereby affecting the levels of PA. However, seipin loss also increases lipolysis in differentiating cells and, in addition, can affect lipid droplet budding and/or morphology. A role for seipin in regulating the cytoskeleton during adipogenesis has also been proposed. The loss of GPAT3, AGPAT2, seipin, or lipin 1 all prevent the normal induction of adipogenic gene expression in cultured models of adipocyte differentiation (red broken arrows). It is not clear how this is mediated but loss of DGAT does not affect adipogenic gene expression, only the synthesis of TG, implying that a lipid intermediate in this pathway or its derivative influences adipogenic gene expression. Both caveolin 1 and PTRF/cavin proteins are essential for the formation or maintenance of caveolae in adipocytes. These proteins are important for FA uptake, cholesterol uptake, cholesterol and TG accumulation in the lipid droplet, the regulation of lipolysis, and signaling by cell-surface growth factor receptors including the insulin receptor. Insulin signaling is important for controlling glucose uptake and lipolysis in the mature adipocyte. However, it may also provide important proadipogenic signals and several insulin-responsive intracellular signaling intermediates play critical roles in adipogenesis.

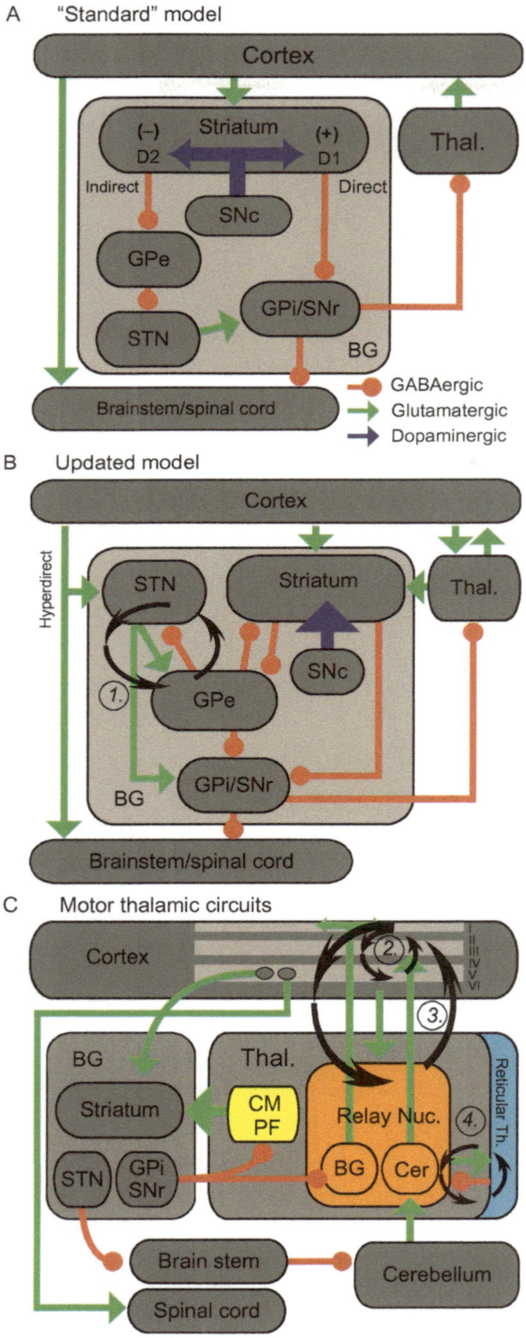
A
"Standard" model
Cortex
Striatum
(–)
D2
(+)
D1
Indirect
Direct
SNc
GPe
STN
GPi/SNr
BG
Thal.
Brainstem/spinal cord
GABAergic
Glutamatergic
Dopaminergic
B
Updated model
Cortex
Hyperdirect
STN
Striatum
Thal.
SNc
1.
GPe
GPi/SNr
BG
Brainstem/spinal cord
C
Motor thalamic circuits
Cortex
I
II
III
IV
V
VI
2.
3.
BG
Striatum
STN
GPi
SNr
Thal.
CM
PF
Relay Nuc.
BG
Cer
4.
Reticular Th.
Brain stem
Spinal cord
Cerebellum

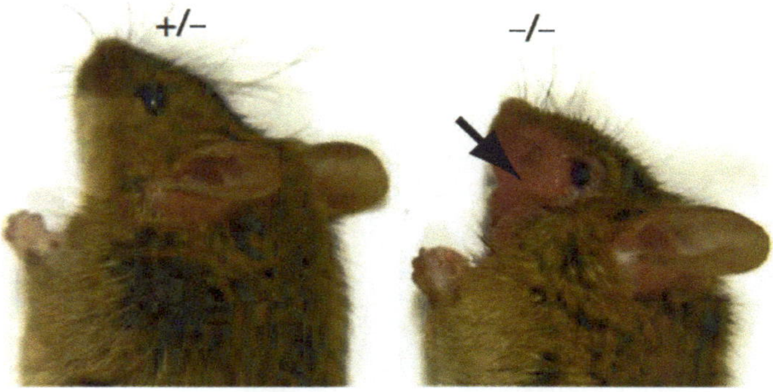

Samuel S. Pappas *et al.*, Figure 3.2 SAPAP3 –/– mice exhibit facial lesions due to self-injurious grooming. *Figure reprinted with permission from Macmillan Publishers Ltd.: [Nature] (Welch et al., 2007), copyright 2007.*

Samuel S. Pappas *et al.*, Figure 3.1 Cortical-BG-thalamic models. (A) The "standard" model. Green arrows, excitatory (glutamatergic) projections; red circles, inhibitory (GABAergic) projections; blue arrows, dopaminergic projections. (+) and (–) indicate excitatory and inhibitory effects of dopamine on MSNs, respectively. Abbreviations are defined in the text. (B) Updated model. Not all connections described in the text are illustrated for clarity. (C) Model emphasizing thalamic circuits. Roman numerals indicate cortical layers. Ovals in layer V indicate distinct populations of IT and PT neurons. Note the distinct projection patterns of the cerebellar and BG-recipient thalamus. In the thalamus, "BG" and "Cer" indicate the BG and cerebellar-recipient regions, respectively. Circular arrows in (B) and (C) indicate candidate oscillation generators. *Figure and caption reprinted with modification from Ellens and Leventhal (2013), copyright 2013, with permission from IOS Press.*

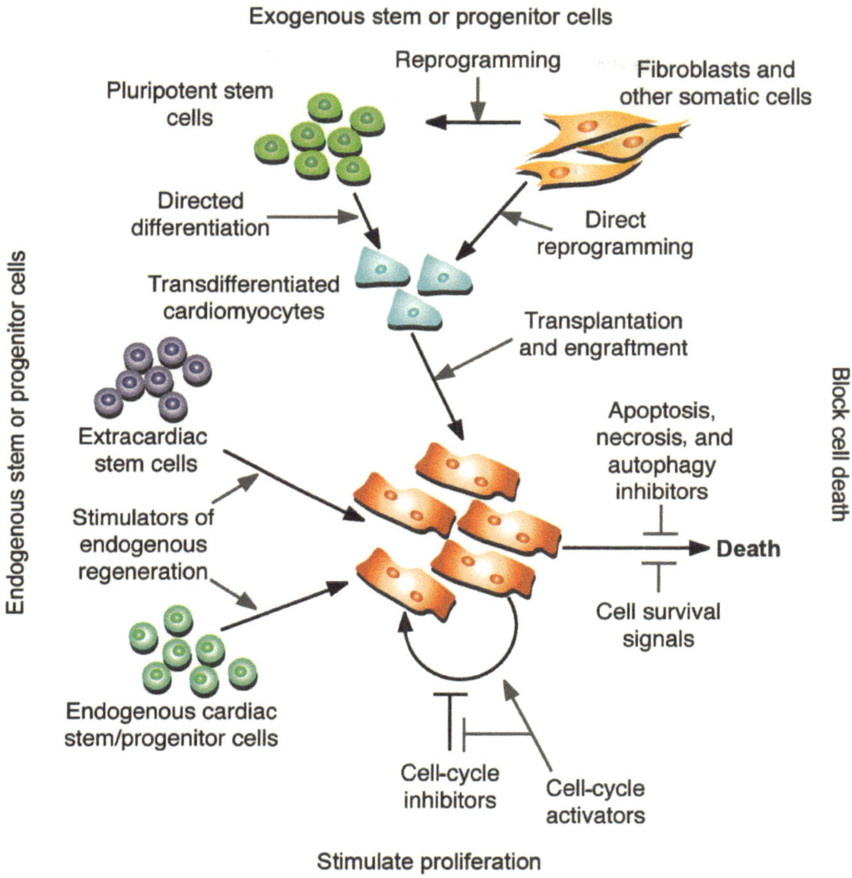

Lorna R. Fiedler *et al.*, Figure 4.1 Strategies to increase cardiac muscle cell number as a therapeutic target. In principle, the limited ability of the heart to replace cardiomyocytes can be improved by reactivating cell division of preexisting cardiomyoctes and/or inhibiting cell death or augmenting survival. Alternatively, new myocytes can be produced from multipotent stem or progenitors that reside within niches in the myocardium, circulating stem cells with cardiac potency, or *ex vivo* cells transplanted into the injured heart. Challenges to regeneration include an endogenous restorative capacity that appears limited by an insufficient number of available stem or progenitor cells, and the need to develop efficient means to produce or deliver exogenous cells. Developmental signals are being investigated for use in enhancing therapeutic regeneration from endogenous and exogenous sources. *Reproduced from Mercola, Ruiz-Losano & Schneider, 2011.*

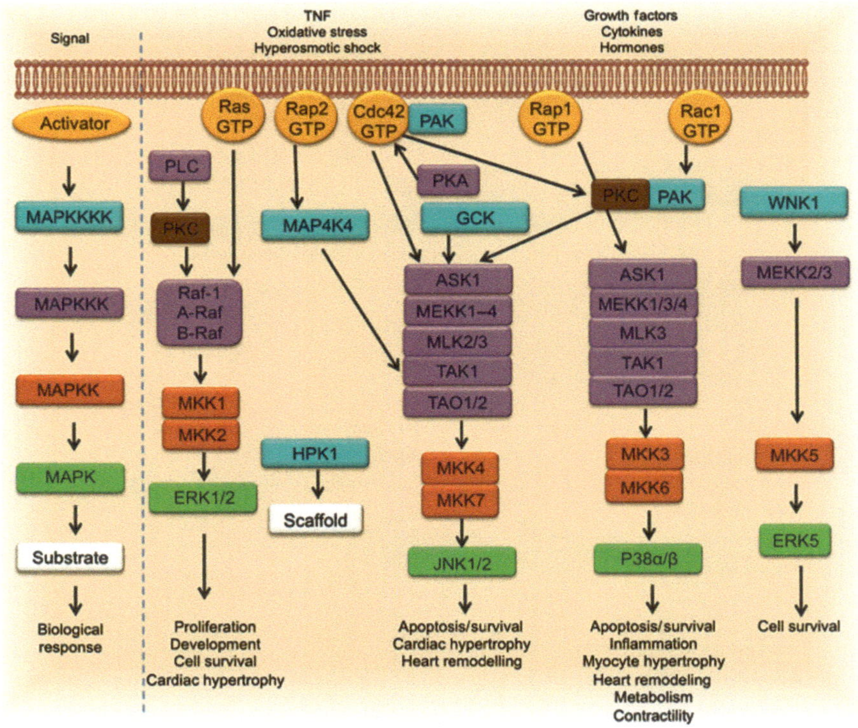

Lorna R. Fiedler *et al.*, Figure 4.2 Schematic representation of signal transduction pathways in the heart by kinases. Signal transduction through a hierarchical MAPK pathway is elicited by stimuli that include cytokines, growth factors, and stress responses such as oxidative stress. Activation of the terminal MAPKs, ERK1/2, JNK1/2/3, p38α/β, or ERK5 leads to a biological response that can range from proliferation, apoptosis, cell motility, osmoregulation, hypertrophy, or metabolism. Thus, kinase signaling couples extracellular signals to cardiac hypertrophy, contractility, remodeling, cell death, and heart failure. In the case of MAP4K4, Rap2-GTP interacts with MAP4K4 through its C-terminal citron homology domain (CHD) and can enhance its activity (Machida et al., 2004). MAP4K4 has also been shown to couple to TAK1 and JNK1/2 in control of cell death. Cardiomyocyte, and nonmyocyte, cell death is a complex process of interactive protein signaling cascades. *Diagram adapted from Coulthard, White, Jones, McDermott, and Burchill (2009).*

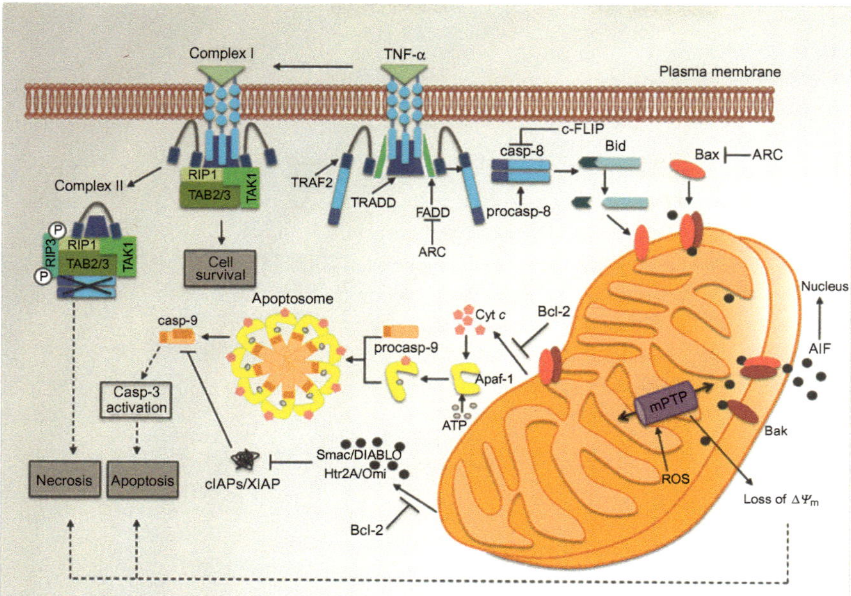

Lorna R. Fiedler *et al.*, Figure 4.3 Illustration of the different branches of apoptosis and necrosis cell-death signaling. Ligand binding (TNF-α) induces recruitment of adaptor proteins and subsequent activation of procaspase-8. Caspase-8 will induce Bid cleavage to tBid that will lead to Bax translocation on the mitochondrial membrane and the formation of mitochondrial pores consisting of Bax/Bak duet. Apoptogens (Smac/DIABLO, Htr2A/Omi, and cyt c) will be released for the inhibition of XIAP/cIAPs and the formation of the apoptosome. Cyt c will account for procaspase-9 activation by stimulating its binding on Apaf-1, which along with ATP, ends in the formation of the apoptosome. Caspase-3 is activated via either caspase-8 directly (extrinsic pathway) or via caspase-9 (intrinsic pathway), which will subsequently target cytoskeletal proteins and other enzymes for apoptosis. AIF presumably translocates in the nucleus from the mitochondria to induce large-scale DNA fragmentation. ARC protein antagonizes both pathways and Bcl-2 antagonizes the release of apoptogens from the mitochondria. During ischemia, depletion of ATP and lower pH sensitizes mPTP opening, an event that occurs upon reperfusion. Mitochondria become overloaded with Ca^{2+} and ROS leading to events, such as dissipation of mitochondrial membrane potential ($\Delta\Psi_m$), mitochondrial swelling, disruption of the outer mitochondrial membrane, and release of intermembrane apoptogens that mark irreversible cell death. Cyclophilin D (CypD or PPIF) has an essential role in protein folding and is involved in the regulation of mPTP (mPTP-dependent necrosis). Alternatively, necrosis can be initiated at the cell surface, when multiprotein complex I-TNFR1 is dissociated to form complex II, to further mediate cell death. Deubiquitination of RIP1 by CYLD (not shown) will lead to recruitment of FADD–procaspase-8 via DD interactions. If procaspase-8 is activated, RIP1 is cleaved by caspase-8 and rendered inactive to signal necrosis. In the case of procaspase-8 inhibition, RIP1 recruits RIP3 instead ensuing a series of phosphorylation events leading to cell death (death receptor-dependent necrosis). In unstimulated cells, complex I comprises TNFR1, TRADD, RIP1, TRAF2, and cIAP1/2, the latter of which polyubiquitinates RIP1 and TRAF2 so that they recruit TAK1 kinase via TAB2/3-binding proteins, which in turn will turn on survival genes via NFκB activation or IKK inhibition (for more details, see main body of review).

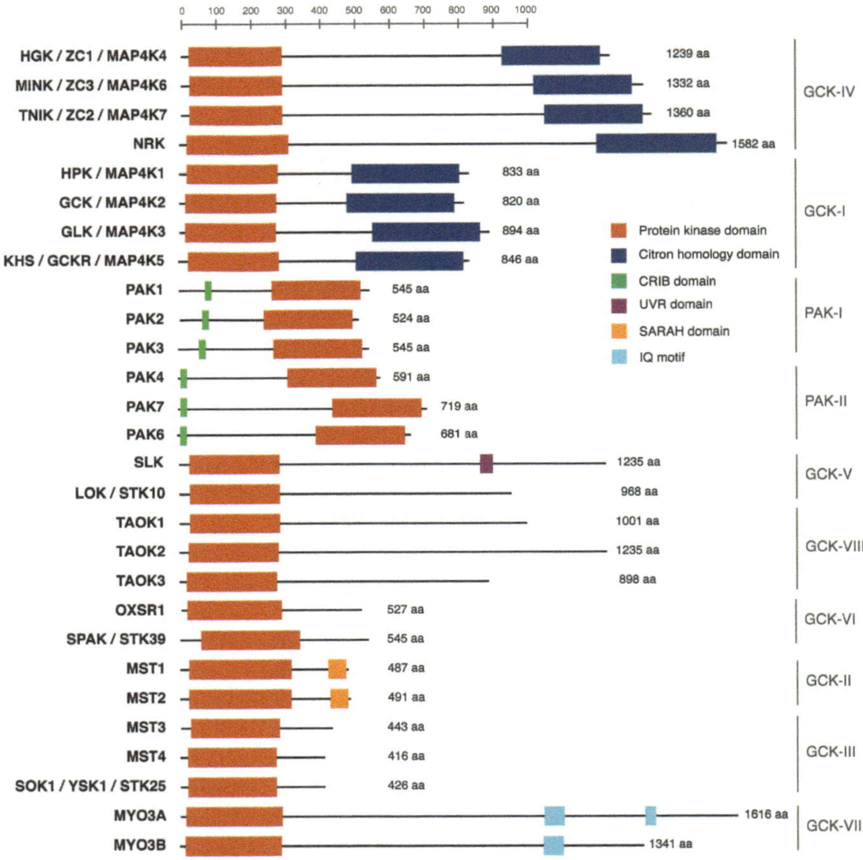

Lorna R. Fiedler *et al.*, Figure 4.4 Human STE20-related kinases. Members of the GCK group possess a characteristic N-terminal kinase domain and C-terminal citron homology (CNH) domain and are divided into eight subgroups on the basis of sequence similarities (I–VIII). A second subgroup of MAP4Ks comprises the p21-activated protein kinases (PAKs) that consist of a C-terminal kinase domain and N-terminal CRIB domain. PAKs are divided into subclasses I and II.

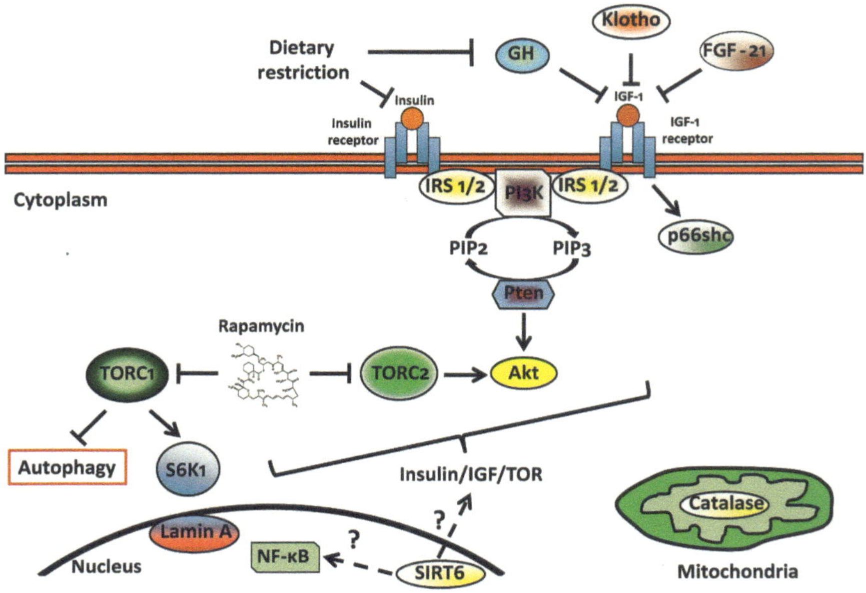

Chen-Yu Liao and Brian K. Kennedy, Figure 5.1 Major molecular pathways that regulate murine lifespan. The schematic represents an overview of the signaling pathways regulating lifespan. NF-κB, nuclear factor-κB; IRS-1/2, insulin receptor substrate-1/2; PI3K, phosphatidylinositol 3-kinase.

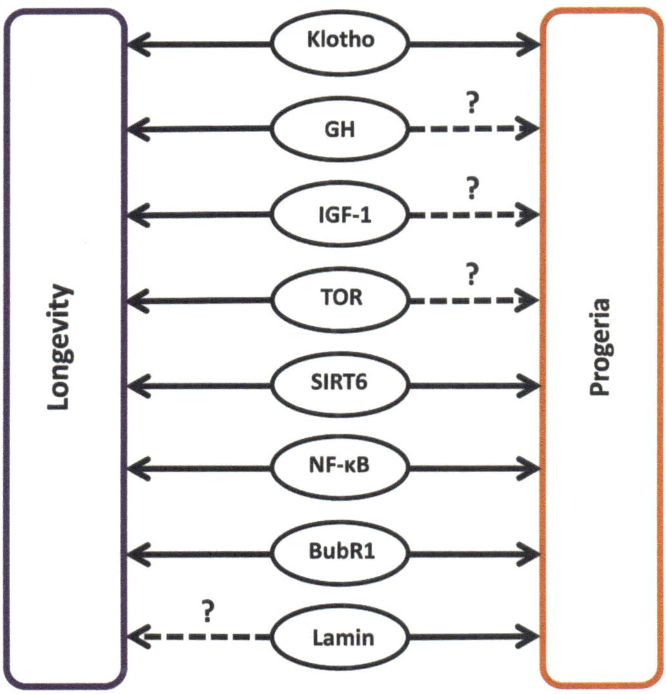

Chen-Yu Liao and Brian K. Kennedy, Figure 5.2 Same pathways that link to longevity and progeria. Pathways that mediate (or are proposed to mediate) both longevity and progeria. The dashed arrows indicate relationships that are yet to be firmly established.

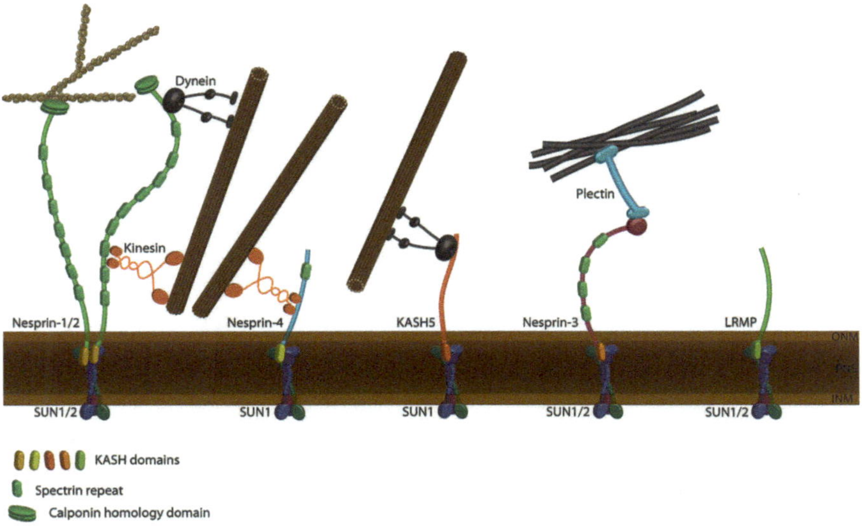

Henning F. Horn, Figure 6.1 Mammalian LINC complex proteins. The six known KASH-domain proteins are shown with their respective SUN1/2 binding partners. Also shown are the cytoskeletal components with which the KASH-domain proteins have been shown to interact. Nesprin-1 and -2 interact with filamentous actin, and also with microtubules via the molecular motor proteins dynein and kinesin. Nesprin-4 interacts with the microtubule network via kinesin, and KASH5 interacts with microtubules via dynein. Nesprin-3 interacts with the intermediate filament network via the adapter protein plectin. No cytoskeletal interactions have been reported for LRMP.

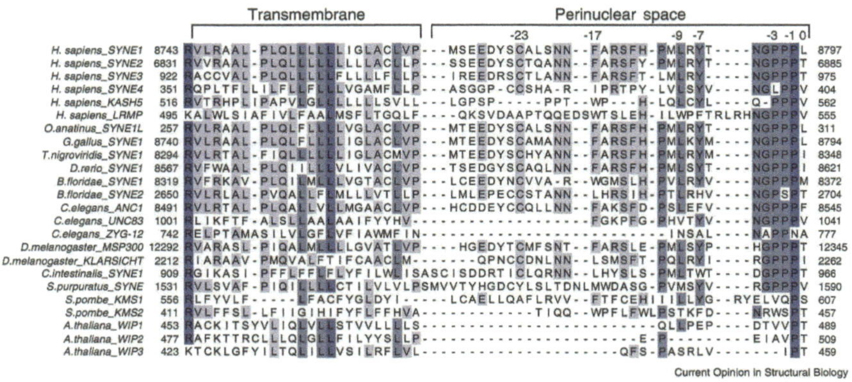

Henning F. Horn, Figure 6.2 Alignment of the KASH and transmembrane domains from a large range of KASH-domain proteins. Organisms range from yeast and plants to humans. Residues that have been identified as important for SUN domain interactions are indicated with numbers. *Figure reproduced with permission from Sosa, Kutay, and Schwartz (2013).*

FSC
www.fsc.org
MIX
Papier | Fördert
gute Waldnutzung
FSC® C083411

Kolibri 360 GmbH
Ferdinand-Jühlke-Straße 7
99095 Erfurt, Deutschland
produktsicherheit@kolibri360.de